P9-DMJ-729

WILDLIFE ECOLOGY

A SERIES OF BOOKS IN AGRICULTURAL SCIENCE

Animal Science

EDITORS

G. W. Salisbury
E. W. Crampton (1957–1970)

2200
16 50
EB

WILDLIFE ECOLOGY

AN ANALYTICAL APPROACH

AARON N. MOEN

Cornell University

with a foreword by Douglas L. Gilbert
Colorado State University

W. H. FREEMAN AND COMPANY

SAN FRANCISCO

Copyright © 1973 by W. H. Freeman and Company

No part of this book may be reproduced by any mechanical,
photographic, or electronic process, or in the form of
a phonographic recording, nor may it be stored in a retrieval
system, transmitted, or otherwise copied for public or
private use without written permission of the publisher.

Printed in the United States of America

1 2 3 4 5 6 7 8 9

Library of Congress Cataloging in Publication Data

Moen, Aaron N   1936–
   Wildlife ecology.

   Bibliography: p.
   1. Zoology—Ecology.   2. Wildlife management.
3. Ruminantia.   I. Title.
QH541.M54   599'.735'045   73-6833
ISBN 0-7167-0826-4

CONTENTS

PART 1

LIFE, INTERACTIONS, AND ECOLOGICAL MODELING

PART 2

THE DISTRIBUTION OF MATTER AND ENERGY IN TIME AND SPACE

PART 3
METABOLISM AND NUTRITION

PART 4

BEHAVIORAL FACTORS IN RELATION TO PRODUCTIVITY

PART 5

ENERGY FLUX AND THE ECOLOGICAL ORGANIZATION OF MATTER

PART 6

PRODUCTIVITY, POPULATIONS, AND DECISION-MAKING

APPENDIXES

FOREWORD

In recent years, environmental problems have created great general concern. Thus, the time has come when a revitalized and more effective approach to the management of natural resources is necessary. This is especially true in light of increased human populations.

In the past, individual abuses of the natural resources have been treated as isolated problems—an approach doomed to failure. Instead, individual abuses can be seen as parts of a larger problem: the increasing pressure of an expanding population on dwindling nature resources. That problem often appears overwhelming. In seeing it, many have given up in despair. But a great problem may be broken down; each part can be attacked separately and perhaps solved. Bit by bit the big problem becomes solvable. The importance of each issue, whether it be protein availablility, harvest of females, or disposal of waste pollutants, depends on the particular role of the issue in the overall environmental structure.

Wild animals, and the management of them are a vital part of the environmental "machine," a part that also is made of smaller parts. Age, sex, and time of year affect the physiology of an individual animal. These, together with nutritional factors, genetic history, and features of the physical environment, combine in the complex system that determines the interactions between an animal, other organisms, and the land.

It is the essence of the wildlife manager's job that he understand the system and be able to work with it. He must understand how an organism fits into the ecosystem. He must understand the effects of the organism on its total environment and the effects of the environment on the organism.

In *Wildlife Ecology: an analytical approach,* Professor Moen has analyzed this natural system. He evaluates each component and welds them together into a unified whole. Although most of the examples deal with white-tailed deer, the concepts are applicable to the other wild ruminants and, indeed, to all organisms.

Professor Moen's creative research and dedication have produced a work in which traditional pieces of wildlife management—numbers and conditions of animals, nutritive values of range plants, behavior patterns—are at last presented as parts of a greater whole. This book should be made available to every wildlife professional, whether technician, manager, biologist, conservation officer, administrator or researcher. It is an important publication and the time for it has come.

Douglas L. Gilbert

Colorado State University
Fort Collins, Colorado
September 1972

PREFACE

Rapid advances in analytical capabilities within the last fifteen years have made it possible for the ecologist to do things within a time dimension that were unheard of a few years ago. The capabilities for rapid analyses pose a threat to the discipline of ecology, however, because there can be a tendency to use numbers, large quantities of them, hoping by some magical means of computer analysis to find some relationships emerge.

The reorganization of numbers within a computer program of storage and computation is nothing more than a rapid bookkeeping system. Computers used in such a way do not usually help much in gaining insight into the mechanisms that are operating in the natural world. They tend to promote a false sense of security.

The real benefits of computer analyses emerge if they are used to extend the analyst's capabilities for analyzing the relationship between one factor or force and another factor or force in the ecosystem. It is important to realize that the human mind must always be ahead of the computer, with the electronic system doing rapid computations that are too numerous and time-consuming to do in any other manner. This suggests that the first models built by analytical ecologists are of necessity very simple ones. Let them be no more complex than the model builder can fully comprehend, insuring that he knows not only the capabilities of his analytical model but also its weakness. A progression of such simple models will result in more complex, working models that represent a *known* portion of the ecosystem.

In this book I have aimed at promoting the building of simple but workable models. They do not require large computer centers for their use; small desk-top computing systems are entirely adequate. In fact, many of the models suggested can be done manually, with the principles of model building illustrated just as well. Thus the book should be of interest to ecology classes in many types of educational institutions, from the small college to the major university. I am convinced that, wherever the student is located, the major factor that will determine his progress in ecology is his ability to think, along with the guidance of a professor who stimulates thinking about meaningful ecological relationships.

Aaron N. Moen

April 1973

ACKNOWLEDGMENTS

The completion of a book is not possible without the help of many people. My own efforts have been made possible through the kind direction and guidance given to me by my parents on their farm in western Minnesota. The opportunities for contact with wild animals and native plants in that area stirred within me an interest to pursue an understanding of the relationships between organism and environment.

My academic career in the field of natural resources began under the guidance of Dr. Max Partch at St. Cloud State College. His enthusiasm for teaching in the field impressed me greatly. Dr. William H. Marshall, of the University of Minnesota, gave me opportunities, freedom, and responsibility as I pursued a Ph.D. The most significant academic work that permitted me to delve into the energy relationships of deer at that time was that of Helenette Silver and her colleagues at the New Hampshire Fish and Game Department and the University of New Hampshire. Without her pioneering efforts in the field of energy metabolism of white-tailed deer, my Ph.D. dissertation could not have started me on the challenging research on the energetics of a free-ranging animal.

I wish to thank the many friends I have made in the field of wildlife management, especially the deer biologists in the State of New York who always provide stimulating interaction as we proceed together to understand this most important resource in New York State. My colleagues at Cornell, especially Dr. Peter Van Soest of the Department of Animal Science, have provided many insights into the animal-environment relationships currently under investigation. Dr. Douglas L. Gilbert, formerly at Cornell and now at Colorado State University, has discussed big-game management with me on many occasions. Dr. Donald Ordway and his staff of aerodynamic engineers have been of great help in our thermal analyses at the BioThermal Laboratory.

Dr. Dwight A. Webster, former head of the Department of Natural Resources, and the administrators of the Agricultural Experiment Station at Cornell have all been most helpful as I established a research program at the BioThermal Laboratory. Funds for research at the Laboratory have been contributed through the Pittman-Robertson Federal Aid program, Project W-124-R, and the New York State Department of Environmental Conservation. Additional funds from the Agricultural Experiment Station at Cornell, the Cornell Research Grants Committee, the National Science Foundation, The Loyalhanna Foundation, and the National Rifle Association have helped support the work at the BioThermal Laboratory.

The staff at the Laboratory has contributed significantly to the work that is described in this text. My respect for the abilities and dedication of my students cannot be fully expressed by acknowledgment but will be manifested by their contributions in the future. I must recognize the help and accomplishments of former students, especially Dr. Keith E. Evans and Dr. Deborah S. Stevens. The work of Nadine L. Jacobsen and Charles T. Robbins, both Ph.D. candidates studying the energy relationships of deer, has provided much insight into the complex animal-environment relationships that are the focus of study at the Laboratory. William Armstrong, laboratory technician, has helped in the design and construction of research equipment and in the care of our experimental deer herd. Richard E. Reynolds, foreman at the Ithaca Game Farm, has contributed much to the program with his help in the construction of the deer pens, maintenance of the facilities, and continual attention to our needs. Eleanor Horwitz offered many fine suggestions on ways to improve the manuscript. I appreciate her efforts to convince me to say things in the simplest way possible.

Students in my courses have raised many stimulating questions. I wish that each one of them could participate actively rather than passively in the educational process of research and discovery.

Finally, the help and encouragement of my wife, Sharon, and of Ronald, Thomas, Daniel, and Lindy cannot be fully expressed in words. It has often been impossible to keep up with some of the domestic duties confronting every husband and father because of the urgency of research according to a biological clock and my own intense interest in the subject. As Tom (age 9) said when I suggested I might write another book, "Oh no, not another five years of that!"

WILDLIFE ECOLOGY

Courtesy of Paul M. Kelsey
New York State Department of Environmental Conservation

PART

1

LIFE, INTERACTIONS, AND ECOLOGICAL MODELING

Life does not occur in a vacuum. It consists of processes with very fundamental characteristics that are related to and affected by the milieu in which they are taking place. The analytical ecologist, regardless of the species he is studying, must not only understand these basic processes but also relate them to the environment of the free-ranging organism. This is a challenge! It is difficult to interject oneself into the life of an organism without disrupting its basic processes. Insight into these processes and organism-environment relationships does, however, provide a starting point for the development of analytical models that represent these processes and relationships in a numerical way. The fund of knowledge available today, along with computer capabilities that permit rapid analyses, provides a potential for understanding that far exceeds anything available in the past.

PRODUCTIVITY GRADIENTS

1-1 THE CONCEPT OF LIFE

Life is a process that involves the expenditure of energy for the redistribution of matter. This fundamental truth applies to life processes at the molecular level, to the whole organism, and to populations. The redistribution of matter or materials at the molecular level is often called nutrient cycling. The movement of nutrients through the ecosystem is a continual process that includes the synthesis of body tissue in the living organism, the transfer of energy from one trophic level to another, and the release of nutrients from the body of an organism upon death and subsequent decomposition.

The whole organism participates in the redistribution of matter as it moves about, selecting a place to eat, rest, and perform its other daily activities. Some organisms move but a short distance during their entire lives, that is, they have a small home range. Others cover hundreds or even thousands of miles; many birds fly thousands of miles each year as they move from wintering grounds to nesting grounds and back to the wintering grounds. Some populations move as a unit. Herds of elk, mule deer, and caribou exhibit a seasonal pattern of movement or migration. Social interaction between individual animals within each herd adds an internal mobility to its external mobility.

The individual animal is the fundamental unit that must meet the energy cost of motion, or of life. The energy cost of life of each individual depends on its growth rate, reproductive condition, amount of daily movement, and other factors unique to that individual. Thus it is logical to analyze the energy relationships between an organism and its environment, determining the effect of different factors on the maintenance, growth, and reproduction of animals with different individual characteristics. How is this done within the basic framework of matter and energy? These two fundamental components of the ecosystem need to be regarded as factors and forces, with an understanding of their *effect* on an organism rather than a mere recognition of their association with an organism.

1-2 THE ROLE OF THE ANALYTICAL ECOLOGIST

The orderly progression of life processes in the natural world is the result of evolutionary development. The ecologist steps into this world and asks first, "What is present?" Natural historians have been doing this for many years, resulting in a very important body of knowledge describing the life histories and characteristics of organisms. Many of the responses of an organism, whether it is responding as an individual or as a member of a population, have been related to the presence of physical forces in its habitat, such as weather, food, space, and others.

The descriptive ecologist proceeds a bit further than the natural historian, relating observed characteristics of the organism to observed characteristics of the habitat in a quantitative way. This has resulted in the formulation of many ecological rules. Bergman's Rule is an example: animals living further north tend to be larger. These types of rules are generally applicable, although exceptions can be found in looking at detailed relationships.

The analytical ecologist asks the question, "Why?" He is interested in the mechanisms operating in the natural world. The recognition of simple relationships such as the condition of the range and the condition of the animal are pursued further by analyzing the requirements of the animal through time and the ability of the range to satisfy these requirements. Scientists working with domestic animals have been doing this for years, but these animals are subject to considerable control by man. Their genetic characteristics can be manipulated, their living space can be limited by fences or pens, their feeding regimen can be controlled and many other conditions can be imposed on them. The animal scientist searches for a combination of forces—feed, cover, and space—that will result in the attainment of a particular production goal. Thus, if a particular diet and feeding regimen results in the desired level of milk production, it may become a management recommendation if there are no other complicating factors.

The analytical ecologist who is concerned with free-ranging organisms does not have such straightforward production goals in mind. His interests lie in the natural world where he has little control. The animals attracting his attention are usually elusive, secretive creatures subject to natural rhythms in activity each day and each year. They feed in accordance with these natural rhythms rather than man's work schedule. Breeding follows from natural stimuli. Their lives are inexorably linked to the rhythmic and arrhythmic fluctuation in their relationships with other organisms, the range, and other natural forces. These natural characteristics present a challenge to the ecologist because he cannot make this natural world conform to his particular desires.

Domestic animals are not under the complete control of man, of course. They exhibit many natural rhythms and other characteristics that are continuations of their natural development. Animal scientists are becoming a bit more "ecological" in their thinking, especially after it has been realized that certain production limits cause complications in other aspects of the animal's biology. High-grain diets, for example, cause problems in the digestive physiology of dairy cattle.

Ecologists, however, can become much more knowledgeable in the field of animal biology if they avail themselves of the vast amount of information that has been accumulated about domestic animals. No animal violates the basic laws of matter and energy, and the literature on domestic animals contains many experimental analogs of natural events. Different diet levels, for example, are analogous to natural variations in the food supply. Different population densities in pastures, pens, or barns are analogous to natural variations in population densities.

The analytical ecologist needs to recognize the dynamic nature of the relationships between organism and environment in the natural world. He must be cognizant of the whole in addition to the relationships between its parts. He must always be ready to recognize the factors and forces that may be affecting an observed relationship. He must be able to establish the importance of different factors and forces present through time, which will enable him to predict the potential impact on the ecosystem that an alteration of these factors and forces would have.

1-3 DEAD OR ALIVE

Biologically, an organism may be classified as either dead or alive. If it is dead, it is subject to decomposition and the recycling of its nutrients through living systems in the future. This is a slow process, and the impact of a single organism on the nutrient cycling picture is usually not very great.

Ecologically, an organism is more than simply "dead or alive," however. If it is "not dead," then it is alive. There is a great possibility for variation in the ecological importance of each individual in the "not dead" or living category. An organism could be merely surviving, consuming resources sufficient only for existence, or it could be at its full productive potential—growing, reproducing, and contributing to population growth. Living organisms determine the growth potential of the population. The dead can contribute only as a part of the slow process of decomposition and nutrient recycling.

This book presents analyses of the factors that affect the living. Life in its fullest complexity cannot be comprehended by the human mind, but some considerations can be made that permit an understanding of some of the basic principles of life processes of organisms in their natural habitat.

1-4 PRODUCTIVITY OF THE INDIVIDUAL

The productivity of a living organism can vary from less than maintenance to maximum productivity as an individual, including both its body weight and size and its reproductive potential (Figure 1-1). An individual animal in a submaintenance condition may be experiencing a negative nitrogen balance with a subsequent loss of protein tissue; it may be in a negative energy balance as fat reserves

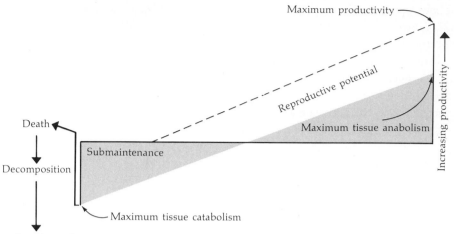

FIGURE 1-1.   The productivity of a living organism can vary from less than maintenance to maximum productivity.

are being mobilized; and it may be losing vitamins, minerals, and water, which results in a general loss of body weight. Such overall patterns of weight changes are exhibited by all free-ranging animals. The general pattern for weight changes in wild ruminants shows clear variations in weight, with losses occurring in the winter in most members of a population, and in the summer also for lactating females.

Lactating females may be in negative productivity as an individual but positive productivity with respect to its suckling offspring. Several mammals exhibit weight losses during lactation, indicating that body reserves are being mobilized to produce milk that is of distinct importance for the survival of the offspring. Ecologically, then, an animal's productivity does not fall into a single category; it may be at two or more points simultaneously on the productivity scale shown in Figure 1-1. The ability of an animal to draw on its own reserves to enhance the survival capabilities of its offspring may be of particular ecological significance because every young animal that reaches reproductive maturity has the potential for contributing more animals to the population.

The young, immature, and rapidly growing animal is in a general state of positive productivity. Protein is being synthesized at a rapid rate, minerals are being deposited as the bones grow and ossify, and the total body mass is usually increasing. This continues beyond the point of reproductive maturity in most species, although there is usually a marked deceleration in the growth rate after an animal reaches reproductive maturity.

The reproductive potential of any animal has a genetic limit, but the actual rate of reproduction is less than that potential because of the effect of natural forces. It is important to consider the ecological potential for any reproductive process, including the production of different types of body tissue, eggs or fetal tissue, and the production of milk. The ability of an animal to cope with the demands of different biochemical processes is dependent not only on the rate of ingestion of nutrients at the time they are required, but also on its ability to translocate nutrients by the mobilization of reserves. Mobilization of protein material with the subsequent resynthesis of amino acids in fetal tissue is one example of translocation; another example is the reverse process, or resorption. Antler growth has been shown to result in the mobilization of phosphorus from the ribs. In deer, elk, and moose, the antlers are shed and the minerals enter a slow recycling process. This process begins when rodents eat the antlers, live, die, and decompose with subsequent resynthesis in plant tissue, once again becoming available to herbivores such as deer. Egg shells are produced by a similar process; some of the calcium deposited in the shell comes from food, and some from metabolic processes that result in bone deposition followed by mobilization for production.

The relationships between ingestion, metabolism, deposition, and the mobilization of nutrients indicate that time is a very important consideration in ecology. Organisms do not just live according to conditions at the moment, but their ability to cope with factors and forces impinging on them at a given moment is dependent on their previous life processes.

1-5 THE NATURAL MOSAIC

The challenge confronting the present student of ecology is that of integrating current information about the components of the ecosystem within a framework that utilizes technological capabilities for a systems analysis. The measurement of environmental parameters in isolation and the calculation of simple correlations between an environmental parameter and an organism's response is seldom, if ever, a meaningful educational experience by itself. Yet, certain general relationships have been described in the literature.

Many examples of parameters measured in isolation can be found in the ecological literature. Temperature has been suggested as a dominant ecological force, and, of course, there are both high and low atmospheric temperatures at which life ceases to exist. However, temperature is not the only changing parameter between, say, the equator and the North Pole. There is a northern limit for most species in the Northern Hemisphere, but other parameters must be taken into consideration. For example, vegetation changes are marked, culminating in the tundra with its unique vegetation forms.

High temperatures have been observed to cause a depression in the reproductive rate of domestic animals. Free-ranging animals can often avoid high-temperature areas, but heat stress can also develop because of excitement or increases in the metabolic rate caused by running. Thus high temperatures alone are not an indication of the potential for heat stress.

Wind has been considered an ecological force by many investigators. Meteorologists describe wind flow in terms of an average velocity over a vertical profile. Aerodynamic analysts consider laminar and turbulent flow characteristics. Ecologists talk of the responses of animals to windbreak effects. Some have reported that animals tend to bed in the areas with higher velocities; others report the opposite. Barometric pressure has also been related to animal behavior, and there are those who think animals can sense not only changes in pressure but also oncoming storms.

Relative humidity has been related to the responses of many organisms. In some cases there has been a positive response observed, in some, a negative one, and, frequently, no relationship is detected. Relative humidity is a function of barometric pressure, air temperature, and the vapor pressure of the air. The ecologist needs to be aware of the interaction between these components of energy and matter before a realistic appraisal of the relationship of an organism to moisture in the air can be made.

Precipitation is an ecological force that has been related to population fluctuations. Conflicting reports can be found in the literature on the success of pheasant hatching during periods of heavy rainfall. Some reports show a depression in the population during a wet spring, although others show an increase in the population. The amount of rainfall affects the eggs and chicks directly; it also affects the growth of vegetation and the insect population associated with this growth. Thus, the amount of rain, the type of rainfall, the timing of the rainfall, and other characteristics of this ecological force are necessary parts of a meaningful analysis that can be used to interpret an organism's responses, both as an individual and as a member of a population.

Snowfall has been given considerable attention by ecologists, since it functions as a mechanical barrier as well as a thermal barrier. Heavy snowfall may enhance the survival of a subnivean animal, since it is an excellent insulator. Snow increases the energy expenditure necessary for movement through it by an active animal, however. It also covers food; however, if it becomes encrusted, animals can walk on it and reach food that would otherwise be unavailable. Thus snow can have both detrimental and beneficial effects, and an analysis of the effect of snow must consider the function of the snow in relation to a given organism.

Light seems to be related to the daily and seasonal cycles of many animals. Some are nocturnal, such as racoon (although they are observed to be active during the day at times) and owls; others are primarily diurnal animals, such as game birds, song birds, and hawks; and still others are crepuscular, becoming most active in the transitional period between daylight and darkness. These differences in the timing of the activity periods of animals may result in almost total non-interaction of species even though they live in the same area. On the other hand, because activity patterns are not absolutely rigid, they may overlap. Thus the amount of interaction that can take place between different species occupying the same area can be represented by a gradient.

Seasonal reproductive cycles appear to have some relationship to the seasonal variations in length of daylight. Some birds are responsive to longer periods of daylight, coming into breeding condition in the spring and laying their eggs in the late spring and early summer. Breeding condition is reached earlier if artificial light is supplied. For example, the provision of artificial light in the pheasant yards at the Ithaca Game Farm on February 15 results in the first egg being laid on about March 10. Wild ruminants, with fairly long gestation periods, respond to shorter days and breeding condition is reached in the fall. Conception generally takes place between October and January, with births following in late spring or summer.

Daylength follows a regular pattern of change, but weather patterns are much more irregular. Spring weather may come much later in some years than in others, and this may delay forage production on the range so that it is not of sufficient quantity and quality to satisfy the requirements of animals after parturition. Effects of variations in weather, plant phenology, and other range conditions on animals are discussed in Part 6.

The thermal effects of radiation as an ecological force have been quite neglected. Livestock shelters have been designed to maximize heat loss to the clear cold sky in the summer. A "cloud" due to respiratory water loss may form above a herd of caribou, reducing the loss of infrared energy from the animals. This may result in little or no metabolic benefit to the animals since they may be moving at the time, causing their heat production to be higher. Thus, the effect of cloud cover, radiant energy, and other thermal forces cannot be determined without a larger analysis of the interaction between these forces and the rest of the animal's characteristics.

Soil type and fertility is an ecological force that permeates the entire food chain, affecting the selection of nesting sites, bedding sites, movement patterns, and the like. For example, deer have been observed to selectively graze on wheat that is growing on fertilized ground. The presence of pheasants seems to be associated with higher calcium levels in the soil. Selenium is a trace element in the soil that is necessary in very small amounts for metabolic processes and is toxic at higher levels. Both desirable and undesirable elements in the soil, along with a complex of other nutritive, physiological, and behavioral factors, affect the productivity of the individual animal. Thus the soil must be considered an integral part of a model that includes considerations of all the factors and forces present in the ecosystem.

Fire is an interesting ecological force. It is a part of the natural scheme of things, since lightning caused forest fires long before man entered the picture. These fires were naturally controlled if they took place in areas that had been burned previously and not enough litter had accumulated to result in a significant fire. Precipitation, which often follows lightning, was also a natural control. Fire causes immediate thermal effects because of high temperatures, and there is a sudden release of nutrients as the organic material is reduced to ash. These nutrients can then find their way into the nutritive processes of plants more quickly than decomposed material. On the other hand, fire can adversely affect the fertility of the soil by destroying organic binding agents in the soil, which results in a greater chance of erosion and an increase in the rate of leaching. Thus, the effects of fire are both short-term and long-term and should be considered to be a part of the whole ecosystem, since they interact with physical components, plants, and animals.

The chemical constituents of the soil are reflected in the chemical characteristics of the forage growing on it. Nutritive quality can be described in chemical terms alone, but this has little relevance in ecology since it is not what is present in the plants that is of importance, but what the animals extract and use from these plants. Thus net energy, rather than gross energy, is an indicator of forage quality. Similarly, it is net protein, rather than crude protein, that is of significance to the organism. Vitamins and minerals are necessary to provide a balanced diet, and the quantity of each that is needed from forage is determined by the requirements of the animals. Adult ruminants, for example, need no vitamin B because this vitamin is synthesized in the rumen. Monogastric animals, however, must rely on external sources for vitamin B. Some minerals act as metabolic inhibitors—selenium has already been mentioned—although some, including selenium, must be present in small but essential amounts. A lack of iron, for example, causes anemia, and this affects the metabolic efficiency of the body by depressing the ability of the blood to carry oxygen. Mercury has been found to affect the viability of pheasant embryos, although it seems to have little effect on the adult birds. Thus one needs to consider not only the mineral and its action in the metabolic machinery of the animal, but also the age of the animal at which this becomes a part of the organism-environment relationships. The time factor may be of real significance in the interaction between factors and forces that affect productivity.

1-6   INTERSPECIES INTERACTION

Several types of relationships can be found between different species in the natural system. Ruminants are dependent on the action of microflora and fauna in the rumen to break down fibrous forage material and produce organic chemicals (volatile fatty acids) that are a source of energy for the host. Heavy infestations of parasites may cause the host's death. Some parasites have a neurological effect on their hosts, causing aberrant behavior that renders the host incapable of coping with other environmental forces. Moose sickness (*Pneumostrongulus tenuis*) is an example of that type of effect. Pathogens may function in much the same way. Predators feed directly on prey species, and they also may affect the prey by harassment, by reducing their feeding time, or by causing other effects on behavior. Predator and prey may also coexist because of changes in the food habits of the predator owing to an abundance of alternate foods. Thus coyotes may feed heavily on fruits and berries, ignoring fawns that might have been a source of food if the berries were unavailable. The density of prey necessary to retain the interest of the predator is worthy of consideration; it is too easy to assume that predators are easily discouraged in the same manner that the human hunter is today. It is likely that the predator has much more at stake, and, consequently, has different thresholds for beginning and ending the utilization of a particular type of prey.

The variability in time and space that can occur in interspecific relations suggests that a comprehensive look must be taken at the entire interspecific community. Predation levels measured between two species alone simply cannot represent the effectiveness of predation in the natural world. Parasite numbers by themselves have little relevance in analytical ecology; it is an understanding of the mechanisms with which the parasite affects the host that is important for an interpretation of the meaning of the number relationship between parasite and host. How can an ecologist escape these basic and fundamental relationships?

1-7 INTRASPECIES INTERACTION

The interaction between members of the same species is frequently very strong, varying from open conflict to gregariousness. There are differences between species, of course, with some species, like the moose, being quite solitary animals that have little association with others of the same species. Other large ruminants, like the bison, are quite gregarious at almost all times of the year. The sociability of song birds, game birds, and waterfowl varies seasonally, with a general intolerance during the reproductive season, and gregariousness during migration, winter flocking, and the like. Some species have a distinct social structure at all times of the year, although the amount of gregariousness varies. The turkey is an example. Competition for food and space affects the amount of intraspecific tolerance; field mice (*Microtus*) in crowded conditions are much less tolerant of each other than when conditions are less crowded. The requirement for space by any species is a function of its basic requirements for matter and energy (food) and its reproductive condition and general behavior pattern at a particular time. Thus the living space required is a dynamic dimension that varies partly because of daily rhythms, partly because of seasonal rhythms, and partly because of local conditions at a particular time.

The areas of the home ranges of species in particular habitats are frequently calculated by ecologists. The size of the home range is dependent on the distribution of energy and matter within a space that the animal can traverse. There must be a maximum space over which an animal can roam and still ingest enough nutrients to supply the necessary energy for those movements. Animals move for many reasons besides foraging, including playing, breeding, escape, and other necessary activities that are a part of daily life. A smaller animal might have a higher requirement for energy and matter if it was a prey species that encountered predators frequently. Its home range might reflect that. Large animals require large areas for living space, and they may migrate seasonally to find sufficient space. Yet elk migrate to winter concentration areas that often do not provide enough forage to meet their current requirements, and the fat reserve is utilized in order to make up the energy deficit. Thus the significance of the home range and its characteristics extends through time; an extensive summer range may be of significant importance to survival on a restricted winter range. It is very important to consider these variations in space and time before the functional organism can be understood in relation to the distribution of energy and matter in its natural habitat.

1-8 REPRODUCTIVE PATTERNS

Reproductive patterns are often a reflection of seasonal variations in light or rainfall, or of the quality of the range, or of some other natural event. For example, the timing of the reproductive period of most birds is related to light, although waterfowl in some areas breed when rains are common and the water level is high. Lower reproductive rates are frequently observed on ranges with a lower soil fertility. Reproduction, however, is not a simple on-off process but can be represented by a gradient with a variable threshold at which breeding commences.

The significance of the number of reproductive attempts per year needs further consideration within a larger ecological model. One parturition may occur, as in deer. Smaller animals, such as *Microtus,* may have several reproductive periods per year. This is of greater interest than merely the number of young born in toto and in each litter. Young animals born at different times of the year enter different worlds: forage is changing, there are seasonal differences in weather, and the growth rate and subsequent productivity of these different litters is going to vary in part because of the different environmental conditions. The same effect may be observed in species of birds that renest if the first nest is destroyed. A successful second hatch after an initial failure may not be the same ecologically even if the same number of chicks are hatched. A late hatch has a different food supply owing to the phenology of the vegetation and insects, and there is less time for a late-born chick to grow and mature prior to the first winter. This may affect its ability to survive the winter and reproduce the following spring. The significance of these variations is of considerable interest to the population ecologist. This is especially true for an analytical ecologist who makes predictions based on organism-environment relationships rather than merely presenting a historical perspective based on large quantities of numbers.

The various relationships discussed in previous sections in this chapter include but a scattering of isolated examples among a large number reported in the literature. It is clear that ecological relationships are complex enough to demand a very systematic and organized approach to their analysis. It is also clear that these analyses must proceed within a comprehensive framework that includes at least the more important variables in the natural system.

The natural world is too complex to represent in its entirety. This text contains samples of analyses concerning the "principal characters" on the ecological stage, representing the main forces that are present, with suggestions of the effects of alternative responses that an animal might exhibit. The situations that are described are not all-inclusive, but the student is asked to consider the larger meaning of these analyses, relating the analysis under consideration to the larger whole. This might be called theoretical ecology, but perhaps a more appropriate term is "theatrical" ecology, with the written text containing profiles of the main characters in the ecological theatre, relating them to the rest of the characters in a compressed form in both space and time.

Just what or who are these principal characters? They are analogs of real organisms in the natural world. The scenes considered do not represent any one particular spot on earth, or any particular individual on earth; rather they represent an environment and an organism that could be present anywhere on earth. Thus the material that follows is not unique to only one situation at the exclusion of all others, but represents the factors and forces present in any ecosystem, with an indication of how these factors and forces might interact with specific organisms by the use of examples that illustrate mechanisms in operation.

1-9 A THEORETICAL MOSAIC

The analytical ecologist cannot present all forces and factors existing in the world. The alternative is to assemble an array of principal characters in the ecological theatre and relate one to another in a realistic manner with no more complexity than can be understood in each and every analysis. This results in an understanding of the role of different factors and forces in the real world, though the analysis may not be for any particular organism at a particular place at a particular time. The characters represent interactions, possible events, and dynamic relationships, making it easier to grasp the fundamental relationships displayed and apply them to particular areas of interest.

Theoretical considerations are hard to grasp conceptually, but they can be made less difficult by the use of specific examples. For example, topography is an ecological factor that affects the distribution of plants and animals, as well as physical components of the ecosystem such as water. We know from geometry that topographic variations reach limits contained within 360°. Practically speaking, topography varies from flat land to steep hills, which can be measured to find the flattest to steepest limits of topography. The analyst can then use a range of values between these limits to analyze the effect of topography on some other ecological factor.

The soil has particular water-holding characteristics. The limits are clear though; the soil may hold none of the water that reaches it, or it may hold all of the water that reaches it. The actual amount retained by the soil is a function of gravity, soil particle size, slope of the substrate, and characteristics of the soil profile. Interactions between water and soil may be reduced in simplest form to specific interactions such as surface runoff, percolation, or some other analysis that considers a minimum of factors. When the simple analysis is understood, additional factors are added to approach greater and greater ecological realism.

Vegetation can be represented theoretically by considering limits to its distribution. There is an upper limit to the growth of trees, which is a function of genetic potential and physical forces such as gravity, wind, and so forth. There is a horizontal limit to the spread of branches; they cannot extend further than the mechanical structure of the woody tissue will support. Thus the principal features of vegetation can be placed within limits that have very fundamental relationships to basic natural laws. If these are recognized as the outer limits, then the effect of additional factors and forces can be added that will indicate why these maximum physical limits are not reached. Such things as life form, strength, mechanical density, optical density, and energy and matter distribution can all be considered and are discussed later in this text.

Weather patterns vary daily in a given locality and are different between different localities. The complexity of wind, cloud density, and other characteristics of the atmosphere is great, but again there are limits that can be recognized for each of those weather parameters. Atmospheric interference reduces the solar radiation striking the earth's surface. Cloud density can be treated as a variable between a minimum of zero—as in a clear sky—and a higher density limit that represents maximum cloud cover such as heavy fog at ground surface. Thus a principal weather character called solar radiation can be described in terms of maximums and minimums that are dependent in part on atmospheric conditions. Thermal radiation can also be treated in a similar way, as can wind, vapor pressure, and temperature. Their patterns are somewhat regular and in each case can be represented by theoretical or observed limits, permitting one to test the ecological effect of variation in each by using a range of values for each parameter within the limits established.

Organisms themselves are principal characters in the ecological theatre, but the analytical ecologist cannot handle all of them in a beginning analysis. Theatrical productions frequently contain a small number of actors and actresses, each of whom represents something larger than himself. Analytical ecology is best approached by using actors too. Each deer considered represents more than an individual animal of a given species; of greater concern is its relationship to the factors and forces in this ecological theatre. It should be related to the forces of topography, water, weather, and food, with an understanding of how it relates to these forces in a simple way before the plot becomes more complex. Further, if fundamental characteristics are considered, the same approach and the same principles apply to any organism that might be considered, permitting an ecologist to apply the principles in his own particular area of interest.

Once the role of a single principal character is understood, two characters might be considered. These could be members of the same species—a doe and a fawn could be related analytically—or they could be of different species. A deer and a wolf, for example, could be considered together with an analysis of their chances for contact, energy flow, energetic efficiencies, or a host of other fundamental considerations. In this situation, the deer represents primary consumers, or animals that feed primarily on vegetation. The wolf represents secondary consumers, or animals that feed on other animals. The basic characteristics of energy and matter are applicable to all species, and the progression of the analysis from the simple to the complex may proceed from consideration of a single animal to two, four, and then larger populations, to communities, and finally to the entire ecological organization.

A frequent approach in ecology has been from the entire ecological organization first, but this presents many difficulties because the dynamic relationships between an organism and its environment are lost in the complexity of the total picture. The result is that students have learned only generalities, "rules," and similar sorts of conclusions that are devoid of the drama associated with the life of the individual. Yet each organism lives or dies as an individual, and its productivity is uniquely its own—a function of the effects of the independent, the compensating, and the additive ecological forces present in its environment. Students will find it exciting to build these analyses from the component parts (the principal characters that they choose), working from the simple "one-act play" to the complex n-dimensional analysis of populations, communities, and systems. The phrase "building the ecological model from the inside out" illustrates the reverse approach taken.

IDEAS FOR CONSIDERATION

Can you grasp the idea of differences between knowledge in empirical, factual form and an understanding of the role of energy and matter, in the form of individual organisms in a dynamic ecosystem?

Eggs and milk are different forms of matter. Ecologically, however, their functions are similar. Both require a female's metabolic activity for synthesis, and both are a part of the perpetuation of a species. What other functions can you identify that are taxonomically different but ecologically similar?

What limits can you describe for various components of the ecosystem? Can you identify the maximum and minimum due to physical laws, and then identify more realistic ecological limits?

What basic, functional similarities do you find between plants and animals? What basic differences are there, in terms of energy and matter utilization and synthesis? How different are primary consumers (herbivores) and secondary consumers (carnivores)? Are these differences more characteristic of variation in habit rather than of basic functions? What about their roles in energy flow? Are you prepared to recognize the role of each as an individual and yet relate each to a position on a trophic level?

Why not assemble life-history data for species of direct interest to you to become acquainted with their "personalities and life styles"? Rather than being simply the extent of your knowledge of them, this information can be used in later analyses.

INTERACTIONS BETWEEN ORGANISMS AND ENVIRONMENT

2-1 FUNCTIONAL RELATIONSHIPS

An understanding of the relationships between an organism and its environment can be attained only when the environmental factors that can be experienced by the organism are considered. This is difficult because it is first necessary for the ecologist to have some knowledge of the neurological and physiological detection abilities of the organism. Sound, for example, should be measured with an instrument that responds to sound energy in the same way that the organism being studied does. Snow depths should be measured in a manner that reflects their effect on the animal. If six inches of snow has no more effect on an animal than three inches, a distinction between the two depths is meaningless. Six inches is not twice three inches in terms of its effect on the animal!

Lower animals differ from man in their response to environmental stimuli. Color vision, for example, is characteristic of man, monkeys, apes, most birds, some domesticated animals, squirrels, and, undoubtedly, others. Deer and other wild ungulates probably detect only shades of grey. Until definite data are obtained on the nature of color vision in an animal, any measurement based on color distinctions could be misleading.

Infrared energy given off by any object warmer than absolute zero ($-273\,°C$) is detected by thermal receptors on some animals. Ticks are sensitive to infrared radiation, and pit vipers detect warm prey with thermal receptors located on the anterior dorsal portion of the skull. Man can detect different levels of infrared radiation with receptors on the skin, but they are not directional nor are they as sensitive as those of ticks and vipers. Thus we must conclude that the environ-

16

ment of lower animals is probably very much different from that of man and is dependent on their own capabilities for detection of environmental stimuli.

This leads to a very basic question: What are the neurological and physiological capabilities of an animal for detecting environmental factors? Can it see? What wavelengths can it distinguish, or is it seeing only varying shades of grey? What is its hearing range? Does the range change with age? What odors can it detect? What are the minimum or threshold levels necessary before detection of a stimulus can occur? What stimuli are not detected neurologically, but affect the animal physiologically?

There is a dearth of information on the sensory and physiological characteristics of wild species. Burton (1970) has published a general summary of the literature on animal senses, including references to several of the more refined sensory capabilities of some species. Information on the more common larger animals of North America is virtually nonexistent. We can conclude from field observation that deer, for example, do have a more sensitive sense of smell than do humans, and they seem to hear better too. Technical capabilities available to the researcher today make it possible to go beyond this superficial level of knowledge of the sensory capabilities of an organism, however. Sensors are available for the measurement of a multiple of physical and chemical environmental characteristics. The value of such sensors depends on how closely the measurement approximates the detection capabilities of the organism being studied.

Techniques are also available for the study of the physiological characteristics of free-ranging species. What are the heart rates of free-ranging animals? What are the red blood cell counts, and how do they vary diurnally? What relationship exists between the number of red blood cells and the ability of the blood to supply oxygen to body tissues? What is the relationship between the heart rate, red and white blood cell counts, oxygen consumption, and the animal's ability to withstand infection by parasites, endure cold weather, escape when pursued by a predator, or cope with other environmental forces?

Consider nuclear radiation, an environmental component that is detected physiologically but not neurally. What dosage can the animal withstand before death occurs, and what is a lethal dose for 50% of the population? What effect does a sublethal dose have on the reproductive capacity of the individual or of a whole population? on its ability to survive predation? What effects do herbicides and other organic poisons have on the survival and production of nontarget species? Do they affect the survival of the individual? suppress reproduction? or both? Do they affect the social hierarchy and territorial behavior of wild animals, thus reducing reproduction even without a change in the physiological condition of an animal?

These are important ecological questions. The effect of organism-environment relationships is not merely one of life or death, but variability in the organism's response to environmental forces. The effect of this variability is reflected in productivity, and variation in productivity results in differences in population levels and the distribution of energy and matter in the entire ecosystem. The excitement of an analytical approach to the study of ecology lies in beginning

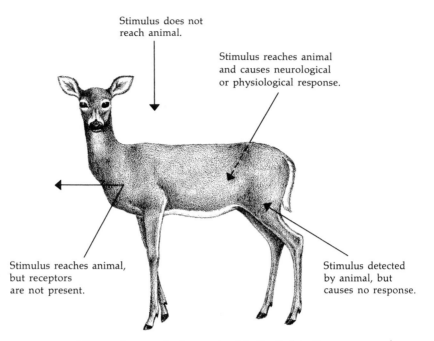

Stimulus does not
reach animal.

Stimulus reaches animal
and causes neurological
or physiological response.

Stimulus reaches animal,
but receptors
are not present.

Stimulus detected
by animal, but
causes no response.

FIGURE 2-1.   The environment of an animal is limited to the energy and
matter that has a neurological or physiological effect on it.

to understand how these things relate rather than in merely seeing the effects
of all these interrelationships buried in a mass of numbers representing the
presence of *n* number of organisms in a particular place.

Meaningful relationships between organism and environment exist only when
a stimulus reaches the organism and has either a neurological or a physiological
effect on it. Environmental forces or stimuli that do not reach the organism or
are not detected by it cannot be considered part of a *functional* organism-environment relationship. All possible organism-environment relationships fall within
four groups (Figure 2-1):

1. Stimulus does not reach the organism. There is no functional relationship.

2. Stimulus reaches the organism but is not detected. Again there is no functional relationship.

3. Stimulus reaches the organism and is detected by the organism, but causes
no response aside from detection. This relationship is of little importance to
either the organism or the ecologist.

4. Stimulus reaches the organism, is detected neurologically or physiologically,
and the organism responds to the stimulus. This is the kind of relationship
that the ecologist must consider.

It must be recognized that there are differences in the type and amount of
stimuli received by any two organisms at a given point in time, simply because
no two organisms can occupy the same space at the same time. Further, no two

organisms will have identical neurological or physiological thresholds at which stimuli are detected and responses made. The individuality of the relationships between an animal and its environment is illustrated in Figure 2-2. Certain conditions exist in the environment: the wind blows from a certain direction, locations of trees are plotted, and thresholds for detection of stimuli are assumed. Note in Figure 2-2A that the visual relationships are dependent on the location of objects in the habitat. If the woodlot located to the right of the animal is so dense that it is essentially an opaque wall, the visual relationship will extend up to it and no further. To the left of the animal, vision is limited by a single shrub

FIGURE 2-2. The environment of each animal is different, depending on its own receptor capabilities and on the distribution of energy and matter in the environment.

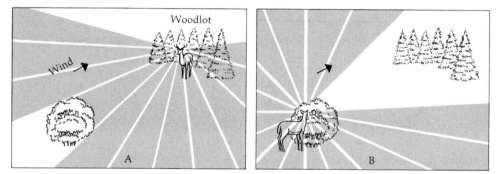

Visual relationships

Sound relationships

Scent relationships

at certain angles but extends for greater distances where there are no obstructions, as indicated by the dots following the lines of sight (__ . . .). The shape of the visual environment in Figure 2-2B is quite different from that in Figure 2-2A. Note that the animal in each of the diagrams would be visually excluded from the other. The indicated wind direction has no effect on these visual relationships.

The patterns of the sound relationships in the environment are considerably different from the patterns of visual relationships (Figure 2-2C). Sound penetrates the woodlot, and the shape of the ellipse representing the sound relationships in 2-2C is partially dependent on the direction of the wind, as sound energy travels slightly further downwind than upwind. If the wind is blowing strongly, the noise of the rustling leaves on the trees will be louder than many other sounds. Thus on a still day the sound pattern for an animal in a given place is more circular and larger in size than on a windy day, assuming that the loudness of the sound is similar in both instances. Note that the animal in Figure 2-2C can detect an animal next to the shrub, but the reverse is not true in Figure 2-2D.

The scent relationships in the environment have a shape that is highly dependent on wind direction (Figure 2-2E). Objects that can be smelled are detected at quite a distance if the source of the odor is downwind. The actual distance depends on the concentration of the detected molecules. This is a function of the strength of the source of the odor and of the velocity and turbulence of the wind.

A comparison of the patterns of the scent component in the environmental relationships in Figure 2-2E and F indicates the effect of wind on the distribution of scent molecules. The animal in Figure 2-2E could detect the animal in Figure 2-2F, but the latter could not detect the former. If the wind were to switch directions, however, a reversal in the pattern of scent relationships would emerge. The animal in Figure 2-2F could detect the one in the position shown in Figure 2-2E, but the reverse would not be true. Thus the scent relationships in the environments of these animals would change without any change in the animal's physiology or behavior.

Analysis of the response of an animal to taste is difficult because of the subjective nature of the response. A common approach is to record the responses to a variety of chemicals. This was done for black-tailed deer; they were presented with salty, sour, sweet, and bitter compounds in water solutions (Crawford and Church, 1971). Their consumption of the experimental solution was compared with their consumption of tap water. These deer, whose ages ranged from 5 to 14 months, showed preferences for some salty solutions (sodium acetate), but not for NaCl. They showed a preference for acetic acid solutions, but not butyric or HCl solutions. Strong preferences were demonstrated for sweet solutions. Males and females responded differently to bitter solutions, but not to the others. These preferential responses, not only between types of taste sensations but between different compound solutions within a single type, demonstrates the complex nature of the taste response.

The sense of taste in birds has been reviewed by Welty (1962). He concludes that it is poorly developed, pointing out that the number of taste buds is low and that they are distributed at the sides and base of the tongue rather than at the tip.

The various senses are all participating in the process of communication between an organism and its environment. Some environmental stimuli are not detected by the senses, but have an effect on physiological processes. Nuclear radiation is an example; the effect of radiation is not "felt" neurologically, but the physiological functions of certain tissues may be effected.

These examples illustrate that *it is the neurological and physiological capabilities of the organism itself that link it to its environment*. It is necessary to think of the environment in terms of its functional relationship to the organism rather than merely as the physical or geographical area in which it lives, if the dynamic relationships between an organism and its environment are to be understood.

The necessity for consideration of the conceptual basis upon which ecologists analyze organism-environment relationships has been heightened by the rapid increase in the capabilities for studying free-ranging animals by radio telemetry techniques. No longer must the field biologist rely only on visual observation of an animal, or on indirect observation of animal activity by tracks, pellets, or other signs. The location of an animal and information about its activity and physiology can be transmitted continuously from the animal to a receiving station.

2-2 THE SCOPE OF THE ENVIRONMENT

OPERATIONAL, POTENTIAL, AND HISTORICAL RELATIONSHIPS. In order to understand what causes an animal to "do" something in a behavioral, psychological, or physiological sense, it is necessary to recognize the stimuli that are in operation at a given time (Mason and Langenheim 1957). If we require only that a stimulus have an effect on the organism, the examples of stimuli that cause the responses given earlier in this chapter fall into the *operational* category, and the habitat components that are not detected by the animal or that do not influence his behavior or physiology in any way are naturally excluded from the environmental relationships. The removal of these habitat components will have no effect on the organism, and the propagation or perpetuation of these nonoperational features is wasted effort.

In addition to the operational components of the environment, there are habitat components that *could become* operational and those that *have been* operational. The former can be called *potential* and the latter *historical* environmental relationships. Thus the very important dimension of time is included in the potential and historical categories, and the various stimuli that elicit a response are recognized at a given point in time.

Several examples of the potential, operational, and historical relationships will illustrate the logic of these categories. Suppose that a deer was standing in an open field exposed to a strong wind on a cold winter day. If a shrub was located 50 yards away from the animal and at right angles to the wind direction, the shrub would not have an effect on the thermal balance of the deer (Figure 2-3A). The deer, however, has the ability to traverse the 50 yards and use the shrub

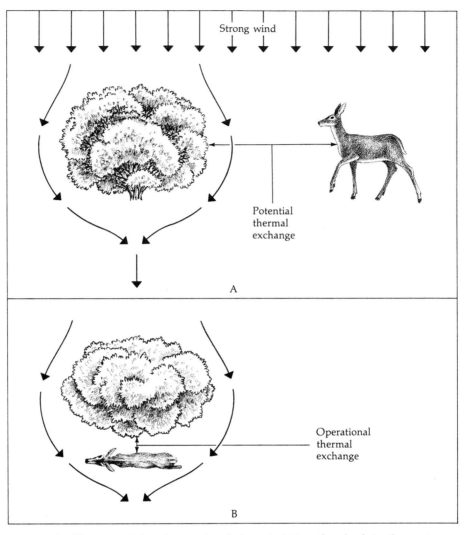

FIGURE 2-3.   The potential and operational characteristics of a shrub in the environment of a deer with respect to thermal exchange.

for protection from the wind. As long as the deer is some distance from the shrub, the shrub can only have a potential effect with respect to air flow and thermal exchange. If the deer moves behind the shrub, placing itself in a more favorable thermal-exchange situation, the shrub is no longer in a potential relationship with the deer but is operational (Figure 2-3B).

Or suppose a male elk has bedded down during the rut, and another male elk is bugling at a distance too great for the sound to be detected. The bugling male elk has only a potential relationship with the resting male; as far as the latter is concerned, the challenger does not even exist (Figure 2-4A).

If the bugling elk moves close enough so the sound can reach the second elk, the latter may, if he is in breeding condition, respond to the challenge. The sound of bugling then takes on a definite operational characteristic (Figure 2-4B). The subdominant elk, determined by the comparative aggressive behavior of the two bulls, then has an encounter in his memory that is a historical relationship; he does not challenge the other bull for the dominant position and control over the harem (Figure 2-4C).

There is a wide range of results that might be produced by the operational relationships between an animal and its environment described in the preceding paragraphs. Movement of a deer to the lee side of a shrub may have little or no effect on the thermal balance of the deer, or it could be critical to the maintenance of homeothermy. The resting male elk may pay little or no attention to the other male, or he may respond very emphatically if in the proper breeding condition.

Aside from genetically determined instinctive behavior that is not dependent on a learning experience by the animal, the historical relationships develop only

FIGURE 2-4. The (A) potential, (B) operational, and (C) historical characteristics of the environmental relationships between two male elk.

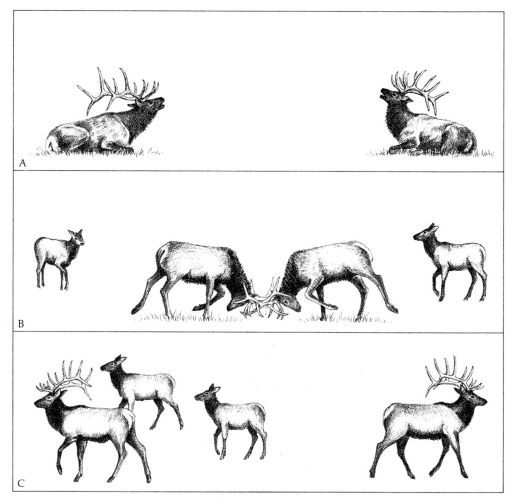

after operational relationships have occurred. At some time in the past, the deer may have experienced a reduction in heat loss when it moved to the lee side of a shrub. That same deer may then "recognize" the effectiveness of the shrub in reducing the velocity of the wind passing its body and could place itself again in the more beneficial situation. This requires some sort of a memory; the animal must have the capacity to be conditioned by a characteristic of the environment before there can be a "history."

The potential-operational-historical series becomes complicated very rapidly for living organisms. When the deer first sees the shrub, the shrub has an operational *visual* relationship with the deer as light waves are received by the retina and are interpreted by the brain. Until the deer moves to the protection of the shrub, however, it has only a potential *thermal* relationship. Thus an object in the environment can have an operational relationship for one type of stimulus-response combination and a potential relationship for another.

The behavior of the resting male elk will depend on its breeding condition. Whether the call of one male elk elicits a strong response in another depends on the hormone balance at that particular time. This is an internal thing, suggesting that in addition to an external environment there is an internal environment that includes stimuli that illicit a response. The role of hormones in determining an animal's response is an important part of the total animal-environment relationship. The ecologist must recognize these internal relationships in order to explain a large number of responses that are observed in the field. The hormone balance may be affected by external influences, such as light. Thus breeding activity might be described in terms of a day-length, hormone-balance, breeding sequence, with physical, physiological, and behavioral factors operating.

TIME.   The idea of time is very simple in a physical sense because of the regular pattern of spatial relationships between the sun and the earth that results in a calendar year and regular periods of day and night. The concept of biological time is considerably more complex. A biological definition of time might simply be that it is "a measure of the intensity of life." Thus hibernating animals may spend four or five months of the year in a very passive condition, and the metabolic requirements over the entire period of hibernation may be equivalent to only a few days of intensive living during the active period of the year. The existence of diurnal and seasonal variations in the physiological condition of an animal indicates that chronological time is inadequate to represent physiological or behavioral events because the organism is not regulated as precisely as the rotation and revolution of the earth.

The age of an animal is another factor affecting its sensory capabilities. The environmental relationships of the developing fetus in the uterus are mostly chemical, and the fetus responds physiologically to the components of its limited environment. Two of the most important processes include the exchange of oxygen and carbon dioxide and the transfer of nutrients across placental membranes. Neurological detection mechanisms of the fetus are quite undeveloped and some—for example, sight—are quite unnecessary.

At birth, the young animal is suddenly exposed to vastly different environmental relationships. Senses are needed, but some neurological capabilities may remain undeveloped for days and even weeks, such as sight in members of the cat family. The newborn bobcat lives in a world filled with scents, sounds, and the touch of physical objects and litter-mates, with no visual relationships for several days. Members of the deer family, on the other hand, are alert and active within a few hours after birth.

As the animal matures, its senses develop and along with them, historical relationships. Thus the old buck thoroughly familiar with its home range is less susceptible to mortality than the newborn fawn. An animal in its prime of life, with maximum neurological and physiological development, has the most complex environmental relationships that it will ever have.

As an animal grows older, the aging process produces a general loss of sensitivity and elasticity of body tissue. Thus hearing is reduced, eyesight becomes weaker, muscular tissue is less capable of sustained contraction, and bones are more brittle. As these changes take place, the capabilities for the detection of stimuli are reduced, along with the animal's ability to react to them. The animal is more subject to predation, mechanical injuries, and other decimating events. The physiological effects of certain environmental components are frequently more severe in an older animal. DDT, for example, accumulates in body tissue and may reach a critical level as the animal grows older.

In summary, the complexity of the environmental relationships depends on the complexity of the nervous system and physiological characteristics of the animal. This does not mean that more complex animals have a greater ability to detect each and every stimulus; man has sensory abilities significantly inferior to those of many lower animals. The total overall complexity of man's environmental relationships is greater because of a highly coordinated nervous system.

ASSOCIATED RELATIONSHIPS. It has been stated that each individual organism has a unique relationship with its environment at any point in time. This environment includes potential, operational, and historical components. The organism does not live by itself in an ecological vacuum, however, since there are many other organisms with their own specific environmental relationships that are associated with and necessary for its survival. Bacteria functioning as decomposers, for example, may not be detected neurologically by humans. They cannot be seen without magnification, yet they are essential for converting organic material to simpler organic substances and inorganic compounds that are cycled through the ecosystem. The end product of decomposition processes (humus) is detectable, and life as we know it on this planet would be impossible without these bacteria.

Consider also the rumen microorganisms in ruminants. The host animal is neurologically unaware of their presence. The microorganisms have no direct physiological link with the host animal, but they are essential to the host who is dependent on the end products of their metabolism. The host even derives some amino acids from their dead bodies as the microorganisms themselves are digested in the small intestine.

Thus there is an ecological complex of environmental relationships, or associations of organism-environment relationships that may be independent, dependent, compensatory, additive, symbiotic, or in some way related. In simple terms, the environment of any organism includes other members of the biota and the functions of these associated relationships are of ecological significance.

HABITAT EVALUATION. A distinction should be made between habitat and environment. The environment has been discussed as a functional thing, something that relates to the organism under consideration. It can only be described in terms of the organism and includes the things that the organism can experience. The habitat, on the other hand, is simply the place in which an organism lives. It is the physical area inhabited by an organism and, on a larger scale, by the species. Thus we can talk about pheasant habitat, deer habitat, wolf habitat, and so forth, simply because those animals live there.

The habitat of an organism includes the organism-environment relationships of all species present there. Thus an area can be both pheasant habitat and fox habitat at the same time, but the environments of these two species are quite different. Indeed, the environments of two individuals of the same species are different.

An evaluation of habitat requires the recognition of those things that relate to the organism under consideration. What is present that the organism responds to? What kinds of responses are made? Does the environmental factor increase productivity? decrease productivity? Does it affect overt behavior? Does it affect some subtle physiological characteristic? Let us examine several characteristics of a habitat and consider how they might relate to the environment of an organism.

Optical density is a measure of the penetration of light through cover. There are absolute limits to this penetration, depending on the degree of transparency or opacity of the material. Vegetative cover ranges between complete transparency (no overhead cover present) and complete visual opaqueness. There is vertical density (Figure 2-5), which can be described in terms of the percentage of the sky that is visible directly overhead. At angles less than 90°, there is a decrease in the transparency and the optical density may reach a maximum even in cover that is quite sparse. This "venetian blind" effect is of interest both optically and thermally.

How does the optical density relate to an animal living in that vegetative cover? If the animal is subject to predators that hunt by sight, then optical density is of particular importance. Optical density is significant to both predator and prey, however, and what the predator can see and what the prey can see are equally important in an analysis of predator-prey relationships. The significance of these visual characteristics is dependent on the escape mechanisms of the prey. If an animal relies on the sighting of a predator as a stimulus for escape behavior, complete transparency might be the best cover for this particular relationship. Some species of ducks, for example, select nest sites in very open places, and may depend on their ability to see to warn them of the presence of predators.

Deer frequently bed down on or near the tops of hills, permitting them to see for some distance. They also bed frequently in small thickets, and this may be advantageous because it is much easier to see out from thickets than it is to see an animal in a thicket, especially if its color blends in well too.

The *mechanical density* of cover can be described in terms of the percent volume occupied by mass of a particular type, but this information is meaningless without an identification of the size and mobility of the organism that might be using the cover. Viruses can go through extremely small holes, penetrating material that appears to be opaque and solid to the human eye. Pheasants can travel easily through the lower few inches of vegetation. Deer can travel through brush that stops many a hunter, even though they may weigh about the same (say, 150 pounds). This brings to mind the importance of posture. A deer walking on all four limbs has a different posture than an upright man, and the difference in the distribution of the weight on the deer skeleton permits it to be much more mobile in dense brush. This clearly indicates that mechanical density can be described in a meaningful way only when the characteristics of the organism are also described.

Thermal density has characteristics in common with optical density. The downward flux of radiation from a plant canopy is less from the zenith than from angles less than $90°$ when the sky is clear. If the sky is cloudy, the environment of an organism is thermally uniform, and the thermal-density effect of vegetative cover is masked by the thermal-density effect of the cloud cover. This indicates that the description of a habitat characteristic may be dependent on the presence or absence of another habitat characteristic with which it may interact. This is true

FIGURE 2-5.   The optical and vertical density of vegetative cover.

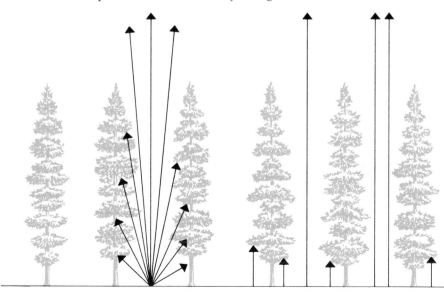

Optical density   Vertical density

for solar and infrared interaction with reference to thermal density too; physical and biological materials have different reflection, transmission, and absorption coefficients for different wavelengths, and these affect the interaction between organism and environment. This is discussed further in Chapters 6 and 13.

The taxonomic structure of a plant community has often been used to describe that community. Thus species lists, abundance lists, frequency distributions, and the like have been used to characterize a plant community. This may be adequate from a humanistic, systematic point of view, but what about the other organisms living in that community? If two plant species have the same chemical characteristics that provide the same nutritive value to an animal, then these two different species are really the same in the nutritive relationships for the animal in question. Perhaps the life form of two different species is very much the same—different maples, for example, look very much alike in general life form—and the optical density, mechanical density, or some other factor that is experienced by an animal may indicate no difference between the different species. In that case, the environment of the animal is much simpler than the taxonomist might make it. Similar plant species may provide the same kind of mechanical support for nest sites. In that case the separation of species would be superfluous. This is discussed further in Chapter 15.

Consider also the source—internal or external—of stimuli that elicit responses from organisms. It may be difficult to separate internal from external stimuli. If a parasite is in the mouth of an animal, it is internal when the mouth is closed and external when the mouth is open. Or consider the intestinal parasite—it really is on a surface continuous with the outside of the animal; and although it might be inside something, it is not in a closed anatomical position, as is a blood parasite. Then too, it is possible for the organ systems to be closed to some organisms, but open to others. Viruses, for example, pass through cell membranes that are impervious to larger organisms, just as a deer exclosure may be closed to deer but open to snowshoe hares.

The role of any organism or species in an ecosystem cannot be categorized in simple terms with respect to other organisms, but can only be described as a functional part of different ecological processes. This philosophy is necessary for the analytical ecologist who recognizes ecological relationships rather than the mere presence of organisms in a natural environment. It is a unique philosophy that at times ignores taxonomic lines, and at other times it may further define the role of particular members of a species as interactions between other species are recognized.

Similarities in function between different species of plants or animals suggest the idea of "ecological equivalents." Taxonomically they might be different, but ecologically they might be similar. This is not true for their every ecological function, however. Different species are ecological equivalents only for particular functions, and it is necessary to think of individual organisms and species as fluid, labile entities that function in different ways in time and space.

The next chapter includes a discussion of the use of models in ecological analyses. It is a simple introduction to model building, with more detail included in the various models that follow in later chapters. The important point to remember is that the simple models are centered on the organism in relation to the environment in a functional way.

LITERATURE CITED IN CHAPTER 2

Burton, R. 1970. *Animal senses*. New York: Taplinger, 183 pp.
Crawford, J. C., and D. C. Church. 1971. Response of black-tailed deer to various chemical taste stimuli. *J. Wildlife Management* **35**(2): 210–215.
Mason, H. L., and J. H. Langenheim. 1957. Language analysis and the concept environment. *Ecology* **38**(2): 325–340.
Welty, J. C. 1962. *The life of birds*. Philadelphia: Saunders, 546 pp.

IDEAS FOR CONSIDERATION

The following books and articles contain information about the relationships between organisms and their environments. Evaluate the data in these papers, assembling some thoughts and examples of specific organism-environment interactions based on the strength of the stimuli and the responses of the organisms.

SELECTED REFERENCES

Amoore, J. E. 1970. *Molecular basis of odor*. Springfield, Illinois: Charles C Thomas, 200 pp.
Arden, G. B., and P. H. Silver. 1962. Visual thresholds and spectral sensitivities of the grey squirrel (*Sciurus carolinensis leucotis*). *J. Physiol.* **163**(3): 540–557.
Bell, F. R. 1959. Preference thresholds for taste discrimination in goats. *J. Agr. Sci.* **52**(1): 125–128.
Bell, F. R. 1959. The sense of taste in domesticated animals. *Vet. Record* **71**(45): 1071–1079.
Bell, F. R., and H. L. Williams. 1959. Threshold values for taste in monozygotic twin calves. *Nature* **183**(4657): 345–346.
Bishop, L. G., and L. Stark. 1965. Pupillary response of the screech owl, *Otus asio*. *Science* **148**(3678): 1750–1752.
Brindley, L. D., and S. Prior. 1968. Effects of age on taste discrimination in the bobwhite quail. *Animal Behaviour* **16**(2/3): 304–307.
Brower, L. P., W. N. Ryerson, L. L. Coppinger, and S. C. Glazier. 1968. Ecological chemistry and the palatability spectrum. *Science* **161**(3848): 1349–1350.
Browning, T. O. 1962. The environments of animals and plants. *J. Theoret. Biol.* **2**(1): 63–68.
Bruce, H. M. 1966. Smell as an exteroceptive factor. *J. Animal Sci.* **25**(supplement): 83–87.
Busnel, R. G., ed. 1963. *Acoustic behavior of animals*. New York: Elsevier, 933 pp.
Case, J. 1966. *Sensory mechanisms*. New York: Macmillan, 113 pp.
Crescitelli, F., B. W. Wilson, and A. L. Lilyblade. 1964. The visual pigments of birds. I. The turkey. *Vision Res.* **4**(5/6): 275–280.
Crescitelli, F., and J. D. Pollack. 1965. Color vision in the antelope ground squirrel. *Science* **150**(3701): 1316–1318.
Davis, L. I. 1964. Biological acoustics and the use of the sound spectrograph. *Southwestern Naturalist* **9**(3): 118–145.
Evans, W. F. 1968. *Communication in the animal world*. New York: Thomas Y. Crowell, 182 pp.

Frank, M., and C. Pfaffmann. 1969. Taste nerve fibers: a random distribution of sensitivities of four tastes. *Science* **164**(3884): 1183–1185.

Friedman, H. 1967. Colour vision in the Virginia opossum. *Nature* **213**(5078): 835–836.

Frings, H., and B. Slocum. 1958. Hearing ranges for several species of birds. *Auk* **75**(1): 99–100.

Frings, H., and M. Frings. 1964. *Animal communication.* New York: Blaisdell, 204 pp.

Goatcher, W. D., and D. C. Church. 1970. Taste responses in ruminants. I. Reactions of sheep to sugars, saccharine, ethanol and salts. *J. Animal Sci.* **30**(5): 777–783.

Goatcher, W. D., and D. C. Church. 1970. Taste responses in ruminants. II. Reactions of sheep to acids, quinine, urea and sodium hydroxide. *J. Animal Sci.* **30**(5): 784–790.

Goatcher, W. D., and D. C. Church. 1970. Taste responses in ruminants. III. Reactions of pygmy goats, normal goats, sheep and cattle to sucrose and sodium chloride. *J. Animal Sci.* **31**(2): 364–372.

Goatcher, W. D., and D. C. Church. 1970. Taste responses in ruminants. IV. Reactions of pygmy goats, normal goats, sheep and cattle to acetic acid and quinine hydrochloride. *J. Animal Sci.* **31**(2): 373–382.

Gottlieb, G. 1965. Components of recognition in ducklings. Auditory cues help develop familial bond. *Nat. Hist.* **74**(2): 12–19.

Hocking, B., and B. L. Mitchell. 1961. Owl vision. *Ibis* **103a**(2): 284–288.

Johnston, J. W., Jr., D. G. Moulton, and A. Turk, eds. 1970. *Communication by chemical signals: advances in chemoreception.* Vol. 1. New York: Appleton-Century-Crofts, 412 pp.

Matthews, L. H., and M. Knight. 1963. *The senses of animals.* New York: Philosophical Library, 240 pp.

Miller, L. K., and L. Irving. 1967. Temperature-related nerve function in warm- and cold-climate muskrats. *Am. J. Physiol.* **213**(5): 1295–1298.

Milne, L., and M. J. Milne. 1962. *The sense of animals and men.* New York: Atheneum, 305 pp.

Moncrieff, R. W. 1967. *The chemical senses.* London: Hill, 760 pp.

Moulton, D. G. 1967. Olfaction in mammals. *Am. Zool.* **7**(3): 421–429.

Pribram, K. H. 1969. The neurophysiology of remembering. *Sci. Am.* **220**(1): 73–86 (Offprint No. 520). *Scientific American* Offprints listed here and for subsequent chapters are available from W. H. Freeman and Company, 660 Market Street, San Francisco 94104, and 58 Kings Road, Reading RG1 3AA, England. Please order by number.

Rahmann, H., M. Rahmann, and J. A. King. 1968. Comparative visual acuity (minimum separable) in five species and subspecies of deer mice (*Peromyscus*). *Physiol. Zool.* **41**(3): 298–312.

Ralls, K. 1967. Auditory sensitivity in mice: *Peromyscus* and *Mus musculus. Animal Behaviour* **15:** 123–128.

Schneider, D. 1969. Insect olfaction: deciphering system for chemical messages. *Science* **163**(3871): 1031–1037.

Smythe, R. H. 1961. *Animal vision: what animals see.* Springfield, Illinois: Charles C Thomas, 250 pp.

Stuart, P. A. 1955. An audibility curve for two ring-necked pheasants. *Ohio J. Sci.* **55**(2): 122–125.

Tansley, K. 1965. *Vision in vertebrates.* London: Chapman and Hall, 132 pp.

Tansley, K., R. M. Copenhaver, and R. D. Gunkel. 1961. Spectral sensitivity curves of diurnal squirrels. *Vision Res.* **1**(1/2): 154–165.

Thorpe, W. H. 1956. The language of birds. *Sci. Am.* **195**(4): 128–138 (Offprint No. 145).

Todd, J. H. 1971. The chemical languages of fishes. *Sci. Am.* **224**(5): 98–108 (Offprint No. 1222).

Tucker, D. 1965. Electrophysiological evidence for olfactory function in birds. *Nature* **207**(4992): 34–36.

Wilson, E. O., and W. H. Bossert. 1963. Chemical communication among animals. *Recent Progr. Hormone Res.* **19**: 673–716.

Zotterman, Y. 1953. Special senses: thermal receptors. *Ann. Rev. Physiol.* **15**: 357–372.

ECOLOGICAL MODELING AND
SIMULATION

Models have been in use in resource analyses as long as man has made attempts to understand, modify, or control part of the ecosystem. The models have often been only mental models, or they may have been written in the form of a project outline or work plan. Models have been increasing in complexity as more biological knowledge is accumulated. Present capabilities for rapid and accurate computations with electronic computers make it possible to build very complex models that include the analyses of many relationships between organism and environment.

How do we begin developing a model? The model starts with the formation of an idea. This conceptual model can be explained verbally to others, and can be described in prose form, outline form, or in a schematic form that illustrates the main components of the model being described in writing. This verbal model can become quantitative by representing the components of the model by numerical quantities. Thus there has been a progression from a conceptual to a verbal to a quantitative model.

This progression in no way guarantees that the model is of any value. The idea might have been poor, words might not adequately express the ideas, and the numbers may be poor representations of the actual strength of the relationships being considered. Its value is dependent on the level of realism associated with the model and the part of the real world that it represents.

The sequence of events in the building of a working model of biological

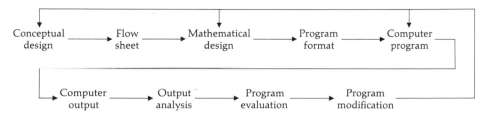

FIGURE 3-1. The sequence of events in the construction of a biological model for computer analysis.

relationships is illustrated in Figure 3-1. The conceptual design comes first; it is a representation of the thoughts or ideas of the model builder. The thoughts are then converted to words in the form of a flow sheet. A flow sheet is a schematic display of the relationships being considered in the model. Representation of these relationships by mathematical equations results in the formation of a mathematical design. The computer enters the picture when the mathematical design has been completed. The computer's main function is the rapid manipulation of the mathematical design, including storage of numerical quantities, execution of mathematical functions, and decision-making with respect to numerical values. The model builder must, of course, use the program format or language that his own particular computer requires.

The computer outputs can be analyzed to see how well they seem to represent the biological investigations under consideration. The program is then evaluated and any necessary modifications made in the program, in the mathematical design, or in the conceptual design.

Corrections in the computer program alone are simply mechanical. They are necessary when a mistake has been made in writing the computer program itself and are analogous to a spelling error. Corrections in the mathematical design or the conceptual design may result in a new model. If an orderly sequence of model building is observed, the new model will usually be larger than the previous one. This is often by intent; the first model that is developed must be within the scope of understanding—perhaps even a little below the full capabilities of the model builder. The use of the simpler models first for generating conclusions will make subsequent models more realistic.

3-1 MATHEMATICAL MODELS

An understanding of the basic characteristics of mathematical models is necessary before their usefulness can be appreciated in resource analyses and management. Models consist of two basic components, including (1) the factors or forces that constitute the model and (2) the relationships between these factors and forces.

A simple numerical example illustrates the basic structure of a model as follows:

$$[(10_1 \times 3) - 6 + 7] \times 10_2 - 50 = 260 \qquad (3\text{-}1)$$

The numbers represent the factors themselves, such as the number of animals, the amount of forage available, or some other biological quantity. The mathematical signs ($\times$, $-$, $+$) represent the relationships between these quantities. Some are multiplicative; others are additive, either positive or negative. The parenthesis and brackets indicate the mathematical order that must be followed in arriving at an answer. The final output (260) represents the cumulative relationship between all factors in the model.

The model represented by equation (3-1) is a very simple one. Complex models can be built only after simpler ones have been assembled and tested, however. The increase in the size of models follows a pattern, as illustrated in Figure 3-2. The number of meaningful conclusions that can be reached is often directly related to the size of the model, assuming that the model contains good inputs that are treated in a biologically reasonable manner. The amount of waste may be quite independent of the size of the model.

The outputs of models that represent biological knowledge are variable owing to both natural biological variation and to error in measurement. The significance of the effects of either natural variations or error in a whole biological system is dependent on the characteristics of the factor(s) itself and on its relationships to other factors. Variations in the numerical factors in equation (3-1) will illustrate this characteristic of models.

Suppose that the factors could vary from their full value shown to one-half of their value shown. If the number 10_1 is divided by 2, the final answer is 110 [equation (3-2)] rather than 260 [equation (3-1)].

FIGURE 3-2. The growth of models, illustrating the larger model size, the increase in conclusions, and the continual output of "waste" information.

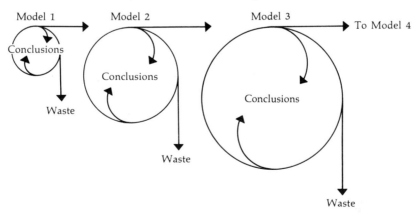

TABLE 3-1 THE EFFECT OF A 50% REDUCTION IN EACH OF THE ORIGINAL
VALUES IN THE MATHEMATICAL MODEL ILLUSTRATED
IN EQUATION (3-1)

Original Value	Original Value ÷2	New Answer	% of Original Answer
3	1.5	110	42
6	3	290	112
7	3.5	225	87
10_2	5	105	40
50	25	285	110

$$\left[\left(\frac{10_1}{2} \times 3\right) - 6 + 7\right] \times 10_2 - 50 = 110 \qquad (3\text{-}2)$$

The new value is 42% of the initial answer. The division of each of the original number components by two results in new outputs ranging from 105 to 290, or 40% to 112% of the original output (Table 3-1).

Natural variations or measurement errors can occur in several factors at one time, however. The end result may be similar to the original, reflecting a compensatory effect, or it may be different owing to an additive or multiplicative effect. If 6/2 and 7/2 are substituted for 6 and 7, respectively, in equation (3-1), the result is 255, or 98% of the original. The two errors compensated for each other. If $10_1/2$ and $10_2/2$ replace the two original values of 10, the new output is 30, or 11% of the original. These errors are multiplicative, and the final output is very different from the original. In both of these cases a "constant" change was introduced; each of the two original values was divided by the same number, that is, 2.

The illustrations clearly show that the importance of any one value is dependent on its relationship to the other values. This is true mathematically and is also true biologically as the different factors and forces act together in a "web of life" that is dynamic and exceedingly complex.

The previous illustration of the use of models to represent ecological relationships cannot fully convey the excitement of computer analyses of these relationships. It is something that must be experienced to be appreciated. The rapid increase in the availability of electronic computing systems ranging in size from desk-top models smaller than a typewriter to central computers linked to time-sharing terminals in remote locations now puts this kind of analysis within the reach of most people in the natural resources field.

It is important to remember, however, that there is no magic in computer analyses. The value of the computer is illustrated by its use in the space program. The basic knowledge necessary for flights to the moon has been available for many years. The computer plays a vital role in the use of this knowledge through the execution of mathematical expressions. It does so with the speed necessary in the decision-making process that is so critical in space flights. The computer is a tool that must be used judiciously in the analyses of different physical or biological relationships. It is essential for the realistic analyses of complex relationships currently under consideration.

3-2   THE ANALYTICAL MODEL IN ECOLOGY

The development of a model that contains ecological dimensions is somewhat more difficult than the development of a mathematical model with numbers that are dimensionless and nonrepresentative of real relationships. Let us proceed to an example illustrating the development of a more realistic model that is both biological and mathematical in content.

Suppose that you are interested in the growth rate of an animal. You might know from past experience that this animal weighs 5 grams at birth and reaches a maximum weight of 105 grams in 50 days. What are the components of that biological relationship that will be of value in the development of a growth model? The two parameters present are weight and time. Given this information, you can assemble a first model that will result in the prediction of weight at any time during the growth period. The information given limits you at this point to the use of a simple model such as linear regression with the components Y, a, b, and X. Biologically, it can be said that $Y = f(a, b, X)$, where Y is the weight at any point in time X, a is the birth weight, and b is the change in weight per day. The numerical equation can be calculated from the above information as follows: $105 = (5 + b50)$, which can be rewritten as $b = (105 - 5) \div 50$, which results in $b = 2$. Thus there is a growth rate of 2 grams per day, and the equation $Y = 5 + 2X$ can be used to calculate the weight at any given day.

The above model is a descriptive one, developed after certain known quantities are available. These quantities are empirical, and the predictive validity of the model is dependent on the similarity between the conditions during the measurement of these quantities and the conditions imposed on an animal whose growth rate is being predicted or assumed from this first model.

Program evaluation and review may result in the conclusion that the model might not be particularly good because time is not the *cause* of growth, but is simply a necessary constituent of the growth process. Other factors, such as metabolic rate, energy availability, protein, vitamins, minerals, activity, genetic characteristics, hormone balance, and others are related to the cause of growth. A second model, more useful for predicting the growth pattern of the animal, might take into consideration the rate of energy metabolism for the synthesis of body tissue. It is obvious that this requires more biological information for a realistic model.

What basic laws of energy and matter are useful in outlining a second model that becomes more realistic as a representative of the growth of the animal? First of all, the body of the animal contains a certain amount of energy and matter at birth and a certain amount at the cessation of growth. These can be measured, and the difference is the net increase in energy or matter over a period of time. Knowledge of the metabolic rate, or rate of energy metabolism during this period of time, permits the determination of the growth rate on an energy base, and ecological conditions that affect the rate of energy metabolism for productive purposes can then be considered in further predictions. In other words, the model is not restricted just to the effect of all metabolic processes through time, but alterations in the metabolic processes that would affect the growth rate can be taken into account.

The same procedure can be followed for protein, minerals, and productive processes such as egg production, milk production, and hair growth. The complexity of such models increases very rapidly, but at this point the student may find it more imposing than it should be because of insufficient supporting information. The story will become more complete in later chapters as the models illustrated become more complex in terms of the number of considerations. They are, however, simple in a way if the function of each component part is described in accord with some physical law relating it to the transformation of matter and energy.

The fact that no parameter can be considered in isolation in ecosystems suggests that tests of statistical significance are meaningful only when considered in the ecosystem context. The importance of variation in any one parameter cannot be determined until the effect of that parameter is considered in relation to other components of the ecosystem that it affects. Differences in snow depth can be used to illustrate this very important concept.

Suppose that snow accumulations on two different areas were compared and found to be statistically different, using a test of the slope of a regression line representing accumulations over time. If this snow never exceeded 10 inches, it would pose no problem to the movement of moose through it and hence would have no ecological significance with respect to the restriction of moose activity. If the distribution of forage was such that none of it, or so little of it, was located in the first 10 inches of the plant community that there was no shortage of food, then the snow would have little or no ecological significance.

The possibility exists that such snow depths would be important for the winter survival of *Microtus,* however. If the temperature was very low, the greater snow depths might be important in maintaining the thermal regime of the mouse. Suppose the critical snow depth for insulation purposes was 8 inches; then the area having that depth would be distinctly better than the other area having less accumulation. The statistical significance of differences in snow accumulations becomes important only after a depth of 8 inches is reached in this particular example. If the temperature never drops much below 0°C, there might be little effect due to differences in snow depths because the critical snow depth for insulation purposes would be small. Thus is can be seen that the importance of differences in snow accumulation is related to a larger number of factors ecologically. Deep snow may be good for *Microtus,* a problem for deer after depths of 20 inches are reached, but irrelevant for moose until 30 or more inches are on the ground.

It is important to understand how any parameter under consideration functions before deciding that variation within or between parameters is or is not important. This requires a simultaneous *n*-dimensional type of thinking that is a necessary component of the thought processes of an analytical ecologist.

IDEAS FOR CONSIDERATION

Consider any biological process. Identify all of the factors and forces that could possibly be related to this process, following a format such as:

$$\left\{ \begin{array}{c} \text{Growth} \\ \text{rate} \end{array} \right\} = \left\{ \begin{array}{c} f(\text{energy metabolism, food availability,} \\ \text{protein quality, vitamins, activity, . . .)} \end{array} \right\}$$

Divide each of these further, for example:

$$\left\{ \begin{array}{c} \text{Protein} \\ \text{quality} \end{array} \right\} = \left\{ \begin{array}{c} f(\text{amino acid spectrum,} \\ \text{productive metabolic processes, . . .)} \end{array} \right\}$$

Are you bound up in a cyclic process of repetition and interrelationships? The system cannot be described in a linear fashion!

Develop simple predictive biological models using your present knowledge, with emphasis on the basic energy and matter relationships that regulate the biological parameters under consideration. Keep them workable!

SELECTED REFERENCES

Milsum, J. H. 1966. *Biological control systems analysis.* New York: McGraw-Hill, 466 pp.
Patten, B. C. 1966. Systems ecology: A course sequence in mathematical ecology. *BioScience* 16(9): 593–598.
Patten, B. C., ed. 1971. *Systems analysis and simulation in ecology.* Vol. 1. New York: Academic Press, 610 pp.
Pielou, E. C. 1969. *An introduction to mathematical ecology.* New York: Wiley-Interscience, 286 pp.

Van Dyne, G. M., ed. 1969. *The ecosystem concept in natural resource management.* New York: Academic Press, 383 pp.

Watt, K. E. F. 1968. *Ecology and resource management: a quantitative approach.* New York: McGraw-Hill, 450 pp.

Courtesy of Dwight A. Webster
Department of Natural Resources, Cornell University

PART

2

THE DISTRIBUTION OF MATTER
AND ENERGY
IN TIME AND SPACE

The ecosystem is composed of energy and matter, and these can be classified according to their physical, chemical, and thermal characteristics. There are physical entities such as soil, water, and air present. These are composed of various forms of energy, including potential energy, kinetic energy, chemical energy, and thermal energy.

Volumes have been devoted to the composition of the earth and its atmosphere from physical, chemical, and thermal viewpoints. The analytical ecologist is interested in relating these characteristics to organisms. These relationships are exceedingly complex. They cannot be described in their entirety; indeed, we know very little about the functional relationships between organism and environment, especially for the game birds and animals that are of interest to a significant segment of the public. On the other hand, there is a considerable amount of information available on the physical, chemical, and thermal characteristics of the biosphere, and the analytical ecologist will find it very useful to synthesize this type of information with the functions of an organism, either plant or animal.

The next three chapters describe some of the more basic physical, chemical, and thermal characteristics of the biosphere, providing some insights into the types of relationships that can be analyzed. These chapters deal with the philosophy of analytical ecology, presenting an increasing amount of detail that will be used in more complex models in later chapters. The philosophy of the analytical approach is most important here, however, and students are urged to build simple models describing some of the physical and chemical characteristics of the soil, water, and the atmosphere. The construction of these models is well suited to the use of small desk-top computing systems. Such models are only the beginning, of course, and students are urged to pursue those models of greatest interest to more meaningful conclusions as their knowledge, abilities, and computing capabilities increase with experience.

SOIL, WATER, AND TOPOGRAPHY

Soil, water, and topographic features combine to form a physical substrate that poses several kinds of forces in the ecosystem. Soil is a mechanical barrier to many organisms—ground squirrels, woodchucks, and other burrowing animals; and the distribution of these animals is partly a function of the distribution of different soil types. This is particularly true on a local scale; soil depths and densities determine the location of burrows. Water provides mechanical support for animals: firm support when frozen, and a moving mechanical force for swimming animals when it is flowing. The flow characteristics of water are a function of topography. The mechanical force with which water scours the substrate is dependent on the amount of sediment or abrasive material being carried by the water.

Soil, water, and topography are all a part of the thermal regime of an organism. Soil has particular thermal characteristics such as conductivity and temperature profiles. Water has a high heat of vaporization, and a considerable amount of heat energy can be dissipated by the evaporation of water. This is particularly important for plants and animals in hot environments. Topography affects the thermal regime of the soil, as a slope facing the direction of the sun's location absorbs more radiant energy than one facing away from the sun. There are often marked differences in vegetation on such slopes, not only on the long slopes of mountains but on the small slopes of local topographic features.

Different types of soil absorb energy at different rates, and the transfer of heat energy within the soil is a function of the conductivity of the soil. This is a function of the soil density, compaction, wetness, and other physical characteristics. Animals may select different soil types for bedding on in response to other thermal conditions. Deep litter, for example, is a warmer substrate than wet, bare soil and is utilized by animals for bedding under conditions that cause a high heat loss. Plants respond to the thermal conditions in the soil, too; cold, wet soil supports less vegetative growth. Bogs are good examples of areas with retarded growth, and areas of permafrost are the most extreme examples.

Soil, water, and topography all play a part in the nutritive relations of both plants and animals. Fertile soil supports the most abundant vegetation if water is adequate, although specific nutrient requirements of different plant species may vary. Too little or too much water affects the health of plants. Topography is related to both water and soil fertility because soil formation is partly a function of the topographic characteristics. Water erosion does not occur on level land, and the development of the soil profile there is different from that on hills because there is more stability in the top layer of decaying humus. Decomposition of the humus results in the release of elements that percolate into the lower layers of the profile.

None of these physical features—soil, water, and topography—can be analyzed in isolation. All are interrelated, and it is necessary to consider their relationships in order to maintain an ecological perspective. The entrance of a single organism into this complex system of interactions complicates things very quickly. Thus we begin by looking at physical characteristics of the ecosystem, assembling a series of simple models that illustrate how these physical characteristics can be considered as factors and forces in an ecological analysis that eventually includes both plants and animals.

4-1 SOIL

Soil formation is a function of the interrelationships between parent material, climatic features, thermal characteristics, biotic influences, and the topographic features of the area. The student of analytical ecology must approach the soil system with insight into the functional relationships between these factors and forces. The soil scientist, specializing in the analysis of at least some of these relationships, has a greater understanding of them than can be presented here, of course. Students with a special interest in soil characteristics in relation to an organism will find several references at the end of this chapter that cover the subject in greater detail. Let us consider here some of the basic physical and chemical characteristics of soil before relating soil characteristics to other functions in the ecosystem.

PHYSICAL CHARACTERISTICS. Mineral soil is a composite of mineral particles and decaying organic matter. Large mineral particles predominate in gravelly or sandy soils. In soils with a finer texture, the few large mineral particles that may be

SOURCE: *Soil Survey Manual* (U.S. Dept. of Agriculture Handbook No. 18, 1951), p. 207.

TABLE 4-1 THE CLASSIFICATION OF SOIL PARTICLES ACCORDING TO U.S. AND INTERNATIONAL SYSTEMS AND THE MECHANICAL ANALYSES OF TWO SOILS USING THE U.S. SYSTEM

Soil Separate	U.S. Department of Agriculture System			International System
	Diameter Limits (mm)	Analyses of Two Typical Soils		Diameter Limits (mm)
		Sandy Loam (%)	Clay Loam (%)	
Very coarse sand	2.00–1.00	3.1	2.2	
Coarse sand	1.00–0.50	10.5	4.0	2.00–0.20
Medium sand	0.50–0.25	8.2	6.3	
Fine sand	0.25–0.10	25.3	8.4	0.20–0.02
Very fine sand	0.10–0.05	22.0	9.6	
Silt	0.05–0.002	21.1	37.2	0.02–0.002
Clay	below 0.002	9.8	32.3	below 0.002

present are embedded in other more finely divided materials. Fine-textured soil that has been deposited by wind contains no large particles. Such a wind-deposited soil is called *loess*.

Soil particles are classified according to size based on diameter (Table 4-1). The particles are not necessarily spherical, of course; the angular characteristics of a particle are a function of the amount of abrasive action it has received. Pebbles in streams are often very round. Soils developing from glacial till may contain rounded particles if the source of the till had been subject to the action of moving water.

Soils containing large percentages of sand are said to be *light* soils. They have a low water-holding capacity and drain rapidly. If there is little organic matter in the sandy soil, it is very loose with no stickiness. Soils with high percentages of silt and clay are *heavy* soils. These soils have a fine texture, with slow drainage and high water-holding capacity. They are also sticky when wet, with poor aeration. Clay expands on wetting, with the release of heat energy. Drying results in the absorption of energy and the contraction that follows results in the formation of hard clods.

The distribution of soil particles of different sizes results in different physical characteristics of a volume of soil, including the particle density, air space, and specific gravity. These characteristics can be measured for particular soils. They can be calculated for a given volume of soil if some assumptions are made about the shapes and distribution patterns of the soil particles. The measurements are of value in describing the soil in geographical areas, and the calculations are of value in assembling a functional model that describes the interaction between factors and forces present.

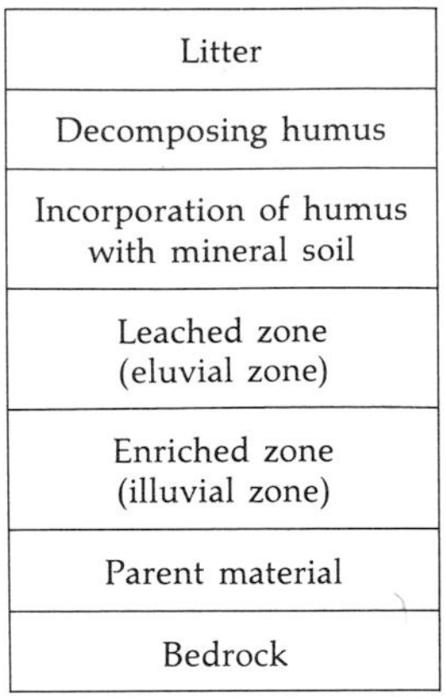

Litter
Decomposing humus
Incorporation of humus with mineral soil
Leached zone (eluvial zone)
Enriched zone (illuvial zone)
Parent material
Bedrock

FIGURE 4-1. Schematic display of zones or horizons in a soil profile. Local characteristics are dependent on the nature of the parent material, climate, topography, vegetation, and time.

SOIL PROFILE.   Soil particles—sand, silt, and clay—are not scattered throughout a developed soil in a random fashion but are organized in a soil profile. This profile consists of the surface soil, subsoil, and parent material. Soil scientists divide the soil profile into layers or horizons that develop as a result of the processes of soil formation (Figure 4-1). The top layer or horizon includes litter and raw humus. Decomposition occurs in the moist litter, resulting in humus that can be incorporated into mineral soil. Incorporation of humus with mineral soil results in a dark-colored mineral horizon. In areas with significant rain fall this dark horizon grades to a lighter-colored leached horizon as water percolates through the soil. This is called an *eluvial zone.* Materials carried by the percolating water are deposited in an *illuvial zone.* The local characteristics of the entire soil profile are dependent on climate, vegetation, parent material, topography (especially drainage patterns), and time. Additional material on the soil system may be found in Buckman and Brady (1969) and Black (1968).

SOIL WATER.   The amount of water in the soil has a definite effect on plant growth. It is also quite a variable physical factor, particularly in soils that permit water absorption at a fairly rapid rate and in areas of intermittent rainfall.

What happens to water at the surface of the soil? Four distributions are possible: (1) evaporation, (2) run-off, (3) absorption by the soil, and (4) surface collection. Evaporation results in the removal of water with little or no effect on soil structure. Run-off water has the potential for changing surface characteristics through erosion, as well as flowing in streams and rivers. Water absorbed by the soil is of particular interest to the analytical ecologist since it has such an important role in the productivity of plants. Surface water, including oceans, lakes, ponds, and intermittent pools, has the potential for changing a terrestrial system to an aquatic one, depending on the time factor.

Soil water has been classified into general categories, including water vapor, hygroscopic water, capillary water, and gravitational water. Water vapor is found in the air spaces between soil particles. Hygroscopic water is found on the surface of soil particles. Capillary water is found in the spaces between soil particles in which the distance is sufficiently small to permit surface tension to hold the liquid water. Gravitational water is found in the larger spaces between soil particles and is drawn away by the force of gravity. Air and water vapor then replace the gravitational water.

Soil water moves in three ways, including capillary adjustment, percolation, and vapor equalization. *Capillary adjustment* results from the adhesion of water to soil particles and the cohesion of water molecules. These forces result in the movement of water upward from the water table. *Percolation* results from the force of gravity; free water moves downward between soil particles because the force of gravity is greater than the surface tension and capillary forces. If there is insufficient water to saturate the profile down to an impervious layer, gravitational forces will be overcome by capillary forces and the downward movement of water will continue owing to capillary action. *Vapor equalization* within the soil results from variation in the vapor pressure within the macropores. The vapor pressure in the macropores is a function of temperature (that is, more water vapor can be present when temperatures are higher), resulting in fluctuation in vapor pressure that reflects changes in the temperature profile in the soil.

The water left after drainage by gravity is the maximum capillary water, and soil under those conditions is at field capacity. As capillary water is removed by evaporation and plant absorption, the point is reached at which plants cannot absorb water fast enough to offset water loss from transpiration. When that inbalance exists, the *wilting point* has been reached. If plants dehydrate beyond the point of recovery, the amount of soil moisture present is called the *permanent wilting percentage*.

These considerations of soil water are general descriptions of soil-water relationships. Consideration of the actual forces that determine these relationships is beyond the scope of the present discussion, although soil characteristics and water absorption are taken up later in the chapter. The student of analytical ecology must comprehend the kinds and the extent of factors and forces present and then proceed with analyses that are no more detailed than his comprehension permits.

CHEMICAL CHARACTERISTICS. Soil fertility varies greatly from one area to another and is a function of the nature of the parent material, the climatic factors in the area, the erosion history, and the plant growth. Plant growth is, of course, determined by the other three factors as well, illustrating once again the interrelationship between physical and biotic factors.

The primary nutrients and organic matter in surface soil are found within certain percentages of abundance in different soil types (Table 4-2). Primary nutrients may be detected chemically in a soil, but the important factor to consider in terms of plant growth is the chemical form of the nutrient in the soil. Nitrogen, for example, may be found in proteins in the soil and in that form is largely unavailable to growing plants. Subsequent decomposition of organic matter with the breakdown of proteins to amino acids and then to the formation of nitrites and nitrates makes the element nitrogen available to plants. Further examples

TABLE 4-2 PRIMARY NUTRIENTS AND ORGANIC MATTER IN SURFACE SOIL
WITH PERCENTAGES FOR TWO REPRESENTATIVE ANALYSES

Constituents	Expected Ranges in %	Representative Analyses	
		Humid-region Soil (%)	Arid-region Soil (%)
Organic matter	0.40–10.00	4.00	3.25
Nitrogen (N)	0.02– 0.50	0.15	0.12
Phosphorus (P)	0.01– 0.20	0.04	0.07
Potassium (K)	0.17– 3.30	1.70	2.00
Calcium (Ca)	0.07– 3.60	0.40	1.00
Magnesium (Mg)	0.12– 1.50	0.30	0.60
Sulfur (S)	0.01– 0.20	0.04	0.08

SOURCE: Adapted from *The Nature and Properties of Soils* by H. O. Buckman and N. C. Brady. Copyright © The Macmillan Company 1960, 1969.

of different forms in which primary nutrients may be found are shown in Buckman and Brady (1969, p. 27).

Organic matter is of significant importance in determining the quality of the soil. It plays a role in the physical structure of the soil by keeping it loose and friable, with enough binding effect on the soil particles to keep the nominal particle size from being too large or too small. The decomposition of organic matter is also an important source of nutrients. Thus organic matter plays both a physical and a chemical role in the dynamic soil system.

The biochemical processes that result in the release of nutrients are very complex and are a part of the life processes of soil bacteria. Soil conditions, including water, thermal, and chemical factors, have a significant effect on the rate of decomposition. Wet soils, for example, have very slow rates of decomposition. Wet soils are always colder, too, because of the high thermal capacity of water and the high heat of vaporization. Soils with a low pH are also characterized by low rates of organic decomposition. The combined action of all of these factors is most pronounced in bog soils, with their characteristic low pH, high organic content, wetness, and subsequent slow plant growth. Black spruce (*Picea marianna*) are found on these soils, and trees just a few feet tall may be 50 or more years old.

BIOLOGICAL CHARACTERISTICS. Soil is a dynamic entity because of the action of biological organisms in synthesizing and decomposing organic materials in the soil along with the mechanical action of larger organisms as they live in the soil. Buckman and Brady (1969) have two excellent chapters on "The Organisms of the Soil" (ch. 5) and "The Organic Matter of Mineral Soils" (ch. 6). The following summary is taken largely from their text.

A classification of the soil biota, according to the roles of organisms as primary consumers, secondary consumers, predators, parasites, decomposers, and so forth, is given in Table 4-3. The metabolic activity of these organisms results in an intake and release of matter and energy within the soil. Gases, liquids, and solid wastes are released, and their body tissue is constantly being replaced in an endless cycle.

The roots of higher plants living above ground make a significant contribution to the organic material in the soil. The roots grow as a result of the translocation of matter from the part of the plant above the ground. The energy of the sun is captured in the process of photosynthesis, stored as chemical energy, translocated in soluble form to the subterranean plant parts, and either metabolized

TABLE 4-3   SOIL BIOTA CLASSIFIED ACCORDING TO THE ROLES OF ANIMALS AND PLANTS

Animals	Macro	Subsisting largely on plant materials	Small mammals—squirrels, gophers, woodchucks, mice, shrew Insects—springtails, ants, beetles, grubs, etc. Millipedes Sowbugs (woodlice) Mites Slugs and snails Earthworms
		Largely predatory	Moles Insects—many ants, beetles, etc. Mites, in some cases Centipedes Spiders
	Micro	Predatory or parasitic or subsisting on plant residues	Nematodes Protozoa Rotifers
Plants	Roots of higher plants		
	Algae	Green Blue-green Diatoms	
	Fungi	Mushroom fungi Yeasts Molds	
	Actinomycetes of many kinds		
	Bacteria	Aerobic Anaerobic	and Autotrophic Heterotrophic

SOURCE: Reprinted with permission of The Macmillan Publishing Co., Inc. from *The Nature and Properties of Soils* by H. O. Buckman and N. C. Brady. Copyright © The Macmillan Company 1960, 1969.

there or stored in an insoluble form where these products become available for decomposition by other subterranean plants, for consumption by animals feeding below ground, or for mobilization and metabolism during the growth of a new plant. This intense biological activity results in a greater abundance of organisms in the root zone or *rhizosphere* than any other part of the profile.

The roots have several functions that result in different soil characteristics. They function as binding agents, holding the mineral soil together. Nutrient uptake occurs in the roots, beginning a nutrient cycle that is completed upon decomposition. Organic acids formed at root surfaces act as solvents. Dead cells slough off the roots and serve as a substrate for microflora.

Almost 700 species of soil fungi have been identified. The most important of these fall into two groups: molds and mushroom fungi. The molds may be either microscopic or macroscopic. They are found in greatest numbers in the surface layers where there is an abundance of organic matter. Some prefer a low *p*H; they are important in the soil development processes in forests where higher acidity is a characteristic of the soil. Mushroom fungi are very important decomposers, attacking such complex compounds as cellulose, lignin, gums, and starch, as well as proteins and sugars. They compliment the action of molds, bacteria, and other decomposers in the soil formation process.

Microorganisms are abundant in the soil. The number of bacteria alone in 1 gram of soil may range from 100,000 to several billion! (Buckman and Brady 1969, p. 15). Other microorganisms include algae and actinomycetes.

Soil bacteria are classified as autotrophic or heterotrophic. Autotrophic bacteria derive energy from the oxidation of mineral constituents of the soil—such as ammonium, sulfur, and iron—and carbon from carbon dioxide. These are the bacteria involved in nitrification, sulfur oxidation, and nitrogen fixation. Their importance is great because of the significance of these processes in the life cycles involving the soil and atmosphere. The heterotrophic bacteria obtain energy and carbon directly from organic matter. Autotrophic and heterotrophic bacteria function together in a dynamic redistribution of matter with the expenditure of energy within the soil.

Actinomycetes are unicellular plants that have about the same diameter as bacteria. They are filamentous, often profusely branched, and produce fruiting bodies similar to those of molds. They are second to bacteria in number in the soil. They function as decomposers, releasing nutrients for absorption by other plants.

Over 60 species of algae have been isolated from soils. Three general groups— blue-green, green, and diatoms—are found, with the most prominent species present in the soil being the same the world over. Soil algae are most abundant in the surface layers (upper one inch) since most of them contain chlorophyll and function much like the higher plants in photosynthesis. Algae are present in subsoils in the form of spores or cysts (resting stages) or in vegetative forms that do not depend on chlorophyll. They do not perform particularly important functions, although their biomass does contribute to the total organic material in the soil. The abundance of different species is apparently related to water and crop characteristics.

Animals, both macroscopic and microscopic, are found in the soil. Rodents (ground squirrels, for example) and insectivorous animals (moles and shrews) live in the ground, along with insects, millepedes, sowbugs, mites, snails, slugs, centipedes, spiders, and earthworms. The action of some of these animals—rodents, moles, and ants, for example—is often confined to particular areas. Some animals tend to be territorial, resulting in a distribution that is dependent on soil conditions and animal behavior.

Many of the animals in the soil begin the process of decomposition by feeding on organic matter, breaking it up mechanically, utilizing parts of it, and excreting parts of it, thus beginning a cycle of breakdown that continues in various forms until the bacteria and fungi have completed the process. Of all soil animals, earthworms have been given the most attention. They ingest soil, grind it, digest the organic matter, and excrete the unused portions (earthworm casts). This process increases the available plant nutrients, as well as aerating the soil and transporting the soil material vertically within the soil profile.

Microscopic animals in the soil include nematodes, protozoa, and rotifers. Nematodes are divided into three feeding types: (1) those that live on decaying organic matter; (2) those that prey on other nematodes and earthworms; and (3) those that are parasites for a part of their life cycle of the roots of higher plants. The first two types fit into the natural biota associated with soil formation. The third type can cause serious crop damage. Protozoa and rotifers are thought to be a part of the organic decomposition process, although their importance is unknown and undoubtedly varies from one soil to another.

4-2 SOIL CLASSIFICATION IN TRANSITION

Soil classification has been undergoing major changes in the last few years. Prior to 1960, soil classification was based largely on how the soils were thought to have formed, that is, the soil genesis. This results in the need for considerable subjective judgement. The soil classification that came into use in the sixties is based on the properties of the soil as identified in the field. This results in the classification of the soils themselves rather than the soil-forming processes. Thus soils of an unknown genesis can be classified—geological and climatological factors are not considered directly—and a greater uniformity is expected since the soil itself is being judged.

This system is based on the use of surface and subsurface diagnostic horizons for soil identification. Soils are identified by general terms followed by more specific terms in the same way that plants and animals are classified. Plants and animals are grouped into phylum, class, order, family, genus, and species. Soils are classified into order, suborder, great group, subgroup, family, series, and type. Many of the root words from which the names of these categories were derived have meanings (for example, the Order, Aridisol comes from the Latin word *Aridus,* which means dry) just as many plant and animal names have meaningful derivations. Some names, however, are nonsense words and others are associated with people or places. This is most true for the soil series.

Students of ecology need to be aware of the changes in soil classification when referring to the literature. The old system was used prior to the sixties, and currently there is a transition to the new system with continued testing of procedures and terms. The new classification system is discussed in detail in the USDA publications (Soil Survey Staff 1960; 1967) and Smith (1963).

4-3   EUTROPHICATION

Surface water constitutes over two-thirds of the total area of the earth's surface. Oceans, lakes, rivers, streams, ponds, and intermittent pools are all a part of this vast surface area, and each has its own particular system characteristics. It is interesting that the huge volumes of water in the oceans should support some of the largest organisms (whales) and some of the tiniest (plankton), as well as the fragile algae on the tide flats. The massive force of water in the oceans is too great for rigidly structured organisms, hence their anatomies are such that they conform to these mechanical forces.

Variations in the physical and chemical characteristics of bodies of water in time is an important ecological consideration. Generally, the smaller bodies of water are subject to the greatest variations, simply because of their smaller mass. It is important for students of ecology to analyze the significance of such variations. Organisms must cope with each variation rather than the average effect of several. No organism lives or dies by the average effect of many factors and forces; rather it is faced with a host of forces that act independently in some cases and interact in others.

The movement of water from the soil to lakes and ponds results in an accumulation of minerals and a subsequent increase in the fertility of the water. This is a natural phenomenon due to geological and biological aging and is called eutrophication. This results in changes in the vegetation patterns. Shore-line vegetation becomes more abundant and plankton density increases. In time, a floating mat of plants may develop along the shore, especially in small ponds where wave action is slight. Organic sediments also accumulate at the bottom of lakes or ponds. Anaerobic conditions frequently develop there in the advanced stages of succession, resulting in a more rapid accumulation of organic material as decomposition subsides. In the most advanced successional stages, the bogs that are formed are transformed into a terrestrial system as open water disappears and vegetation covers the basin completely.

The process of eutrophication just described is very slow geologically. The activities of man increase the rate. Agricultural practices such as plowing expose topsoil that is more subject to erosion than is undisturbed soil held in place by natural vegetation. Recent developments in tillage practices result in less disturbance of the topsoil by utilizing the concept of minimum tillage. Corn, for example, can be planted in sod with resulting high yields. Such a practice requires the use of selective herbicides since corn is not able to compete with forbs and grasses without some help.

Agricultural practices are often considered the cause of problems relating to environmental quality and pollution. The use of fertilizers, herbicides, pesticides, and other synthetic compounds was challenged by environmentalists in the sixties and seventies. It is true that agricultural practices result in a faster rate of redistribution of matter and energy than would occur in undisturbed ecosystems. Arguments related to ecologically isolated observations have often resulted in insufficient analyses of the total ecology, however. The use of fertilizers has been condemned without regard for the fact that their use on land with a high production potential results in increased yields from fewer acres, permitting a reversion of marginal land to nonagricultural uses. Thus lands with steeper slopes that should not be tilled can be left to natural succession, and this may result in a net reduction in erosion and mineral relocation. Herbicides and pesticides also contribute to increased yields, and this can result in wiser use of the total land and water resource. This is discussed further in Aldrich (1972).

It is not the author's intent to elaborate on the benefits and detriments of agricultural practices, ecological theories, environmental philosophies, and so forth, here. It is my firm conviction that meaningful insights into the effects, both short- and long-term, of all of man's activities can only be gained after careful attention has been given to the functional relationships between organism and environment. This is possible only through analytical procedures that evaluate the relationships between energy and matter through time in both the physical and biological components of the ecosystem. There is a grave danger in looking at isolated effects without a consideration of the causes as well as other related effects in the total ecological complex. Thus I suggest that students in analytical ecology turn their attention to understanding life processes of plants and animals in relation to the physical processes of energy and matter redistribution in the earth-atmosphere interface.

4-4   BIOGEOCHEMICAL CYCLES

Life is supported on the planet Earth by the continual cycling of matter through the release of energy. The sole source (almost) of this energy is the sun. The other component necessary for these cycles to continue (the component apparently lacking on other planets in our solar system) is liquid water.

Cycles of interest to the ecologist include energy, water, oxygen, carbon, nitrogen, and mineral cycles. These are depicted in Figure 4-2 and discussed further in a series of articles in the September 1970 issue of *Scientific American*. These articles describe the general pathways of energy and matter as transformation involving the sun, atmosphere, earth, and organisms takes place. They are guides to specific analyses that can be completed by students in analytical ecology of limited systems, such as aquaria, terraria, chambers, and so forth. The concept of life as a continual redistribution of matter with the expenditure of energy falls within these cycles. The remaining chapters in this book deal with more specific interactions between energy and matter. Those discussed are of interest to many ecologists, and those not included in this text (of which there are myriads) can be analyzed with the same type of modeling approach. The

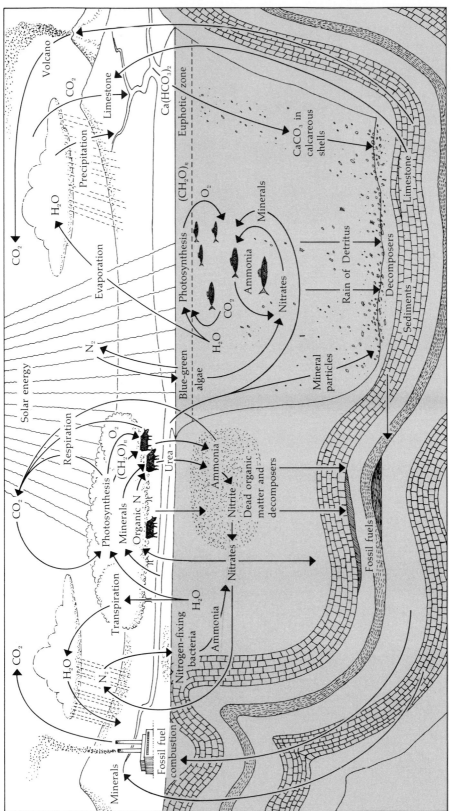

FIGURE 4-2. Major cycles of the biosphere. (From "The Biosphere" by G. Evelyn Hutchinson. Copyright © 1970 by Scientific American, Inc. All rights reserved.)

student will note that there is a greater emphasis on single organism—environment relationships within a simple format than on large scale, but very general, analyses that include populations, trophic levels, biomasses, or gross ecological units. There are many exciting relationships between an individual organism and its environment, and comparisons between different kinds of individuals, such as male and female, pregnant and nonpregnant, large and small, young and old, and so forth, are of great interest because of the different roles that individuals play. At some time in the ecological future, the sythesis of these types of analyses of individual relationships (autecology) will result in a better understanding of community ecology (synecology).

LITERATURE CITED IN CHAPTER 4

Aldrich, S. R. 1972. Some effects of crop-production technology on environmental quality. *BioScience* **22**(2): 90–95.

Black, C. A. 1968. *Soil-plant relationships.* New York: Wiley, 792 pp.

Bolin, B. 1970. The carbon cycle. *Sci. Am.* **223**(3): 124–132 (Offprint No. 1193).

Buckman, H. O., and N. C. Brady, 1969. *The nature and properties of soils.* New York: Macmillan, 653 pp.

Cloud, P., and A. Gibor. 1970. The oxygen cycle. *Sci. Am.* **223**(3): 110–123 (Offprint No. 1192).

Deevey, E. S., Jr. 1970. Mineral cycles. *Sci. Am.* **223**(3): 148–158 (Offprint No. 1195).

Delwiche, C. C. 1970. The nitrogen cycle. *Sci. Am.* **223**(3): 136–146 (Offprint No. 1194).

Hutchinson, G. E. 1970. The biosphere. *Sci. Am.* **223**(3): 44–53 (Offprint No. 1188).

Oort, A. H. 1970. The energy cycle of the earth. *Sci. Am.* **223**(3): 54–63 (Offprint No. 1189).

Penman, H. L. 1970. The water cycle. *Sci. Am.* **223**(3): 98–108 (Offprint No. 1191).

Smith, G. D. 1963. Objectives and basic assumptions of the new soil classification system. *Soil Sci.* **96:** 6–16.

Soil Survey Staff, Soil Conservation Service. 1960. *Soil classification: a comprehensive system. 7th approximation.* USDA.

Soil Survey Staff, Soil Conservation Service. 1967. *Supplement to soil classification system. 7th approximation.* USDA.

USDA. 1951. *Soil survey manual.* USDA Handbook No. 18.

Woodwell, G. M. 1970. The energy cycle of the biosphere. *Sci. Am.* **223**(3): 64–74 (Offprint No. 1190).

IDEAS FOR CONSIDERATION

Use a Soil Conservation Service soils map to identify the different soil types in a local area. Characterize each of these soil types according to the physical and chemical characteristics that affect the interactions of topography, water, and the soil, and consider the possible relationships between these soil types and the plant communities found on them. Can you develop a model that is useful for predicting the type (not necessarily species) of plants you would expect under these conditions?

Determine the air space (by the water-displacement method) in containers filled with spheres of different sizes. Measure the air space for containers filled with uniformly sized spheres and spheres of various diameters. Do measured values agree with those predicted using geometric formulas? Can you relate this technique to soil with different ratios of sand, silt, and clay?

SELECTED REFERENCES

Bay, R. R. 1958. Occurrence and depth of frozen soil. *J. Soil Water Conserv.* **13**(5): 232–233.

Bay, R. R. 1967. Ground water and vegetation in two peat bogs in northern Minnesota. *Ecology* **48**(2): 308–310.

Cary, J. W. 1971. Energy levels of water in a community of plants as influenced by soil moisture. *Ecology* **52**(4): 710–715.

Chiang, H. C., and D. G. Baker, 1968. Utilization of soil temperature data for ecological work. *Ecology* **49**(6): 1155–1160.

Fisher, S. G., and G. E. Likens. 1972. Stream ecosystem: organic energy budget. *BioScience* **22**(1): 33–35.

Fried, M., and H. Broeshart. 1967. *The soil-plant system in relation to inorganic nutrition.* New York: Academic Press, 358 pp.

Gersper, P. L., and N. Holowaychuk. 1971. Some effects of stem flow from forest canopy trees on chemical properties of soils. *Ecology* **52**(4): 691–702.

Hillel, D. 1971. *Soil and water: physical principles and processes.* New York: Academic Press, 288 pp.

Hudson, N. 1971. *Soil conservation.* Ithaca, New York: Cornell University Press, 320 pp.

Kucera, C. L., and D. R. Kirkham. 1971. Soil respiration studies in tallgrass prairie in Missouri. *Ecology* **52**(5): 912–915.

WEATHER IN RELATION TO PHYSICAL CHARACTERISTICS

All of the matter and energy in the physical and biological system in which we live has, theoretically, potential ecological significance. Of greatest interest to the ecologist is the matter and energy distribution within the earth-atmosphere interface, or biosphere, which supports life. Most of this energy originates outside of the atmosphere, such as the radiant energy from the sun. Some comes from within the earth; geothermal energy is an example. The analytical ecologist is interested in the physical and biological interaction between matter and energy in the biosphere. Before analyzing some of these interactions, let us review both the relationships between the sun and earth and the weather in relation to physical characteristics at the earth's surface.

5-1 THE DISTRIBUTION OF SUNLIGHT

One of the more precise physical relationships in the universe is the position of the sun relative to the earth. This precision is manifested by the regularity with which the earth revolves around the sun and rotates on its axis. The calendar does not represent this relationship perfectly; the addition of a day in February every four years is a correction factor for this.

The ecological role of sunlight is significant. Different day lengths affect the growth and reproduction of plants and animals. Sunlight generates different thermal relationships in the atmosphere that result in weather patterns. It is a source of energy for the process of photosynthesis upon which all life depends. It alters the thermal regime at the earth's surface, causing changes in physical conditions as well as in the behavior of plants and animals. Many more effects

57

can also be recognized, of course. Before considering its effect on plants and animals, let us consider how it is related to physical characteristics on the earth, such as topography over daily and seasonal time periods.

The times at which the sun rises and sets for different latitudes are available in tabular form. Differences within a one-hour time zone can be corrected for in order to determine solar time. Such tables are not suitable for direct entry into a computer program. Sunrise and sunset times can be calculated on the basis of the spatial relationships between the sun and earth, however. An equation can be used to store this information in a computing system, which can then calculate the times at which the sun rises and sets, the length of daylight, and other solar considerations in an ecological analysis.

The sunrise time, sunset time, hours of daylight, altitude of the sun at solar noon, and the solar insolation in langleys per minute for any slope aspect, slope angle, and time can be calculated from inputs of date (Julian Calendar), latitude, slope aspect, slope angle, time of day (0 to 2400 hours), and transmittance of the atmosphere (Robbins, unpublished data, BioThermal Laboratory).

A similar type of program that calculates additional factors pertaining to the spectral characteristics of radiation has been described by McCullough and Porter (1971). Their program generates clear-day direct and diffuse components of natural terrestrial radiation for any time of the day, elevation, terrestrial latitude, and time of year. It computes radiation spectra for large zenith angles in which atmospheric curvature and refraction are important, irradiation patterns at latitudes $>66°$ $30'$ (polar zones), the variation of the diffused spectral components of solar radiation with the elevation and reflectivity of the underlying surface, and diffused ultraviolet radiation spectra where ozone absorption of the scattered radiation must be accounted for. These outputs are generated from eight inputs, including time of day, day of year, latitude, longitude, sea level, meteorological range (visibility), atmospheric pressure (i.e., elevation), reflectivity of the underlying earth's surface, and total precipitable water vapor.

There are data in the literature that are useful in evaluating the outputs of these two programs. Sellers (1965), for example, shows the distribution of solar radiation throughout the United States at different times of the year (Figure 5-1). Variations in the radiation values shown are due to differences in latitude, altitude, distribution of water, and topographical effects on weather patterns, and so forth. Note the large difference between January and June. Other data are found in textbooks on meteorology and in numerous research reports in meteorological journals.

The outputs from these two programs are not perfect representations of these parameters at every point on the earth's surface. The programs are useful for setting an ecological "stage" with a few principal characters. The interactions between sunrise and sunset time, length of daylight, animal movements, breeding conditions, plant productivity, thermal energy balance, and other significant components of the ecosystem might be considered initially. Since the equipment necessary to make these fairly complex programs workable is seldom available to students, it is suggested that simpler approximations be made that can be

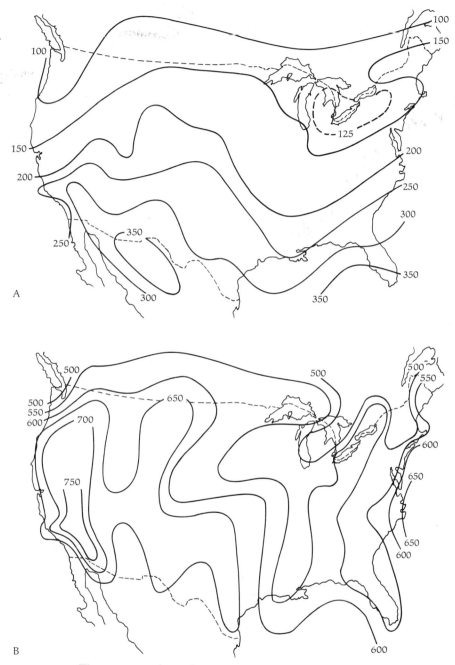

FIGURE 5-1. The average solar radiation on a horizontal surface in the United States in (A) January and (B) July. The units are langleys per day. (From Sellers 1965.)

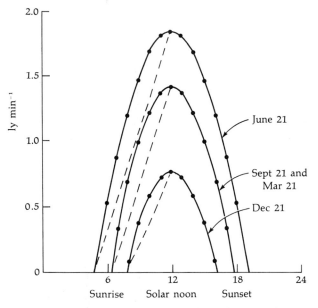

FIGURE 5-2. Sunrise, sunset, and solar radiation values for a latitude of 42°26′ N (Ithaca, N.Y.). The dashed lines are linear approximations of the radiation curve (see text).

handled by desk-top computing systems or calculators, or can even be done without the aid of a machine. For example, solar radiation curves can be approximated by a linear regression equation (Figure 5-2). This obviously results in potentially large errors. The student of analytical ecology is urged to consider the interreactions between solar radiation and an organism, however, and the linear approximation is quite adequate for working out the mechanisms of a predictive solar radiation curve in relation to the biological response of an organism. In other words, extensive detail in part of the total organism-environment relationship is undesirable until all parts of the relationship have been considered.

THE ATMOSPHERE. The atmosphere is divided into several general regions (Figure 5-3). These include the troposphere, the stratosphere, the mesosphere, the thermosphere, and the exosphere. The *troposphere* contains the "weather" and shows a general decrease in temperature with height. Its upper limit is characterized by the maximum height of most clouds and storms. The *stratosphere* has a fairly complex temperature structure that varies geographically and seasonally. It also has wind regimes that vary seasonally and affect weather conditions at the earth's surface.

Above the stratosphere is the *mesosphere*, characterized by cold temperatures and very low atmospheric pressure. Above the mesosphere is the *thermosphere* in which theoretical radiant temperatures rise with height, although artificial satellites do not acquire such temperatures because of the rarefied air (Barry and Chorley 1970). The thermosphere includes the region in which ultraviolet radiation and cosmic rays cause ionization; this region is called the *ionosphere*. The outer limits of the atmosphere grade into a region called the *exosphere,* or outer space, with its almost total lack of atoms. The earth's magnetic field becomes more important than gravity in the distribution of atomic particles in the exosphere. A more detailed discussion of the characteristics of the atmosphere may be found in Barry and Chorley (1970).

All of the layers of the atmosphere are of interest to the ecologist since together they form the total blanket of air in the biosphere. The atmospheric components serve particular functions, including the filtering of radiant energy from the sun, insulation from heat loss at the earth's surface, and stabilization of weather and climate owing to the heat capacity of the air. Several cycles are present that relate to the movement of matter between an organism and its environment. These include the water cycle, the carbon cycle, the nitrogen cycle, the phosphorus cycle, and others.

Gases in the atmosphere include nitrogen, oxygen, argon, carbon dioxide, and water vapor, along with traces of several other elements. Oxygen is necessary for most forms of life, but other forms exist only in an anaerobic environment. In terms of quantity, carbon dioxide is the most variable of these gases. Considerations have been made for the enrichment of the atmosphere (carbon dioxide fertilization) to promote plant growth. These are discussed in Chapter 15. Water vapor condenses in the atmosphere, resulting in precipitation. Its effect on the distribution of plants and animals is obvious on both a small and a large scale. Precipitation limits visibility, too, so observations of animals are more difficult and field work is less efficiently carried out in rain or snow. Consequently, information on animals in storms is lacking, yet these extreme conditions may affect their survival and productivity.

The ionosphere is affected by magnetic storms on the sun, with an increase

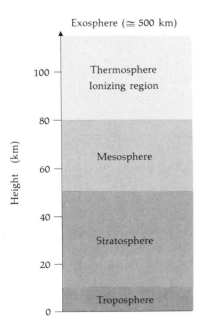

FIGURE 5-3. Atmospheric layers; each grades into the next without a sharp line of demarcation between them.

in the number of free electrons present during periods of intense activity. Population cycles have been attributed to these storms and subsequent electrical activity in the ionosphere, although there is considerable controversy over the validity of such correlations. Personal observations by the author indicate that white-tailed deer seem to respond to unidentified factors, with an increase in nervousness, activity, and physiological parameters such as heart rate. The possible interactions between the energy from the sun, cosmic rays, ionization in the thermosphere, electrical activity at ground level, and physiological processes in plants and animals need further analyses through basic investigations of atmospheric physics, physiology, neurology, and behavior.

5-2 ATMOSPHERIC WATER

Atmospheric water is found in gaseous form or droplet form. The quantity of water present in gaseous form is expressed as vapor pressure. When the vapor pressure is maximum for the temperature of the atmosphere, the air is said to be saturated. Thus the maximum vapor pressure of the air is called its saturation pressure, and the actual quantity of vapor present (vapor pressure) is a function of the temperature of the atmosphere. Relative humidity is equal to the vapor pressure divided by saturation pressure times 100. Students are reminded that relative humidity is *temperature dependent,* a fact often overlooked in ecological analysis. This will be discussed again in Chapter 6.

Clouds form as condensation takes place around hygroscopic nuclei in the atmosphere. These particles can be dust, smoke, sulphur dioxide, salts (NaCl), or similar microscopic substances (Barry and Chorley 1970). Clouds have distinct characteristics that are classified in an internationally adopted system according to the shape, structure, vertical height, and altitude of the cloud. Cloud types can be identified through a "keying out" process similar to a taxonomic key for plants and animals (Table 5-1). Clouds reflect and absorb solar radiation and are good absorbers and emitters of infrared radiation. They are also the centers of considerable electrical activity, as well as the source of precipitation.

5-3 PRECIPITATION

Precipitation consists of all liquid and frozen forms of water, including rain, sleet, snow, hail, dew, hoarfrost, fog-drip, and rime (Barry and Chorley 1970). Rain and snow are the only forms that contribute significantly to the precipitation totals, but some of the other forms, such as dew and hail, can have significant ecological impact. The analytical ecologist is interested in the functional relationships between precipitation and organisms, and this demands an understanding of the energy and matter relationships between them rather than the mere correlation of observed responses of organisms in different precipitation regimes.

RAIN. A basic analysis of the functional relationships between precipitation and organisms permits these mechanisms to be related to other interactions

TABLE 5-1 A KEY FOR THE IDENTIFICATION OF CLOUD FORMATIONS
(Beginning with couplet number one, select the most appropriate choice
and go to the numbered couplet indicated until a cloud formation has
been identified.)

1 Clouds piled up, puffy, currents 2
   Clouds formed without vertical movement 3
2 Clouds, puffy, changing shape . . . . cumulus, 13
   Strong vertical development . . . . cumulonimbus, 15
3 Cloud veils or sheets 4
   More or less broken 7
4 Large halo present . . . . .cirrostratus
   No halo effect 5
5 Sun visible through veil. . . . . altostratus, 12
   Sun not visible, heavy veil 6
6 Low uniform sheet, no rain . . . . .stratus
   Low heavy sheet, rain streaks . . . . nimbostratus
7 High ice clouds, usually thin wispy streaks . . . . cirrus, 16
   Heavier clouds, patchy or irregular masses 8
8 Patches or layers of puffy or roll-like gray or whitish
   clouds, corona often 9
   Irregular masses in a rolling or puffy layer, gray with darker
   shading . . . . stratocumulus, 18
9 White to light gray roll-like, or roll-like in combination with patchy
   or wispy, high ice clouds . . . . cirrocumulus
   Patches or layers of puffy or roll-like gray or whitish clouds, corona
   often, middle water and ice clouds . . . . .altocumulus, 10
10 White to gray roll-like middle water and ice clouds 11
   Patchy to nearly continuous gray middle water and ice
   clouds . . . . .altocumulus perlucidus
11 White to light gray clouds, roll-like or less distinctly
   so . . . . .altocumulus translucidus
   Gray, roll-like clouds . . . . altocumulus translucidus undulatus
12 Darker clouds, but sun appears to be behind frosted
   glass . . . . altostratus opacus
   Lighter clouds, sun appears to be behind frosted
   glass . . . . altostratus translucidus
13 Irregular patches broken by strong winds . . . . .cumulus fractus
   Bulky patches, white to gray 14
14 White to light gray, low water clouds, may be
   towering. . . . . cumulus humilis
   Darker gray low water clouds, becoming more
   dense. . . . . .cumulus congestus
15 Low vertical water clouds to towering water to ice clouds, often
   anvil shaped . . . . .cumulonimbus capillatus
   Dense clouds with vertical development indicated by ventral
   projections, seldom seen . . . . cumulonimbus capillatus mammatus
16 Thin wispy streaks, broken pattern 17
   Heavier streaks, almost patches, white, high ice clouds . . . . . cirrus densus
17 Streaks broken, not spreading over sky . . . . .cirrus filosus
   Streaks broken, spreading over sky . . . . cirrus cinus
18 Low water clouds, light gray. . . . . stratocumulus translucidus
   Low water clouds, dark gray . . . . stratocumulus opacus

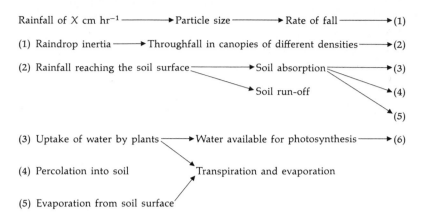

Rainfall of X cm hr^{-1} ⟶ Particle size ⟶ Rate of fall ⟶ (1)

(1) Raindrop inertia ⟶ Throughfall in canopies of different densities ⟶ (2)

(2) Rainfall reaching the soil surface ⟶ Soil absorption ⟶ (3)

Soil run-off

(4)

(5)

(3) Uptake of water by plants ⟶ Water available for photosynthesis ⟶ (6)

(4) Percolation into soil   Transpiration and evaporation

(5) Evaporation from soil surface

(6) Gross energy produced by plants ⟶ Net energy for production by consumers

FIGURE 5-4. A flow sheet showing the relationships between rainfall per hour, raindrop characteristics, soil absorption, plant productivity, and consumer energy for production.

between organisms and environment. This is illustrated in Figure 5-4, in which the amount of rainfall in cm hr^{-1} is the beginning point for calculating the mass transport of water within a plant community, relating this movement to the feedback interactions between evaporation from the surface of the soil and evapotranspiration and to plant production.

FIGURE 5-5. There is a predictable relationship between rainfall intensity and the diameter of the raindrops. (Data from Barry and Chorley 1970.)

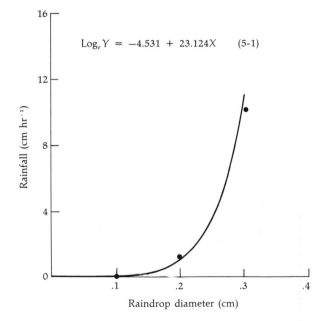

$$\text{Log}_e Y = -4.531 + 23.124X \qquad (5\text{-}1)$$

How can the flow sheet in Figure 5-4 be useful? What kinds of relationships exist that make such an analysis work? Some basic data have been reported by meteorologists that can be used to develop a series of calculations representing the relationships between rainfall in cm hr $^{-1}$ and subsequent soil water, plant uptake of water, evapotranspiration, and plant production. For example, there is a relationship between precipitation intensity and raindrop size; a greater amount of rainfall per unit time is more dependent on the size of the raindrops than on the number of raindrops. Barry and Chorley (1970) point out that the most frequent diameter of drops in a rainfall of 0.1 cm hr^{-1} is 0.1 cm; in 1.3 cm hr^{-1} it is 0.2 cm; and in 10.2 cm hr^{-1} it is 0.3 cm. These data can be expressed in equation form [equation (5-1) in Figure 5-5]. This relates to the number/volume ratio of drops, too, as a drop having a diameter of 0.1 cm has a volume of 0.000524 cc, and 19,083,970 drops are necessary for 1 cm of water to cover an area of one square meter. This is a volume of 10 liters.

Once the particle size has been established, the rate at which the drops fall can be calculated from data in Taylor (1954). He presents data that show how a larger drop falls at a faster rate than a smaller one, and this relationship, plotted in Figure 5-6, can be approximated by equation (5-2) relating the $X:Y$ values.

The rate of fall can be used to calculate the inertia of the raindrops as they strike the soil surface (Table 5-2; their weight was calculated from data shown

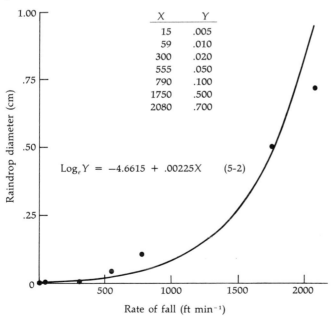

FIGURE 5-6.   A first approximation of the relationship between raindrop diameter and the rate at which an individual raindrop falls. (Data from Taylor 1954, Table 7-1, p. 155.)

X	Y
15	.005
59	.010
300	.020
555	.050
790	.100
1750	.500
2080	.700

$$\text{Log}_e Y = -4.6615 + .00225X \qquad (5\text{-}2)$$

TABLE 5-2 THE INERTIA OF INDIVIDUAL RAINDROPS AT DIFFERENT RAINFALL INTENSITY

Rainfall $(cm\ hr^{-1})$	Raindrop Diameter (cm)	Raindrop Weight $(\times\ 10^{-3}\ g)$	Rate of Fall $(ft\ min^{-1})$	Inertia: Raindrop Weight $(g) \times$ Rate of Fall $(ft\ min^{-1})$
0.1	0.096	0.46	1031.96	0.478
0.2	0.126	1.05	1152.33	1.207
0.3	0.144	1.56	1210.09	1.891
0.4	0.156	1.99	1246.95	2.479
0.5	0.166	2.40	1273.57	3.050
1.0	0.196	3.94	1347.37	5.312
1.5	0.213	5.06	1385.46	7.010
2.0	0.226	6.04	1410.63	8.526
2.5	0.236	6.88	1429.22	9.836
3.0	0.243	7.51	1443.85	10.847
4.0	0.256	8.78	1466.00	12.878
5.0	0.265	9.74	1482.45	14.445
6.0	0.273	10.65	1495.46	15.932
7.0	0.280	11.49	1506.17	17.312
8.0	0.286	12.25	1515.24	18.560
9.0	0.291	12.90	1523.08	19.652
10.0	0.296	13.58	1529.99	20.776

Note: The values in this table and all subsequent tables containing calculations may vary according to the computing system used and the arrangement of program steps in relation to rounding of numbers.

in Figure 5-5; specific gravity of water = 1.0), and this inertia is of major importance in determining how much of the rainfall can penetrate a plant canopy (throughfall), how much reaches the soil surface, and the mechanical impact it has on the soil surface. The inertia can be related to throughfall if the mechanical strength of the plant canopy is known. The inertia can be related to soil disturbance if the mechanical strength of the soil surface is known. This depends on the distribution of particle size and density and the cohesiveness of the soil material.

RAINFALL IN RELATION TO SOIL AND TOPOGRAPHY. The results shown in Table 5-2 illustrate relationships between the amount of rainfall per hour and the inertia with which the drops strike a surface. The equations representing the factors and forces in the model were combined into a computing program that results in the outputs shown in the table. It is an example of a "rainfall per hour" to "inertia of the drops" conversion.

Let us expand the previous model to include soil absorption characteristics. Suppose that a rainfall of 1.3 cm hr^{-1} is occurring. These raindrops strike the surface of the soil at a velocity of 1372.14 ft min^{-1} (see Table 5-2). Suppose that the soil was on a slope of 45°, and its characteristics were such that there was no resistance to the flow of water across the soil surface and none of the water was absorbed (Figure 5-7). This would result in a run-off equal to the rainfall per hour, once steady-state conditions were reached. Note that an absorption coefficient of 0 was used in this example. A value at the other extreme could be used here, too (i.e., 1.00), and then all of the water striking the surface of the soil would be absorbed. The run-off would then be 0, of course.

What affects the absorption coefficient of a soil? One factor is the particle size. Soils contain mixed particle sizes, of course, but before going into the details of real soils, let us consider a theoretical soil with homogenous spherical particles. If these particles were arranged as in Figure 5-8, the pore space is a constant 48% of the soil volume.

Natural soils, with particles of varying sizes and forms, are much more complex than the illustration. Calculations of the pore space of real soils can be made if enough is known about the geometry and hygroscopic characteristics of different soil types. The alternative method for determining pore space is measurement by displacement.

The absorption characteristics of a soil interact with its resistance to water flow. If the resistance to flow is high, the water will remain on the soil for a longer time, increasing the possibility for absorption. Vegetation also affects the run-off and absorption characteristics of a soil. Decaying vegetation is absorbant, and a vegetative canopy breaks larger raindrops up into smaller ones, reducing the rate at which they fall and their inertia. Thus there is a whole series of interactions that need to be considered for the modeling of rainfall, erosion, and other mechanical effects of rain.

FIGURE 5-7.   A very simple model illustrating the relationship between rainfall, absorption, resistance to flow, and run-off.

$$Q_{run-off} = rainfall - (absorption\ coefficient \times rainfall)$$

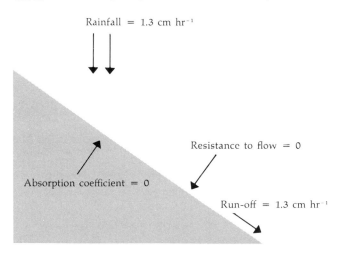

Rainfall = 1.3 cm hr^{-1}

Resistance to flow = 0

Absorption coefficient = 0

Run-off = 1.3 cm hr^{-1}

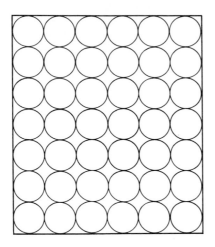

FIGURE 5-8. Spherical particles arranged in the simplest geometry.

The main point thus far is the illustration of the logic of the model-building process, showing how to proceed from the very simple to the more complex, but limiting this complexity to that which is fully understood. Real values are not necessary for building the initial model since many limits can be recognized in the natural world. Resistance to water flow, for example, must vary from no resistance to complete resistance. The effect of this variation can be analyzed within those limits, and if there is a significant ecological effect, further analyses are necessary.

snow.   Snow is an extremely important ecological force that performs many different functions. It reduces visibility when it falls; it is a mechanical barrier, a good insulator, a source of soil moisture, and a source of run-off water. Its many different functions as a physical material, coupled with its many different inter- actions with organisms, compels the analytical ecologist to consider it an important component of the ecosystem. It is an interesting component of analytical models because it has so many different functions. Let us consider its physical charac- teristics first and relate it to organisms in later chapters.

Ice-crystals are formed in the atmosphere at temperatures below freezing by the sublimation of water vapor on hygroscopic nuclei. The ice crystals take various shapes, dependent on atmospheric conditions at the time of crystallization. New- fallen snow generally has a low density (0.05–0.10 g cm^{-3}) owing to the dendritic structure of the crystals. Atmospheric temperature and wind are the two primary factors that alter the density of new-fallen snow. Snow density will increase an average of 0.0065 g cm^{-3} for each 1°C increase in surface air temperature at the time of deposition.

Reported density of new-fallen snow varies from 0.06 for calm conditions to 0.34 for snow deposited during gale winds. Snow density increases to 0.2– 0.4 g cm^{-3} as the age of the snowpack increases. As each new layer of snow is deposited, its upper surface is subjected to the weathering effects of radiation, rain, and wind, and the action of percolating water and diffusing water vapor.

The original delicate crystals become coarse grains; a developed snowpack shows distinct layers characteristic of individual snow-storm deposits and weathering effects (U.S. Army 1956; Nakaya 1954).

CONDUCTIVITY. Factors affecting the thermal conductivity of snow are: (1) the structural and crystalline character of the snowpack, (2) the degree of compaction, (3) the extent of ice planes, (4) the wetness, and (5) the temperature of the snow. Experimental work shows that the thermal properties of snow (specific heat, conductivity, and diffusivity) can be predicted from snow density measurements (Table 5-3).

Heat transfer in a natural snowpack is complicated by the simultaneous occurrence of many different heat-exchange processes. The water vapor condenses and yields its heat of vaporization ($\approx$0.600 kcal g^{-1}) upon reaching a cold surface. Rain or meltwater freezes within the subfreezing layers and adds the heat of fusion (0.08 kcal g^{-1}). These two processes tend to change and influence the conductivity and diffusivity of the snow throughout the pack and influence the heat transfer rates (U.S. Army 1956).

Temperature gradients in the snowpack are more pronounced in the winter than in the spring. When the snowpack reaches an isothermal condition at 0°C, the heat energy is dissipated in melting the snow.

5-4 SNOW COVER IN RELATION TO KINETIC ENERGY

WINDPACK. The wind profile that develops over different surfaces plays an important part in the characteristics of wind-packed snow (Figure 5-9). Wind flow over a snow surface has a sharp profile with high velocities very near the snow surface. The inertia of drifting snow results in a more densely packed snow layer. Vegetation interrupts the flow of air, so two separate profiles are formed. The characteristics of these wind profiles, one in and one above the vegetation, depend on the height and the density of the vegetation. The profile within the vegetation has much lower velocities resulting in a less dense snowpack. The relationships between plants, animals, and wind flow are discussed further in Chapter 6.

TABLE 5-3 THE THERMAL PROPERTIES OF THE SNOWPACK ARE RELATED TO THE DENSITY OF THE SNOW

Density ($g\ cm^{-3}$)	Specific Heat ($cal\ cm^{-3}\ °C^{-1}$)	Conductivity ($cal\ cm^{-2}\ °C^{-1}\ cm^{-1}\ sec^{-1}$)	Diffusivity ($°C\ cm^{-2}\ sec^{-1}$)
1.000 (water)	1.0000	0.00130	0.00130
0.900 (ice)	0.4500	0.00535	0.01190
0.500	0.2500	0.00205	0.00820
0.350	0.1755	0.00087	0.00494
0.250	0.1250	0.00042	0.00336
0.050	0.0250	0.00002	0.00080

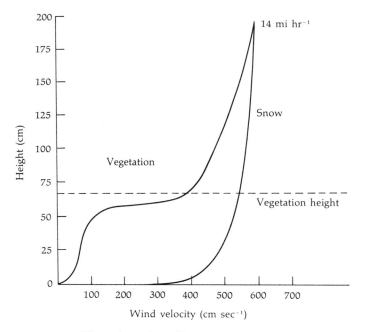

FIGURE 5-9.   Vertical wind profiles over grass that is 60–70 cm in height and over a snow surface.

THE EFFECT OF WINDBREAKS.   The effect of windbreaks on the distribution of snow is well known, but the turbulent flow generated by windbreaks is often described only in general terms. This is because of the difficulty in visually observing the wind and measuring the turbulence patterns. Experiments in the Thermal Environment Simulation Tunnel (TEST) in the BioThermal Laboratory, Cornell University, utilize a bubble generator that extrudes tiny neutrally bouyant bubbles that follow the movement of air as it passes through and around vegetation (Fig. 5–10).

The movement of air on the lee side of a canopy has some interesting characteristics. First, it is highly turbulent. The air is moving in three directions, including reverse flow illustrated by the trace of bubbles posterior to the deer. Second, there is a general downward trend in the flow of air behind the trees. The deer is located at a point where the wind direction is primarily vertical rather than horizontal. Third, the velocity and direction changes abruptly as the air approaches the surface, and the profile over snow shown in Figure 5-9 begins to develop.

Snowflakes follow the air flow, resulting in drift formation that is unlike snow accumulation in zero wind. Note the rounded upper surface of the drift shown in the cross-sectional drawing in Figure 5-11. The drift has an inversion that follows the pattern of air flow shown in Figure 5-10. Wind flow in the field is more complex than in the TEST, but the basic principles remain the same.

The significance of this wind flow, which is basically a kinetic-energy distribution, is discussed later. One important point to emphasize here is that cup anemometers are very unsatisfactory for measuring wind velocities in the region

FIGURE 5-10.   The movement of air on the lee side of an experimental windbreak in the Thermal Environment Simulation Tunnel.

around the deer shown in Figure 5-10 because these instruments do not respond to attack angles greater than 70° [see Hetzler, Willis, and George (1967)]. Thus a cup anemometer in a horizontal orientation just posterior to the deer at about the height of its tail would record zero velocity, but the deer would be experiencing the vertical air flow. The analytical ecologist must be aware of the importance of using instruments that measure the functional relationships between animal and environment if meaningful interpretations are to be made.

FIGURE 5-11.   Cross-section of a snowdrift behind a natural wind-break, North Lansing, New York. (Moen, unpublished data.)

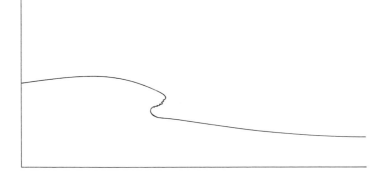

SNOWFALL INTERCEPTION.   The interception of falling snow by vegetative cover is an important factor when predicting snow accumulation on the ground, just as it is in predicting rainfall at the soil surface. The amount of interception varies greatly, depending on the type and density of the vegetation cover and the magnitude, intensity, and frequency of storms. High winds and intense solar radiation reduce the amount of snow trapped in the canopy. A moderately dense coniferous forest, in an area with an annual precipitation of 30–50 inches, may intercept 15%–30% of the total winter precipitation. Equation (5-3) was developed for estimating the amount of interception in a coniferous tree stand in the north-western United States (U.S. Army 1956).

$$\left\{ \begin{array}{l} \text{Percentage of} \\ \text{interception} \end{array} \right\} = \left\{ \begin{array}{l} 0.36 \times \text{percentage} \\ \text{of canopy cover} \end{array} \right\} \tag{5-3}$$

THE ROLE OF SNOW IN A PRECIPITATION-CANOPY-SUBSTRATE MODEL.   The amount of information about the structural, mechanical, and thermal characteristics of snow is quite adequate for assembling initial models of the role of snow in the ecology of different organisms. The equations for snow interception by different canopies and for conductivity of snow of different densities are important for determining the amount that reaches the ground's surface and its role as a mechanical barrier and an insulator against heat loss. Students are urged to develop models that permit calculations of these functions of snow. In the absence of real data that can be used to describe some of the functions, limits can still be recognized that serve the purpose of making first approximations. For example, the aging process of snow results in a continual change in snow density, and this in turn affects conductivity. The aging process is very dependent on radiation, wind, air temperature, and precipitation, and these are not readily predicted. Their effect, however, can be approximated since the lowest snow density possible is one limit and maximum density—ice—is the other. An initial model containing changes in snow density could include an equation that describes these changes purely as a function of time. The simplest format is a linear regression equation. Initial analyses can then begin with the philosophy "What if . . . ?", and the analytical ecologist uses the outputs from such considerations in determining the *effect* of such changes on the organism(s) in question. Once this has been done, it is desirable to go back to the first approximations and improve on them so they become more and more representative of real situations in the natural world.

THE EFFECT OF SNOW DISTRIBUTION ON ANIMALS.   The distribution of snow is a reflection of the distribution of kinetic energy in time and space. Animals are subject to these patterns and must be able to cope with them if they are to survive. High-density snowpacks can have opposite effects on animal life. They may pose a mechanical barrier, increasing the metabolic energy necessary for the animal to move through the snow. This is a cost to the animal that must be compensated for either by food ingested and metabolized or by reserves that have been built up during more favorable periods. Or the snowpack may be dense enough to support the weight of the animal, facilitating its movement. This results in the conservation of energy and it may also place the animal within reach of a food supply that would otherwise be too high.

The mechanical characteristics of the snowpack have a direct effect on the thermal benefit it provides such birds as grouse, which roost in the snow on cold winter nights. If the birds can penetrate the snow, they will be in a thermal regime that is very different from one above the snow surface.

Snow accumulation may directly influence the amount of available food for a wild animal. For example, the vertical distribution of potential food for deer that browse in an upland hardwood community in the Connecticut Hill Game Management Area south of Ithaca, New York, is such that one foot of snow renders 97% of it available. The zone of invading plants between a hemlock stand and an abandoned field have quite a different vertical distribution of food. One foot of snow in that habitat reduced the food supply by 25%, and two feet of snow reduced it by 40%. The marked differences between these two types of habitats indicate that the effect of one foot of snow is dependent on the vertical distribution of food, so snow depths of one foot in each of the two stands are not ecological equivalents.

The interrelationships between snowpack characteristics, forage production and availability, and animal requirements can be illustrated in a flow sheet as in Figure 5-12. The arrows indicate that there is a relationship between two or more factors, and the expression of these relationships in mathematical form will convert the picture-type model to a working model. For example, forage production over a vertical height of six feet can be expressed quantitatively. Snow depth can be expressed in relation to the removal of forage from the "available" category. Animals will eat some or all of the available forage, depending on their energy requirement. The energy requirement is in part a function of the snow depth, since it takes more energy to move through deep snow than through shallow snow.

These kinds of relationships are discussed in later chapters, especially in relation to the concept of carrying capacity. Students are urged to think about the relationships shown in Figure 5-12, progressing toward a numerical model that might be made up of dimensionless numbers at first, and then of measurements in different habitats.

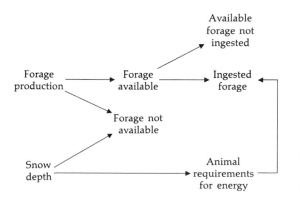

FIGURE 5-12. A flow sheet indicating the existence of interrelationships between forage availability, snow depth, and forage ingested.

Since the model-building process is such a significant part of analytical ecology, it is essential that students grasp the logic and philosophy sufficiently well to begin building meaningful models themselves. Let us turn our attention in the next chapter to weather and the processes of thermal exchange, with additional models that illustrate the process of model building and at the same time provide information on the distribution of thermal energy in the real world. Keep in mind that the process of analytical ecology results in an analysis that may be compared to dramatic art, moving from simple one-act ecological plays, each with just a few principal characters, to the more comprehensive productions that approach greater and greater realism.

LITERATURE CITED IN CHAPTER 5

Barry, R. G., and R. J. Chorley. 1970. *Atmosphere, weather, and climate.* New York: Holt, Rinehart and Winston, 320 pp.

Hetzler, R. E., W. O. Willis, and E. J. George. 1967. Cup anemometer behavior with respect to attack angle variation of the relative wind. *Trans. Am. Soc. Agr. Eng.* **10**(3): 376–377.

McCullough, E. C., and W. P. Porter. 1971. Computing clear day solar radiation spectra for the terrestrial ecological environment. *Ecology* **52**(6): 1008–1015.

Nakaya, U. 1954. *Snow crystals: natural and artificial.* Cambridge: Harvard University Press, 510 pp.

Sellers, W. D. 1965. *Physical climatology.* Chicago: University of Chicago Press, 272 pp.

Taylor, G. F. 1954. *Elementary meteorology.* New York: Prentice-Hall, 364 pp.

U.S. Army. 1956. *Snow hydrology.* Portland, Oregon: N. Pacific Div., Corps of Engineers, U.S. Army, 437 pp.

SELECTED REFERENCES

Day, J. A. 1966. *The science of weather.* Reading, Massachusetts: Addison-Wesley, 214 pp.

Geiger, R. 1965. *The climate near the ground.* Cambridge: Harvard University Press, 611 pp.

Munn, R. E. 1966. *Descriptive micrometeorology.* New York: Academic Press, 245 pp.

Smithsonian Meteorological Tables. 1951. 6th rev. ed. Publication No. 4014, 527 pp.

Sutcliffe, R. C. 1966. *Weather and climate.* London: Weidenfeld and Nicolson, 206 pp.

Wang, J. 1963. *Agricultural meteorology.* Milwaukee: Pacemaker Press, 693 pp.

Willet, H. C., and F. Sanders. 1959. *Descriptive meteorology.* New York: Academic Press, 355 pp.

WEATHER AND THE PROCESSES
OF THERMAL
EXCHANGE

Weather has long been an important consideration of ecologists. It is important for the analytical ecologist to remember that he is interested in the *effects* of weather on organisms. An analysis of functional relationships between weather and organism causes the analytical ecologist to consider the distribution of thermal energy, since it is thermal energy that is the most common bond in these relationships. Analyses of thermal energy relationships between an organism and its environment are centered on the four basic modes of heat transfer. These occur within the organism, in the interface between the organism and its environment (a thermal boundary layer), and between the environment and the thermal surface of an organism. All organisms are continually exchanging thermal or heat energy with their environments by these four modes. The rates of thermal energy exchange depend on the thermal characteristics of the organism and its environment and the interaction between different modes of heat transfer between the two.

6-1 THE FOUR MODES OF HEAT TRANSFER

Thermal energy is exchanged between animal and environment by radiation, conduction, convection, and evaporation. A conceptual picture of the complex nature of heat exchange for a free-ranging animal is useful for recognizing components of the energy regime (Figure 6-1). It is impossible to describe mathematically all of the dynamic thermal relationships between an organism and its environment, but research on the thermal energy exchange of both plants and animals has provided insight into the mechanisms involved.

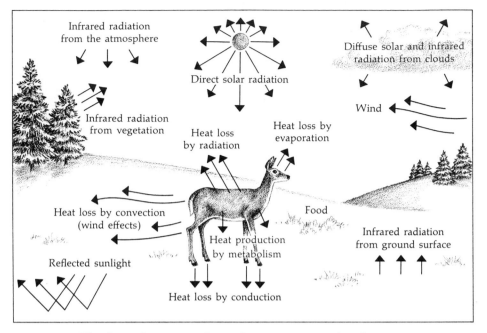

FIGURE 6-1.   The thermal energy exchange between an animal and its environment includes radiation, conduction, convection, and evaporation as the four basic modes of heat transfer.

6-2 RADIANT ENERGY EXCHANGE

Radiant energy exchange occurs between two surfaces as each surface emits energy at wavelengths that are dependent on the temperature of the emitting surface. Radiant energy travels through those media that are transparent, neither reflecting nor absorbing the radiation. A complete vacuum presents no obstructions to radiant energy exchange. A discussion of the basic characteristics of radiant heat exchange follows.

THE ELECTROMAGNETIC SPECTRUM.   The electromagnetic spectrum includes wavelengths as long as hundreds of miles and as short as 1×10^{-10} cm (0.0000000001 cm). Between these two extremes there are, in order of decreasing wavelength, the radio waves that are received by standard radio sets, the short waves, the infrared portion (perceived as heat), visible light, ultraviolet rays, x rays, and gamma rays. The radiant energy that is important in the maintenance of homeothermy includes the wavelengths in the visible portion and the longer wavelengths in the infrared portion of the electromagnetic spectrum.

Both the amount of radiation from an object and the wavelengths emitted are functions of the temperature of the object. The Stefan-Boltzmann law states that the amount of radiation emitted from a black body is directly proportional to the fourth power of the absolute temperature (°K) of the object. A black body is an object that absorbs all the radiant energy that reaches its surface. If a surface is not a black body, the amount of energy that can be absorbed is expressed as a coefficient ranging from zero to one. The absorption coefficient for long-wave radiation is also equal to the emissivity of that surface, since it is equally as good an emitter of long-wave radiation as it is an absorber. This relationship can be expressed with equation (6-1).

$$Q_r = \epsilon \sigma T_s^4 \qquad (6\text{-}1)$$

where:

$Q_r =$ radiant energy emitted in kcal m^{-2} hr^{-1}
$\epsilon =$ emissivity (range from zero to one)
$\sigma =$ Stefan-Boltzmann constant $= 4.93 \times 10^{-8}$ kcal m^{-2} hr^{-1}
$T_s =$ surface temperature of the object in °K

The Wien displacement law states that the wavelength (λ) of maximum intensity that is emitted from the surface of a black body is inversely proportional to the absolute temperature of the body [equation (6-2)].

$$\lambda \text{ max } (\mu) = 2897 \ T^{-1} \qquad (6\text{-}2)$$

Thus a very hot surface emits shorter wavelengths, while a cooler surface emits longer wavelengths.

SOLAR RADIATION. The wavelength of maximum emission from the sun is 0.5μ, which is in the visible portion of the spectrum (Sellers 1965). Sellers points out that 99% of the sun's radiation is in the wavelength range of 0.15μ to 4.0μ, including 9% in the ultraviolet ($<0.4\mu$), 45% in the visible (0.4μ to 0.74μ), and 46% in the infrared ($>0.74\mu$).

The amount of energy reaching a surface perpendicular to the rays of the sun at the outer limits of the earth's atmosphere is called the solar constant. Textbooks published prior to the mid-fifties included a value for the solar constant of 1.94 calories per square centimeter per minute. This value has been revised by Johnson (1954) to 2.00 calories per square centimeter per minute. It actually varies by about 1.5% because of differences in the total energy emanating from the sun and because of variation in the distance between the earth and the sun.

Not all of the sun's energy reaches the surface of the earth. Some of it is reflected into space or absorbed by dust particles and moisture in the atmosphere. On a clear day, a high percentage of the solar radiation is transmitted through the atmosphere. With a completely overcast sky, no direct solar radiation penetrates the cloud cover.

There are three possible pathways for radiant energy to take once it reaches a plant or animal. It may be reflected from the surface, it may be absorbed by the surface, or it may be transmitted through the material (Figure 6-2). Energy that is reflected from the surface is of no thermal benefit to an animal or plant. Reflected energy in the visible portion of the spectrum is detected as shades of gray or as color, depending on the receptors of the organism detecting the light energy. Transmitted solar energy is of no value to an animal, and all animals

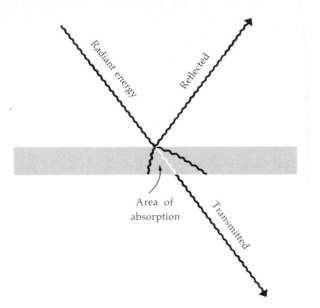

FIGURE 6-2. The three pathways for radiant energy reaching a plant or an animal.

except the smallest protozoans are essentially opaque. The leaves of plants, however, transmit some solar energy. Absorbed energy becomes a part of the thermal and physiological regime of an organism, and the quantity and distribution of absorbed radiant energy is of interest to the physiologist and the ecologist.

The amount of energy absorbed by the hair surface of a mammal depends on the spectral characteristics of the hair and the angle at which the solar energy strikes the surface. The absorption coefficients for cattle have been measured by Riemerschmid and Elder (1945) and are shown in Figure 6-3. White coats absorb less and reflect more solar energy; black coats absorb the most solar energy. The greatest amount of energy is absorbed when the solar radiation strikes perpendicular to the surface; no absorption takes place when the rays are parallel to the surface. Inclination of the hair, the smoothness or curliness of the coat, and seasonal changes in the characteristics of the coat have little effect on the absorptivity. Since animals are not plane surfaces but have a complicated geometry, the absorption characteristics of a whole animal include all angles from 0 to 90 degrees. The distribution of solar radiation on the surface of an animal is called the solar radiation profile and is discussed later in the chapter.

INFRARED RADIATION. Radiation of wavelengths longer than those in the visible portion of the spectrum is called infrared radiation. The same pathways of reflection, absorption, or transmission are followed as for solar radiation. All objects emit infrared radiation according to the Stefan-Boltzmann law [equation (6-1)]. The energy of the shorter wavelengths characteristic of the visible portion of the spectrum that is absorbed by an object is reradiated as infrared energy according to the Stefan-Boltzmann law.

ATMOSPHERIC TRANSMISSION AND ABSORPTION CHARACTERISTICS. The clear atmosphere is nearly transparent to the wavelengths within the visible portion of the electromagnetic spectrum. Emissivity and absorption characteristics of the atmosphere in the infrared portion of the electromagnetic spectrum are dependent on the composition of the atmosphere. Water vapor and carbon dioxide have high emissivities and absorptivities at certain wavelengths, so the transmission of infrared energy through the atmosphere at those wavelengths is very low. Cloud cover has essentially 100% absorptivity. Some wavelengths are not absorbed by the atmosphere and hence pass through the "atmospheric windows" (Figure 6-4).

Solar radiation that is transmitted through the atmosphere during daylight is partially absorbed by the earth's surface, and the heat energy is reradiated at longer wavelengths. Some of these wavelengths are transmitted through the atmospheric windows and their heat energy is dissipated into space.

If the atmosphere were completely transparent to radiation, the earth would be considerably warmer during the day and colder during the night. The blanketing effect of the atmosphere results in the maintenance of relatively stable climates. The surface of the moon has fluctuations in temperature from about 240°F to −260°F because there is no atmosphere to buffer the radiant heat exchange.

FIGURE 6-3.   The absorption of solar radiation by three breeds of cattle at different angles of incidence. (From "The absorbtivity for solar radiation of different coloured hairy coats of cattle" by G. Riemerschmid and J. S. Elder. *Onderstepoort J. Vet. Sci. Animal Ind.* 1945. Reproduced under Copyright Authority 4614 of 23.11.1971 of the Government Printer of the Republic of South Africa.)

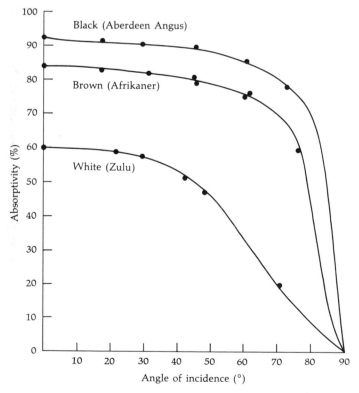

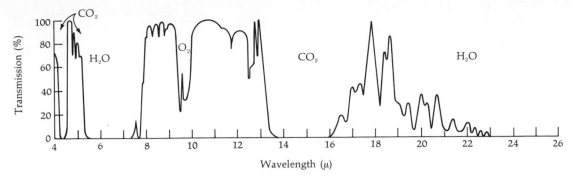

FIGURE 6-4. Atmospheric transmission in the infrared portion of the electromagnetic spectrum. (From Gates 1962.)

The blanketing effect of the earth's atmosphere is often called the "greenhouse effect." A greenhouse transmits solar radiation through glass, and this radiant energy is absorbed by the objects inside. This energy is reradiated at longer wavelengths, but since glass is not highly transparent to infrared waves, some of the energy is reflected back from the glass and retained inside the greenhouse. This effect is also obvious in an automobile when the windows are closed on a hot, sunny day.

ATMOSPHERIC EMISSION. The amount of infrared radiation emitted from the atmosphere during different weather conditions has been measured by meteorologists, and radiation charts for estimating atmospheric radiation from certain meteorological parameters have been constructed. Empirical equations have also been used to estimate the radiation exchange. They have been criticized because they are based on the assumption that outgoing radiation at any given point is determined chiefly by thermal conditions at the surface (Sellers 1965). If empirical equations are applied to specific atmospheric conditions, such as a clear sky, the results are quite reasonable owing to the "screen effect" discussed by Swinbank (1963). The screen effect results from radiation exchange within the atmosphere that limits the effective radiating height of an air column to a few feet. Swinbank presents data that show this to be the case at higher altitudes also, indicating that the total height of the atmospheric column is not particularly important. Empirical equations for given habitats and atmospheric conditions are presented later in this chapter (see Figure 6-5 and Table 6-1).

Infrared energy flux between the atmosphere, overhead vegetative cover, and the earth's surface can be divided into downward and upward components. The difference between the two is called the *net* radiation, and the sum of the two is the *total* radiation. Before considering the ecological implications of these two, let us look at some downward and upward flux measurements under different sky conditions and in different habitats.

Extensive field measurements by the author in both Minnesota and New York indicate that the amount of radiant energy flux under clear skies at night can be predicted with considerable precision if the atmospheric temperature is known. This method was used by Swinbank (1963) also.

Radiant energy flux under clear skies at night in open fields in western Minnesota near Kensington, eastern Minnesota near Bethel, and western New York in Lansing in both winter and spring is shown in Figure 6-5. Some significant conclusions can be drawn from this figure. The downward radiation during clear nights is obviously less than the upward radiation; the clear night sky is cold. The upward radiation—a function of surface (snow or vegetation) temperature and emissivity—is very closely related to air temperature. This is shown in the regression equation for upward flux where the slope of the line is 1.03, or essentially 1.0, and the intercept is −0.049, or just about zero. Thus, radiation from a snow surface can be approximated by the use of air temperature for T (in °K) in equation (6-1). This assumes that the snow-surface temperature is equal to air temperature, which is a good approximation up to 0°C. The data for upward radiation in Figure 6-5 and Table 6-1 clearly indicate this relationship. Less reflective and more absorbent surfaces, such as soil, will show greater temperature differences, especially during the day and in the early part of the evening before radiational cooling has occurred.

The relationship between the radiant temperature of the sky and air temperature is shown in Figure 6-6. The regression equations and correlation coefficients for the four different periods of measurement are listed in Table 6-1. Note the

FIGURE 6-5. Downward and upward radiation flux in open fields under clear night skies. (From Moen and Evans 1971.)

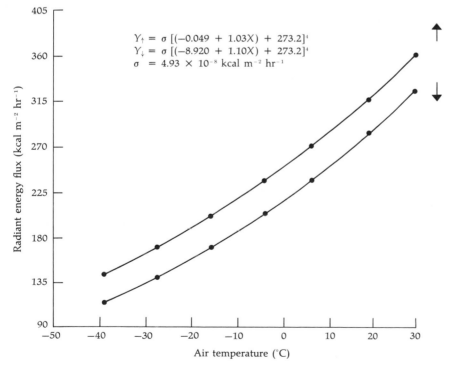

$$Y_\uparrow = \sigma\,[(-0.049 + 1.03X) + 273.2]^4$$
$$Y_\downarrow = \sigma\,[(-8.920 + 1.10X) + 273.2]^4$$
$$\sigma = 4.93 \times 10^{-8}\ \text{kcal m}^{-2}\ \text{hr}^{-1}$$

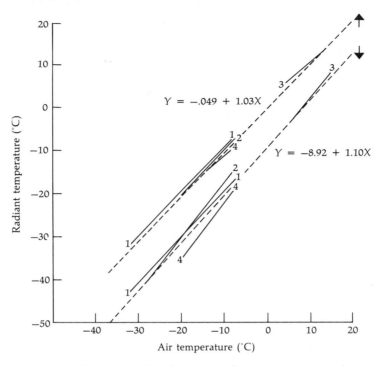

FIGURE 6-6.   The relationship between radiant temperature and air temperature under clear skies at night.

TABLE 6-1   REGRESSION EQUATIONS AND CORRELATION COEFFICIENTS FOR THE TEMPERATURE MEASUREMENTS IN FIGURE 6-6

Location	Upward or Downward	Formula	Correlation Coefficient
Kensington, Minn. (winter)	Upward	$Y = -0.025 + 1.015X$	0.994
	Downward	$Y = -9.202 + 1.039X$	0.968
Bethel, Minn. (winter)	Upward	$Y = 0.490 + 1.059X$	0.993
	Downward	$Y = -4.611 + 1.264X$	0.935
Lansing, N.Y. (spring)	Upward	$Y = 1.065 + 0.947X$	0.995
	Downward	$Y = -10.736 + 1.288X$	0.994
Lansing, N.Y. (winter)	Upward	$Y = -1.568 + 0.948X$	0.984
	Downward	$Y = -8.888 + 1.285X$	0.973
All measurements	Upward	$Y = -0.0491 + 1.030X$	0.999
	Downward	$Y = -8.9150 + 1.103X$	0.995

Radiant Temperature spans the Upward or Downward, Formula columns.

SOURCE: Moen and Evans 1971.

greater variation in the radiant temperature of the sky than in the radiant temperature of the snow or plant surfaces. All measurements were made under clear skies, but differences in the atmospheric temperature profile and in the vapor pressure of the atmosphere contribute to variation in the downward radiation flux. Additional field measurements by the author show that, under cloudy skies, the radiant temperature of the atmosphere is very nearly equal to air temperature.

What is the ecological significance of net and total radiation flux? First of all, the application of these terms to the *atmospheric energy balance* is of interest when the energy balance of the earth is being considered. In that context, they are meteorological terms and not ecological terms. To illustrate, the downward radiation flux under a clear sky with an air temperature of $-30°C$ is 141 kcal m^{-2} hr^{-1} (see Figure 6-5), the upward flux is 169 kcal m^{-2} hr^{-1}, the total is 310 kcal m^{-2} hr^{-1}, and the net is -28 kcal m^{-2} hr^{-1}. At an air temperature of $+20°C$, the downward flux is 366 kcal m^{-2} hr^{-1}, the upward is 325 kcal m^{-2} hr^{-1}, the total is 691 kcal m^{-2} hr^{-1}, and the net is -41 kcal m^{-2} hr^{-1}. Note that there is a larger negative balance at the higher temperatures, but the total flux is more than two times greater at $+20°$ than at $-30°C$. The total radiant energy flux strikes the surface of plants and animals, and the additional radiation at warmer temperatures results in a greater amount of absorbed radiation by the organisms. Geometric considerations necessary to calculate the absorbed thermal radiation are very complex. The ecologist is interested in the exchange of heat between organisms and environment, just as the meteorologist is interested in the exchange of heat between the earth, the atmosphere, and the exosphere. Thus a careful distinction must be made between the net energy balance of the earth and the net energy balance of an organism living on the earth.

The amount of infrared energy from three different cover types under clear nocturnal skies in the winter is distinctly different. Of the three types indicated in Figure 6-7, the least amount of downward radiation comes from the clear sky and the most from the cedar (*Thuja occidentalis*) cover. The differences between them are related to the density of the overhead cover. The density, however, is not the density viewed from the ground vertically through the canopy at a number of points; rather it is the effective thermal density as measured from a single location in the stand. At angles of less than 90°, the effective density of the stand increases in a manner similar to a "venetian blind effect." For example, the cedar canopy in the Cedar Creek Natural History Area in Minnesota occluded from 50% to 80% of the sky within a 35° field of view above the radiometer. In the upland hardwood stand, only 10% to 50% of the sky was occluded in a 35° field of view above the radiometer, but the downward radiation was midway between the downward flux in cedar cover and from the clear sky in an open field. This illustrates that, at increasing angles from the zenith, a canopy becomes optically and thermally more opaque. Since the radiometer senses almost the entire hemisphere, the measured downward flux is considerably higher than the canopy density viewed vertically would indicate. The effect of this radiant energy on the snow is apparent in the formation of icy crusts in open fields and the looser, less crusty snow found in forest stands.

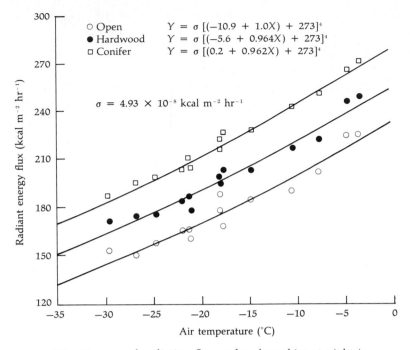

FIGURE 6-7.   Downward radiation flux under clear skies at night in three cover types on the Cedar Creek Natural History Area, Minnesota. (From Moen 1968, *J. Wildlife Management.*)

What limits are imposed on the amount of radiation from vegetation and snow in the canopy? Complete obstruction of the sky results in a maximum amount of downward radiation, equal to the flux from a completely overcast sky when the vertical temperature profile from the ground surface into the cloud cover is isothermal. An animal in this circumstance is exposed to a homogeneous radiant environment in all directions, very similar to a chamber with uniform temperature distribution. The actual thermal benefits derived from this radiant energy cannot be determined until the interaction between radiation and convection in the insulating layer of hair has been analyzed.

RADIATION FROM PHYSICAL AND BIOLOGICAL OBJECTS.   The amount of thermal radiation from physical and biological objects is a function of their emissivity (ϵ) and temperature (°K). The emissivity is an important part of the equation when it is applied to biological organisms that have a limited surface area. Potential error in radiation measurements that may be introduced by making false assumptions about emissivity of plants is discussed in Fuchs and Tanner (1966). Porter (1969) discusses the importance of considering emissivity in making measurements in a chamber. The general rule to follow is that there are larger differences between apparent radiant temperature and real radiant temperature as the difference between the radiant temperatures of the target organism and the environment becomes greater. The importance of radiant energy exchange in the entire thermal balance also increases under these conditions of greater temperature differences.

EMISSIVITY AND REFLECTIVITY OF SNOW.   The snow surface is composed of small ice crystals, making it extremely rough. The rough snow surface is almost a perfect black body for the absorption and emission of long-wave radiation. Since the temperature of snow is limited to a maximum of $0°C$, the maximum intensity of radiation that may be emitted is 27.45 ly hr^{-1} (calories cm^{-2} hr^{-1}) or 274.48 kcal m^{-2} hr^{-1}, calculated from equation (6-1).

Snow is a good reflector of radiant energy in the visible portion of the electromagnetic spectrum. Freshly fallen snow may have an albedo of 75%–95%, although snow several days old may reflect 40%–70% of the short-wave radiation (Sellers 1965).

The infrared emissivities of most biological materials are close to 1.0. Several hair surfaces have been tested, with measured emissivities ranging from 0.92 to 1.0 for several species (Table 6-2). An emissivity of 1.0 is a satisfactory first approximation at this point. The error will be quite small in the range of temperatures experienced by animals in natural habitats.

RADIANT TEMPERATURE IN RELATION TO AIR TEMPERATURE.   There is a predictable relationship between the radiant temperature (T_r) of an animal and the air temperature (T_a) if environmental radiation striking the animal's surface and wind flow over the surface are known. This is illustrated in Figure 6-8 for deer and sharp-tailed grouse. Note that the radiant temperature of the animal's surface drops as the air temperature drops but in both cases the difference between T_r and T_a is greater at colder temperatures. This is an important consideration in later thermal analyses. For now, remember that as the air temperature drops, the radiant surface temperature of an animal also drops, but at a slower rate. This results in a relatively warmer radiant surface when exposed to a colder air temperature.

Variations in wind velocity and environmental radiation affect the relationship between T_r and T_a. The effects of different wind velocities across the surfaces of deer and grouse simulators are shown in Figure 6-9 and 6-10, respectively. At $-20°C$, the radiant temperature of deer is about $-6°C$ when the velocity of the wind is less than 1 mi hr^{-1}, but $-14°C$ when the velocity is 14 mi hr^{-1}. This drop of eight degrees in the surface temperature is due to the increased convection losses at the higher velocities. The nonlinear effect of wind is also observed when a change from 0 to 8 mi hr^{-1} at $-20°C$ reduces the radiant temperature by 5.9 degrees, but a change from 8 to 14 mi hr^{-1} reduces it further by only 2.4 degrees (Figure 6-9).

The dashed line in Figure 6-9 illustrates radiant temperature in relation to air temperature for a wind velocity of 0 mi hr^{-1} when the sky acts as a cold heat sink. The radiant temperature is depressed when there is no wind blowing across

TABLE 6-2 INFRARED EMISSIVITIES OF DIFFERENT HAIR OR FEATHER SURFACES

Species	Condition of Pelage	Emissivity
Willow Ptarmigan† *(Lagopus lagopus)*	On frozen carcass	.98
Snowshoe hare† *(Lepus americanus)*	On frozen carcass	.99
Cottontail rabbit*	Dorsal sample	0.97–0.98
(Sylvilagus floridanus)	Ventral sample	0.92–0.93
Barren ground caribou† *(Rangifer arcticus)*	Frozen, off carcass	1.00
Sea otter† *(Enhydra lutris)*	Tanned	0.98
Grey wolf† *(Canis lupus)*	Tanned	0.99
Beaver† *(Castor canadensis)*	Tanned	0.99
Beaver† *(Castor canadensis)*	Dry	1.00
Lynx† *(Lynx canadensis)*	Tanned	1.00
Red fox† *(Vulpes fulva)*	Tanned	0.98–1.00
Marten† *(Martes americana)*	Tanned	1.00
Bobcat† *(Lynx rufus)*	Tanned	1.00
Flying squirrel*	Dorsal sample	0.95
(Glaucomys volans)	Ventral sample	0.95–0.99
Woodchuck* *(Marmota monax)*	Dorsal sample	0.98
Red squirrel*	Dorsal sample	0.95–0.98
(Tamiasciurus hudsonicus)	Ventral sample	0.97–1.00
Grey squirrel*	Dorsal sample	0.99
(Sciurus carolinensis)	Ventral sample	0.99
Mole* *(Scalopus aquaticus)*	Dorsal sample	0.97
Deer Mouse* *(Peromyscus sp.)*	Ventral sample	0.94

†Data from Hammel 1956.
*Data from Birkebak, Birkebak, and Warner 1963.

the hairy surface. At wind velocities of 1 mi hr^{-1} or more, the effect of the cold sky disappears since the air temperature is the dominant thermal factor. This can be attributed to advection, which is basically the process of convection discussed in Section 6-4.

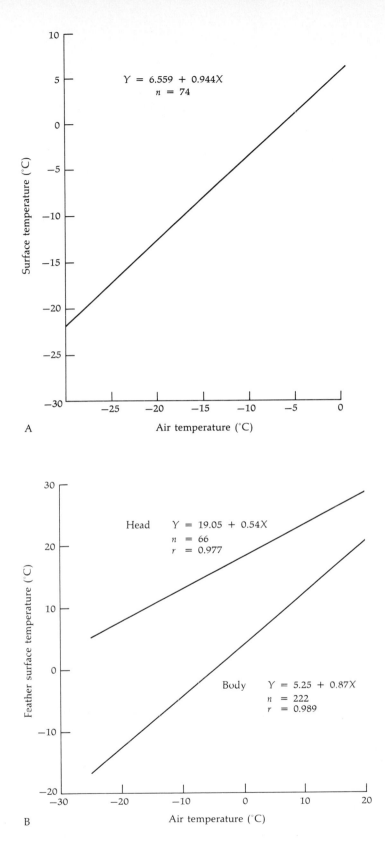

FIGURE 6-8. Radiant surface temperature related to air temperature for (A) deer and (B) grouse. (Data on deer are from Moen 1968 and on grouse from Evans 1971.)

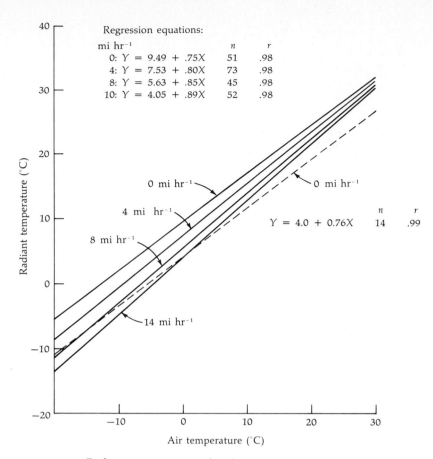

FIGURE 6-9. Radiant temperature related to air temperature for the white-tailed deer simulator in the TMST with two levels of environmental radiation. Solid line: T_r when $T_e = T_a$ in test chamber. Dashed line: T_r when $T_e = T_a - 10$ in test chamber. (Additional data are given in Appendix 4.)

The radiant temperature of an animal's surface rises when the animal is exposed to higher levels of environmental radiation. The surface temperature of an adult male pheasant, for example, was 45°C when its back was exposed at right angles to a bright afternoon sun. No wind was blowing past the surface at the time. Since the body temperature of a pheasant is about 40°C, the radiant surface was warmer than the body itself. When a cloud shaded the sun and a slight wind blew over the pheasant's surface, the radiant temperature dropped to 20°C, which was about equal to the air temperature, in a manner of seconds. This indicates the variability of the animal's radiant temperature, which changes with changes in wind, radiation, temperature, and other thermal factors. Hair and feathers do not merely provide insulation from the cold, but from the heat as well. They tend to ameliorate the thermal regime of the body proper by buffering the thermal variations.

The effect of direct solar radiation on the radiant temperature over the entire surface of a plant or animal is not uniform because both the color of this surface

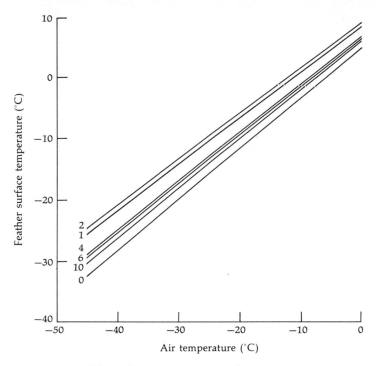

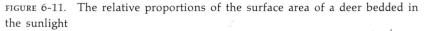

FIGURE 6-10. Effect of air temperature and wind velocity on feather surface temperatures. Prediction formulas are listed in Appendix 4. (Evans 1971.)

and the angle of the rays striking the surface are important in determining just how much solar energy is absorbed. For example, a deer bedded in the sun (Figure 6-11) might have 40% of its surface exposed to direct solar radiation, 80% exposed to indirect solar radiation, and 80% exposed to infrared radiation. These differ-

FIGURE 6-11. The relative proportions of the surface area of a deer bedded in the sunlight

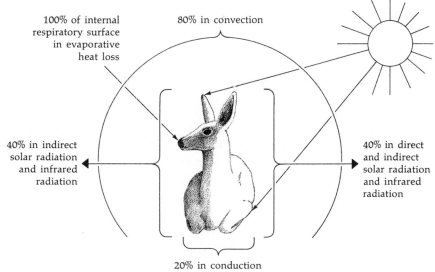

ences result in different radiant temperatures, but they act in combination with the distribution of tissue metabolism beneath the hair surface as well as with the distribution of blood. The radiant temperature distribution of an animal exposed to only infrared radiation at night is much simpler since the hair surface is almost a black body and infrared energy is much more uniformly distributed in the environment than is solar energy. The biological characteristics of tissue metabolism and blood flow still contribute to variations, however.

Radiant temperatures of leaf surfaces can be measured remotely with an infrared thermometer just as animal surfaces are measured. Radiant temperature and air temperature are much more similar for plants than for animals because only a small amount of metabolic heat is released inside the leaf. In homeothermic animals, however, there is heat flow from the many exothermic reactions inside the animal to the outside through a layer of insulating hair or feathers. Thus the surface temperature of an animal is dependent on blood flow beneath the skin, local tissue metabolism, and on the quality of the insulating pelage. In a plant, the surface temperature is dependent primarily on the interactions between thermal parameters alone, with virtually no input from metabolic reactions within the leaf.

The considerations described above are indicative of the complexity facing the ecologist who looks at the way things function in the real world. There is so much interaction that it is difficult and frustrating to talk about isolated things because they never function in isolation! Yet, in writing a general text for students of ecology, it is necessary to cover things one at a time, synthesizing more and more as knowledge and understanding accumulate.

6-3 INSTRUMENTATION FOR MEASURING RADIATION

The basic design of equipment for measuring radiation in the visible and infrared portion of the spectrum is quite simple. An economical radiometer (Figure 6-12), designed by Suomi and Kuhn (1958) and described in greater detail by Tanner, Businger, and Kuhn (1960), has the necessary components for measuring radiation (Figure 6-13).

The sensing element is the basic component of an instrument that measures radiation. This sensor may have spectral characteristics that measure only solar or long-wave thermal radiation. A black paint such as Minnesota Mining Nextel Velvet Coating (101-C10 Black) has a quoted emmissivity of 0.9 or better at wavelengths from 2 to 35 microns. Thus it is almost a black body in the infrared portion of the spectrum. White or silver paints have low absorptivities and high reflectivities in the visible portion of the spectrum. These are used on the sensing elements of instruments that separate solar from long-wave radiation.

A thermometer, thermocouple, or thermistor may be used to measure the temperature of the sensing element. Commercial instruments usually have thermocouples or thermistors as temperature sensors. They respond to changes more quickly than thermometers do, an advantage if the radiation flux varies over short

FIGURE 6-12. An economical radiometer in a maple-basswood stand.

FIGURE 6-13. The components of an economical homemade radiometer. The dimensions suggested can be altered if necessary.

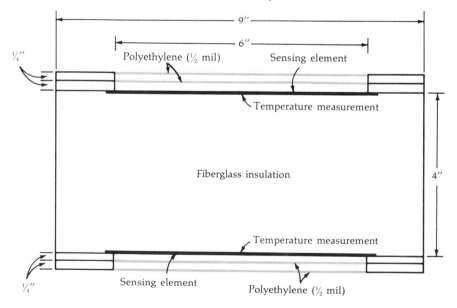

time spans. Thermometers do not require a power source, making them more convenient for field use.

Another component is the shield over the sensing element. The shield may act as a filter that permits only certain wavelengths to reach the sensing element. Polyethylene, for example, is used in the economical radiometer for measuring long-wave radiation because it has a high transmission coefficient in the infrared. Glass would not be suitable because it is not transparent in the infrared. The shield on the economical radiometer is flat. On some instruments it is hemispherical, which is better for measuring solar radiation because the reflectivity of a flat surface increases as the angle of incidence decreases.

Another function of the shield is to protect the sensing element from wind. A sensing element warmed by the sun is cooled if wind blows over its surface. This results in an underestimation of the radiant energy. Actually, free convection occurs at the surface of even an enclosed sensor, but this is fairly negligible with proper instrument design. Some instruments are "ventilated," with a fan providing a constant wind effect that is corrected for in the electronic circuitry. The insulation in the economical radiometer reduces heat flow between the top and bottom of the sensing elements. Any commercial insulation with a known thermal conductivity can be used. A correction for one-dimensional heat flow through the insulation from one sensor to the other is made.

The equations for calculating radiation flux with the economical radiometer are shown in Appendix 2. Good sources for the design of other radiometers are Gates (1962) and Platt and Griffiths (1964).

The accuracy of different instruments is an important consideration before selecting one for field or laboratory use. The accuracy required is related to the use of the data, so selection should not be made on the basis of stated accuracies alone. The economical radiometer is sufficiently accurate for measuring radiation flux in different habitats at night, especially if several instruments are used and the results are averaged, since the application of radiation data to animals is not a particularly precise procedure.

6-4   CONVECTION

Heat energy may be removed from the surface of an object by fluid (air) flowing over the surface. This process is called convection. Two types of convection occur: free convection (also called natural convection) and forced convection. Free convection occurs when temperature differences in the boundary layer of air surrounding an object cause a movement of the air in response to changes in air density. Forced convection occurs when external pressure differences cause wind to blow past the object. Before considering convection in relation to biological organisms, let us consider air flow over different surfaces and past objects such as windbreaks and animals.

AIR MOVEMENT.   The air surrounding an organism has certain physical and thermal characteristics that should be understood before the convection processes are described. One of these is the presence of a velocity boundary layer (Figure

6-14), which develops because of surface friction. At the surface, the velocity of air flow is, theoretically, zero. The velocity increases at greater distances from the surface, and the point at which maximum velocity is reached marks the beginning of the free air stream. The depth of the boundary layer depends primarily on the roughness of the surface.

MEAN VELOCITY WIND PROFILES.   The mean horizontal velocities can be calculated with equation (6-3) (from Sellers, 1965, with modification).

$$U_z = (u_*/k) \ln (Z/Z_0) \qquad\qquad (6\text{-}3)$$

where:

U_z = wind velocity at height Z
u_* = friction velocity
k = von Karman constant = 0.4
Z = height in cm
Z_0 = roughness parameter of surface; the height at which velocity is zero

The reduction in wind velocity owing to friction will be greater over a rough, vegetation-covered surface than over a smooth, snow surface. If the wind velocities were the same, say, 622 cm sec^{-1} (13.9 mi hr^{-1}), at a height of 2 meters over both types of surfaces, the velocity at a height of 75 cm (or around the top of the back of an average deer) would be about 433 cm sec^{-1} (9.7 mi hr^{-1}) over

FIGURE 6-14.   Predicted vertical profiles of mean wind velocity over grass that is 60–70 cm in height and over a snow surface. The profiles within the vegetation are based on data on corn. (From Stevens and Moen 1970.)

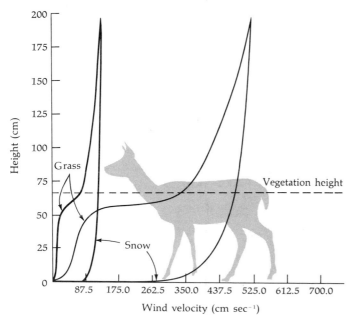

grass that is 60–70 cm high, and 572 cm sec^{-1} (12.8 mi hr^{-1}) over a snow field (see Figure 6-14). The shape of the velocity profile within the vegetation depends on the life-form and density of the vegetation. The profile within the vegetation in Figure 6-14 is approximated for grass from data for air flow through a corn field (Ordway 1969). Similar profiles are shown by Plate and Quraishi (1965) for corn and wheat.

Mean velocity profiles give the impression that wind flow is layered or *laminar*. Actually, wind flow in the field is three-dimensional. There is air movement on a horizontal plane in one direction (U_u), perpendicular to U_u but in the same plane (U_w), and vertical (U_v), or perpendicular to the plane of U_u and U_w.

Three-dimensional wind flow is caused by friction between air molecules and the ground surface, causing turbulence, or a mixing of the air. Turbulent flow can be described in terms of scale and intensity. Turbulence scale is a measure of the size of the turbulent wind mass, and turbulence intensity is the magnitude of fluctuations relative to the average or mean wind speed. In the field, vegetation and objects on the ground impede the wind flow, reducing mean wind velocity and increasing the turbulence. The amount of turbulence created depends on the physical characteristics of the vegetation and of the ground surface.

Animals in natural habitats are not exposed to a single wind velocity but to a range of velocities in all three dimensions that are a function of the physical characteristics of the environment. This was illustrated in Figure 5-10, in which tiny aerodynamically stable bubbles follow the flow of air through a model windbreak. Note the downward wind flow over the back of the deer, the reversal in wind direction behind the deer, and the development of the profile on the right edge of the photograph.

An individual animal also presents an obstruction to wind, creating small wind patterns that are dependent on its body shape and posture. Analyses of the patterns of wind flow around a model white-tailed deer in a bedded posture indicate that the air flows smoothly around the windward side of the animal, with a turbulent zone on the lee side of the animal (Figure 6-15A). A quail causes a spiraling effect when facing directly into the wind (Figure 6-15B).

6-5   CONVECTIVE HEAT LOSS

The amount of heat that is removed from an object by convection is a function of the factors expressed in equation (6-4).

$$Q_c = h_c At(T_s - T_a) \tag{6-4}$$

where

Q_c = calories of heat transferred by convection
h_c = convection coefficient
A = area
t = time
T_s = temperature of the surface of the convector
T_a = temperature of the air (fluid)

FIGURE 6-15.   Patterns of air flow around (A) a model deer in a bedded posture and (B) a quail in the TEST. Note the turbulent area that develops on the leeward side of the animal.

The equations for calculating convection coefficients for flat plates and cylinders in free and forced convection are shown in Gates (1962). The equation for forced convection h_c across a flat plate is:

$$h_c = 5.73 \times 10^{-3} \sqrt{U/L} \qquad (6\text{-}5)$$

where

h_c = convection coefficient in cal cm^{-2} min^{-1} °C^{-1}
U = velocity in cm sec^{-1}
L = length of the flat plate in cm

The equation expressing h_c for cylinders is:

$$h_c = 6.17 \times 10^{-3} \frac{U^{1/3}}{D^{2/3}} \qquad (6\text{-}6)$$

where

h_c = convection coefficient in cal cm^{-2} min^{-1} °C^{-1}
U = wind velocity in cm sec^{-1}
D = diameter of the cylinder in cm

The effect of differences in both velocities and diameters is shown in Figure 6-16. There is a higher convection loss per unit area for small cylinders than for large ones. Also, the effect of a change in diameter of smaller cylinders from 2 to 10 cm is several times greater than for a change in the diameter of larger cylinders from 20 to 100 cm. The effect of changes in wind velocity is greatest at the low air speeds; the rate of convection loss for all cylinders rises more steeply in the first $\frac{1}{2}$ to 4 mi hr^{-1}, but then begins to level off.

Two conclusions can be reached from Figure 6-16: (1) small cylinders are more efficient convectors than large ones, and (2) low air velocities have a greater

FIGURE 6-16. Convective heat loss from cylinders of different diameters for each °C temperature difference between the surface of the cylinder and the air.

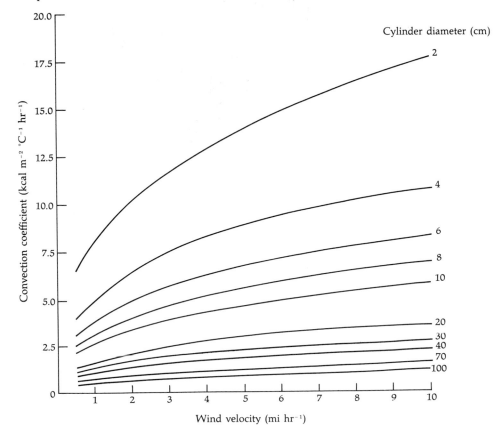

relative effect on convective heat loss than do high air velocities. These conclusions are of ecological significance because animals are, geometrically, collections of imperfect cylinders and cones. Cylindrical hairs are very efficient convectors because of their very small diameter. Thus convective forces can be very effective in removing the radiant heat energy absorbed by the hairs and the heat that is conducted along the shafts of the hairs from the skin surface through the coat.

The convection coefficients expressed in the thermal engineering literature are usually expressed as two-dimensional parameters, dependent on the size of the object and the velocity of the wind. Convection coefficients for animals are n-dimensional, dependent on such factors as size, wind velocity, orientation, hair density, turbulence, radiation absorbed, temperature, and others. These are discussed in Chapter 13.

The relatively greater effect of the lower wind velocities is of ecological interest because so many organisms live in the lower vegetation-filled zone marked by low but highly variable wind velocities owing to the effect of the vegetation on wind flow.

6-6   CONDUCTION

The transfer of heat by conduction results from the exchange of energy when oscillating molecules collide, with a higher rate of exchange during more rapid oscillations. Energy dissipation by conduction is from the higher temperatures resulting from more rapidly oscillating molecules to the lower temperatures. The perfect conductor permits the complete movement of heat energy through the conducting medium, and the perfect insulator prevents all movement of heat energy through the medium. The thermal conductivity (k) of a material is an expression of the rate of heat flow by conduction through the material under a specified set of conditions.

The basic expression used to determine the amount of heat flow by conduction is:

$$Q_k = \frac{kA\,t\,(T_1 - T_2)}{d} \qquad (6\text{-}7)$$

where:

Q_k = calories of heat transferred by conduction
k = thermal conduction coefficient
A = area
t = time
T_1 = temperature of first surface
T_2 = temperature of second surface
d = distance between the surfaces

Thus the amount of heat flow by conduction depends on the thermal conductivity of the medium, the area over which it is taking place, the temperature gradients, and the depth of the conducting medium. If the thermal conductivity coefficient (k) increases, more heat will be conducted. If either the area or the time increases, heat flow by conduction increases. If the temperature difference between two points increases, the gradient is steeper and conduction increases. If the depth of the insulating material increases, the amount of heat conducted is reduced.

The conductivity coefficient increases as the temperature of the conductive medium rises. The change is quite small for insulation material. For air, the k value at $-40°C$ is about $\frac{3}{4}$ of the value at $50°C$. There is a linear relationship between the conductivity of air and air temperature, so k can be expressed as:

$$Y = 2.066 + 0.00648X \qquad (6\text{-}8)$$

where

$Y = k$ in $(kcal\ m^{-2}\ hr^{-1})(°C/cm)^{-1}$
$X = T_a$ in $°C$

This is a useful equation in programing the correction factors when using the economical radiometer (see Appendix 2).

Conduction through the hair layer includes heat flow through the hair shafts themselves and through the air trapped between the hairs. Since air is a better insulator than hair, the important function of hair is the stabilization of the trapped air. This trapped air is a more important thermal barrier than the hair shafts themselves (Herrington 1951). This was demonstrated by Hammel (1953) when he replaced air with freon and determined that the new conductivity coefficient was more dependent on the conductivity of freon than on the hair itself. The same relationship is true for feather surfaces.

6-7   HEAT LOSS BY EVAPORATION

Heat is lost from the surface of a plant or animal by evaporation because energy is absorbed as liquid water is changed to a gaseous state. At $100°C$ the heat of vaporization is 540 kcal per gram, and at $0°C$ it is 595 per gram. There is a linear relationship between air temperature and the heat of vaporization of water (Figure 6-17), which can be expressed by the linear regression equation shown in the figure. Thus if one knows the amount of water evaporated from an organism regardless of the transport mechanism, the amount of energy removed by vaporization can be determined. This has been one of the ways in which Q_e has been measured; subjects are weighed before and after a period of time, and the difference between the first and second weights, minus the weight loss due to urine, feces, and respired CO_2, is the approximate weight of water lost by evaporation.

Evaporative heat loss comes from two sources: evaporation from the skin surface in the form of perspiration from the animal, and evaporation from the lungs and linings of the nasal passages. The relative importance of each source in the total heat loss depends on the characteristics of the animal and the energy characteristics of the atmosphere.

Ecologists have long recognized the importance of water vapor in the distribu-

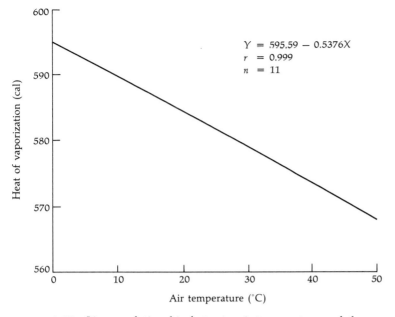

$Y = 595.59 - 0.5376X$
$r = 0.999$
$n = 11$

FIGURE 6-17. Linear relationship between air temperature and the heat of vaporization of water.

tion and activities of plants and animals. The technique of correlating a biological response with a measured meteorological parameter such as relative humidity has often been used. The next few pages call attention to the concept of heat loss by evaporation, and examples show the errors possible when basic relationships are overlooked.

Relative humidity is the amount of moisture in the air at a particular temperature and atmospheric pressure relative to the moisture content possible under these conditions at saturation. Relative humidity is easily determined (although not necessarily accurately determined), and it is expressed in a well-known unit of measurement—percent. The term relative, however, indicates that it is dependent on another factor, which is temperature. The dependence of relative humidity on temperature results in a "sliding" measurement that changes as either the vapor pressure or the temperature changes. Thus, valid comparisons between relative humidity values can only be made when the air temperatures during measurements are equal.

A common method of determining the relative humidity is by comparing the readings of a dry-bulb and a wet-bulb thermometer. The dry-bulb reading represents the air temperature, and the wet-bulb reading is a reflection of the evaporating power of the air as water evaporates from a wick over the bulb of a thermometer, cooling the bulb. Meteorological tables are used to find the percent relative humidity when the dry-bulb temperature, the wet-bulb temperature, and the difference between the two readings are known (Table 6-3).

TABLE 6-3   A PART OF A METEOROLOGICAL TABLE USED IN DETERMINING RELATIVE HUMIDITY AND VAPOR PRESSURE

Air Temperature (T_a)	Saturation Pressure	Depression of Wet-bulb Thermometer $(T_a - T_w)$				
		4	5	6	7	8
39	.237	68	60	52	43	37
40	.247	68	61	53	46	38
41	.256	69	62	54	47	40
⋮						
67	.661	80	76	71	67	62
68	.684	81	76	72	67	63
69	.707	81	77	72	68	64

An examination of the characteristics of relative-humidity measurements can lead to several interesting conclusions. Suppose, for example, that the air temperature measured by the dry-bulb thermometer were 68°F and the wet-bulb reading were 62°F. The relative humidity would be 72%, from Table 6-3. If the air temperature were 40°F and the wet-bulb reading 34°F, the relative humidity would be 53%.

The saturation pressure of the air at a temperature of 68°F is .684 inches of mercury, and at an air temperature of 40°F, .247 inches. When the air is saturated, the relative humidity is 100%, of course. When the relative humidity is 72% and the air temperature 68°F, the actual vapor pressure is .684 × .72 = .492 inches, and the vapor-pressure deficit (VPD), or the difference between the actual vapor pressure and the saturation pressure, is .192. At an air temperature of 40°F and a relative humidity of 53%, the vapor pressure is .247 × .53 = .131 inches, and the vapor-pressure deficit is .116 inches. The VPD is a meaningful parameter since it indicates the amount of additional water vapor that can be absorbed by the air up to saturation, and therefore determines the potential amount of evaporative heat loss.

Note in the above example that the relative humidity was 19% higher at 68°F than at 40°F, and the vapor pressure was .361 inches higher. Yet the VPD was also higher: .192 compared with .116 inches! This indicates that the air could hold additional moisture even with a higher relative humidity; the potential for evaporative heat loss would be greater at a higher relative humidity under these conditions.

The usual accuracy of mercury thermometers in sling psychrometers is ±1°F. The actual temperature of the air when the thermometer indicated 68°F could be anywhere from 67° to 69°F. Assuming that the measured temperatures were 68° and 62°F as in the example above, the actual temperature could be 69° and 61°F at one extreme, or 67° and 63°F at the other. The relative humidity values are 64% and 80%, the vapor pressures .452 and .529 inches, and the VPDs are

.255 and .132 in inches for these two examples, respectively. A similar procedure for the 40° and 34°F examples results in 41° and 33°F at one extreme and 39° and 35°F at the other. The relative humidities are 40% and 68%, vapor pressures, .102 and .161, and VPDs are .154 and .076 inches. Note that there is about a two-fold difference in VPDs in the two examples given that can be attributed to *thermometer error alone!*

One final example illustrating the kind of problems associated with the application of relative-humidity data. Suppose that the relative humidity were 50% and air temperatures ranged from 20° to 100°F. Air-temperature increments, vapor pressures, and VPDs representing the cooling power of the atmosphere at different air temperatures are shown in Figure 6-18. Note that it is distinctly nonlinear although the relative humidity is a constant 50% in this example.

The foregoing examples indicate the need for using meaningful parameters in studying the relationship between an animal and its environment. Relative-humidity data are inadequate for the analyses of evaporative heat loss from animals. Allen et al. (1964) found that the moisture content of cattle coats was not related to the measured relative humidity, but it was related to the vapor

FIGURE 6-18. Vapor pressure deficits at different air temperatures with a *constant* relative humidity of 50%.

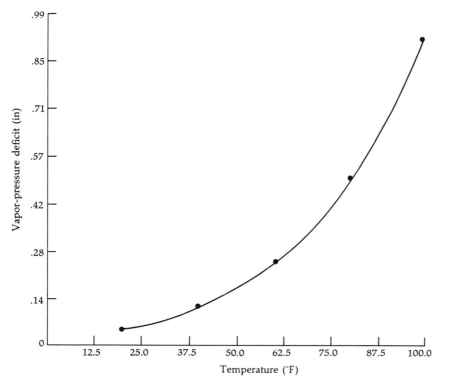

pressure of the air. Relative humidity is only an index, and its use alone may result in fallacious conclusions about evaporation losses.

The use of VPDs in ecological analyses is facilitated by the use of an equation for the calculation of VPD from inputs of T_a in relative humidity (%) (Table 6-4). The nonlinearity of this relationship is accounted for by the log transformation. The use of equations for these kinds of calculations facilitates the evaluation of the importance of one environmental factor in relation to others under consideration over a wide range of values. If the VPD were found to be important, the accuracy of the equation in Table 6-4 may need to be improved.

At low temperatures, the amount of heat lost by evaporation appears to be fairly constant. Evaporative heat loss by sheep was constant until ambient temperatures reached 30°C (Blaxter et al. 1959). This is reasonable since surface-water loss is slight when the surface of the animal is cool. Also, the vapor pressure of the air is low at colder temperatures, so the VPD cannot be very large. Wind, however, constantly brings new air over the source of evaporating water, so even air with a low VPD may still not become saturated. Thus heat loss by evaporation can continue even with low VPDs when there is wind.

The calculation of heat loss by evaporation is very complex because of the interaction between thermal factors, mass transport of water vapor and air, and surface characteristics. Sturkie (1965) discusses evaporative heat loss from birds,

TABLE 6-4   TABULATED AND CALCULATED SATURATION PRESSURES

T_a	Tabular Saturation Pressure (mm of Hg)*	Calculated Saturation Pressure†	Deviation	% Deviation
0	4.579	4.855	+0.276	6.0
5	6.543	6.619	+0.076	1.2
10	9.209	9.025	−0.184	2.0
15	12.788	12.305	−0.483	3.8
20	17.535	16.777	−0.758	4.3
25	23.756	22.874	−0.882	3.7
30	31.824	31.187	−0.637	2.0
35	42.175	42.521	+0.346	0.8
40	55.324	57.974	+2.650	4.8

* Handbook of Chemistry and Physics.
† The vapor pressure can be expressed by the following equation:

$$\log VP = (1.580 + 0.062\,T_a);\ r = 0.999;\ n = 9$$

The vapor-pressure deficit can then be calculated by the following equation:

$$VPD = e^{(1.580\,+\,0.062T_a)}\left(\frac{100 - \%R.H.}{100}\right)$$

and others have considered evaporation losses from plants. One thing is clear to anyone who has considered this question—relative humidity is an entirely inadequate parameter for ecological considerations, and continued presentation of relative-humidity data alone is of no value.

6-8 CONCLUSION

The apparent distribution of matter and energy in relation to physical and biological objects is complex, especially because of the changing distributions due to daily and seasonal cycles such as daylength and seasons, local variations due to weather, long-term changes due to geological aging, and the mobility of organisms that traverse energy flux and different types of matter as they go about daily activities. There is a basic simplicity to the distribution, however, and large quantities of tabular information can often be reduced to mathematical equations that can be successfully used to analyze relationships between physical and biological factors if the model or system is analyzed in a functional way.

Part 2 has included descriptions of simple models that illustrate the manner in which analytical ecology is approached. The emphasis in this chapter on the thermal characteristics of weather is because of the tendency of ecologists merely to compare weather data with biological data, when in reality there is thermal interaction. This interaction cannot be studied in an ecologically meaningful way until the metabolic, nutritive, and behavioral characteristics of organisms are understood. Parts 3 and 4 include considerations of physiology and behavior, with an emphasis on wild ruminants. The principles are applicable to any species, and students of analytical ecology are urged to develop equations expressing physiological and behavioral characteristics of other species. This will result in new information on a variety of species.

LITERATURE CITED IN CHAPTER 6

Allen, T. E., J. W. Bennett, S. M. Donegan, and J. C. D. Hutchinson. 1964. Moisture in the coats of sweating cattle. *Proc. Australian Soc. Animal Prod.* **5:** 167–172.

Birkebak, R. C.; R. C. Birkebak, and D. W. Warner. 1963. Total emittance of animal integuments. Meeting of the American Society of Mechanical Engineers. Nov. 17–22. Paper No. 63-WA-20, 4 pp.

Blaxter, K. L., N. McC. Graham, F. W. Wainman, and D. G. Armstrong. 1959. Environmental temperature, energy metabolism and heat regulation in sheep. II. The partition of heat losses in closely clipped sheep. *J. Agr. Sci.* **52**(1): 25–40.

Evans, K. E. 1971. Energetics of sharp-tailed grouse (*Pediocetes phasianellus*) during winter in western South Dakota. Ph. D. dissertation, Cornell University, 169 pp.

Fuchs, M., and C. B. Tanner, 1966. Infrared thermometry of vegetation. *Agron. J.* **58:** 597–601.

Gates, D. M. 1962. *Energy exchange in the biosphere.* New York: Harper & Row, 151 pp.

Hammel, H. T. 1953. A study of the role of fur in the physiology of heat regulation in mammals. Ph. D. dissertation, Cornell University, 105 pp.

Hammel, H. T. 1956. Infrared emissivities of some arctic fauna. *J. Mammal.* **37**(3): 375–377.

Herrington, L. P. 1951. The role of the piliary system in mammals and its relation to the thermal environment. *Ann. N.Y. Acad. Sci.* **53**: 600–607.

Johnson, F. S. 1954. The solar constant. *J. Meteorol.* **11**: 431–439.

Moen, A. N. 1968. Surface temperatures and radiant heat loss from white-tailed deer. *J. Wildlife Management* **32**(2): 338–344.

Moen, A. N., and K. E. Evans. 1971. The distribution of energy in relation to snow cover in wildlife habitat. In *Proceedings of symposium on the snow and ice in relation to wildlife and recreation,* ed. A. O. Haugen. Ames, Iowa: Iowa State University, pp. 147–162.

Ordway, D. E. 1969. An aerodynamicist's analysis of the odum cylinder approach to net CO_2 exchange. *Photosynthetica* **3**(2): 199–209.

Plate, E. J., and A. A. Quraishi. 1965. Modeling of velocity distributions inside and above tall crops. *J. Appl. Meteorol.* **4**(3): 400–408.

Platt, R. B., and J. F. Griffiths. 1964. *Environmental measurement and interpretation.* New York: Reinhold, 235 pp.

Porter, W. P. 1969. Thermal radiation in metabolic chambers. *Science* **166**(3901): 115–117.

Riemerschmid, G., and J. S. Elder. 1945. The absorptivity for solar radiation of different coloured hairy coats of cattle. *Onderstepoort J. Vet. Sci. Animal Ind.* **20**(2): 223–234.

Sellers, W. D. 1965. *Physical climatology.* Chicago: University of Chicago Press, 272 pp.

Stevens, D. S. 1972. Thermal energy exchange and the maintenance of homeothermy in white-tailed deer. Ph. D. dissertation, Cornell University, 231 pp.

Stevens, D. S., and A. N. Moen. 1970. Functional aspects of wind as an ecological and thermal force. *Trans. North Am. Wildlife Nat. Resources Conf.* **35**: 106–114.

Sturkie, P. D. 1965. *Avian physiology.* Ithaca, New York: Cornell University Press, 766 pp.

Suomi, V. E., and P. M. Kuhn. 1958. An economical net radiometer. *Tellus* **10**(1): 160–163.

Swinbank, W. C. 1963. Long-wave radiation from clear skies. *Quart. J. Roy. Meteorol. Soc.* **89**: 339–348.

Tanner, C. B., J. A. Businger, and P. M. Kuhn. 1960. The economical net radiometer. *J. Geophys. Res.* **65**(11): 3657–3667.

Weast, R. C., ed. 1967. *Handbook of chemistry and physics.* 48th ed. Cleveland: The Chemical Rubber Co.

SELECTED REFERENCES

Gebhart, B. 1971. *Heat transfer.* 2d ed. New York: McGraw-Hill, 596 pp.

Kuhn, P. M., V. E. Suomi, and G. L. Darkow. 1959. Soundings of terrestrial radiation flux over Wisconsin. *Monthly Weather Rev.* **87**(4): 129–135.

Oke, T. R. 1970. The temperature profile near the ground on calm clear nights. *Quart. J. Roy. Meteorol. Soc.* **96**(407): 14–23.

Reifsnyder, W. E., and H. W. Lull. 1965. *Radiant energy in relation to forests.* Techical Bulletin No. 1344, USDA Forest Service.

Stoll, A. M., and J. D. Hardy. 1955. Thermal radiation measurements in summer and winter Alaskan climates. *Trans. Am. Geophys. Union* **36**(2): 213–225.

Viskanta, R. 1966. Radiation transfer and interaction of convection with radiation heat transfer. In *Advances in heat transfer,* ed. T. F. Irvine, Jr., and J. P. Hartnet. New York: Academic Press, pp. 175–251.

Courtesy of Dwight A. Webster
Department of Natural Resources, Cornell University

PART

METABOLISM AND
NUTRITION

The three basic requirements of animals are food, cover, and space. This idea has been repeated many times in courses in wildlife ecology and management. The requirements for food have frequently been discussed at a level confined to food habits, the average amount required per animal per day, and other similar considerations. Each animal, however, ingests food, digests it, and uses it metabolically to meet its current needs. These needs change, depending on the activity and productivity of each animal. Thus the need for food is based on dynamic biological processes ranging from biochemical reactions at the molecular level to forage production on different ranges at the community or ecosystem level. An understanding of the basic biological processes helps in an evaluation of the importance of different components in the total picture of nutritive ecology.

7

ENERGY

METABOLISM

All living organisms must utilize energy to maintain their life processes. Organisms are subject to the first law of thermodynamics, which states that energy can be neither created nor destroyed, but only changed in form. Not all of the energy that reaches the surface of an organism or is ingested by it is of direct benefit because some of it is in a form that is not directly useful to the organism. The amount of energy utilized compared with the amount actually available to an organism is an expression of the energetic efficiency of the organism.

The conversion of energy from one form to another in living organisms is called metabolism. These energy conversions or metabolic processes occur at different rates in different tissues within the body. The rate of metabolism in an active animal is high and in an inactive animal, low.

If cells accumulate more rapidly than they are destroyed, growth occurs. If cells are destroyed more rapidly, the animal loses weight. Growth results from the assimilation of the products of digestion and, in ruminants, rumen fermentation into new body tissue. Amino acids, for example, are synthesized into bacterial protein in the rumen. The bacteria are then broken down into amino acids for assimilation into body tissue in the ruminant. During catabolism, carbohydrate and fat molecules are broken down into simpler carbon dioxide and water molecules. Protein molecules are broken down into carbon dioxide, water, and the nitrogen compound urea, which is excreted in the urine. As a result of this process, energy stored in the large molecules is released and becomes available to the animal.

Metabolic energy is expressed in units called calories. One gram-calorie is approximately equal to the amount of heat required to raise the temperature of one gram of water from 14.5° to 15.5°C. A more useful unit in physiological work is the kilocalorie, which is one thousand times larger than the gram-calorie.

7-1 BASAL METABOLISM AND ASSOCIATED TERMINOLOGY

Basal metabolism has been defined as the minimal energy cost when an animal is at rest in a thermoneutral environment and in a post-absorptive condition (Brody 1945). The measurement does not represent the minimum rate of metabolism needed to support life. In humans, for example, measurements indicate that the metabolic rate is lower during sleep than during rest. The post-absorptive condition is necessary to reduce as much as possible any heat production that can be attributed to the heat of fermentation of food or the heat of nutrient metabolism.

The energy required to maintain life at the basal metabolic rate provides for circulation, excretion, secretion, respiration activities, and the maintenance of muscle tone. Crampton and Harris (1969) estimate that 75% of the energy of basal metabolism is spent in maintaining muscle tone and body temperature, with 25% being used in circulation, excretion, secretion, and respiration.

USES FOR MEASUREMENTS OF BASAL METABOLIC RATE.   The basal metabolic rate (BMR) is useful as a base line with which comparisons may be made when the animal is physically active, on different diets, diseased, infected with parasites, pregnant, lactating, or in some condition that requires expenditure of energy. Thus it is useful in diagnostic work and for evaluating the effect of activity and production on the energy requirements of free-ranging animals.

CONDITIONS FOR MEASUREMENT.   The standard conditions of thermoneutrality, post-absorptive digestion, a lying posture, and a calm psychological state are necessary for establishing a base line for energy metabolism because animals react physiologically to heat or cold stress, digestive processes, changes in activity, and psychological stress. Thermoneutral conditions are required because the metabolic rate increases when an animal is subject to either heat stress or cold stress. Air temperatures of 20° to 25°C are usually considered to be in the thermoneutral range of most large animals, although this range is dependent on physiological, behavioral, and environmental factors (Moen 1968). Air temperature is a useful index to the thermal conditions in a metabolism chamber, but in the field environment many other factors must also be considered. These are discussed in Chapters 6, 13, and 14.

A post-absorptive digestive state is necessary as a standard condition because the digestion of different diets affects the rate of heat production. Ritzman and Benedict (1931) found that sheep reach a post-absorptive state in 34 to 48 hours although foodstuffs may still remain in the intestinal tract. The level of feeding affects the amount of time that an elevated heat production persists; sheep that had been on a high level of feeding had a higher heat production for a seven-day

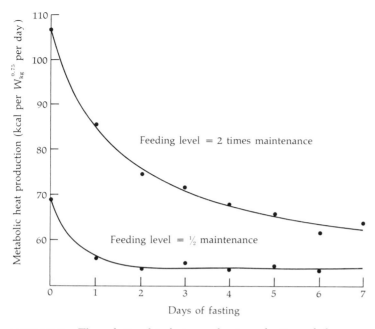

FIGURE 7-1. The relationship between heat production of sheep and the number of days of fasting. (Data from Marston 1948.)

fasting period than those that had been on a lower level of feeding (Figure 7-1).

The lying posture is an easy requirement to meet when working with humans, but domestic and wild animals lie down and stand up at their own discretion. The only practical way to solve this problem is by mathematically correcting the standing measurements to the comparable lying-posture level, eliminating the muscular effect of supporting the body during basal-metabolism measurements. The data may be left uncorrected and expressed as fasting metabolism, although variation in results will then be partly due to differences in the proportion of time spent standing and lying in the metabolism chamber.

A calm psychological state can be attained only after an animal has received the proper amount of training in the chamber. This may be accomplished in a few days or weeks for domestic animals. Wild ruminants generally require longer periods of training before they can be confined in a chamber. Even with training, as many sensory stimuli as possible should be excluded from the chamber environment. This reduces the psychological tension of the animal and may reduce the number of changes in posture.

The results of metabolic measurements are not uniform for all individuals within a species. There are inherent differences between individuals that contribute to this variation. There are also differences in the presumed standard conditions that can be attributed to experimental error. Metabolic measurements under standard conditions are actually somewhere between true basal metabolism and some upper limit.

The value of basal metabolic measurements and other physiological parameters will increase as wildlife research advances from a descriptive field phase to a more basic diagnostic and experimental phase that includes the entire laboratory-pen-field spectrum. It is important to establish a base line so that the effect of experimental variables can be analyzed.

ASSOCIATED TERMINOLOGY.   Heat production can increase in the absence of voluntary muscular activity. There may be a physiological response to cold or a psychological response to environmental stimuli. Hoar (1966) credits Giaja with the term "summit metabolism," which is the highest metabolic rate attainable at normal body temperature without voluntary muscle activity.

Since the term basal metabolism implies a sort of minimum, Brody (1945) used the term "resting metabolism" when the metabolism is not post-absorptive and "fasting metabolism" when the resting metabolism is approximately post-absorptive. Hoar (1966) suggests that the term "standard metabolism" should be a preferred term for use by comparative physiologists. Fasting heat production may be preferred for ruminants with the length of the fasting period specified, since it is difficult to determine just when they reach a post-absorptive state (Crampton and Harris 1969).

Standard metabolism or fasting heat production are perhaps better terms since they imply that the measurements were made under standard conditions or at a certain time after eating without attaching exact significance to the resulting level of measurement. The term basal, however, is so firmly established with reference to mammals that its replacement by a more appropriate term is highly unlikely.

7-2 MEASUREMENTS OF BASAL METABOLIC RATE

DIRECT METHODS.   Direct measurements of the metabolic rate involve the measurement of the actual heat production by the animal. Lavoisier, in the latter part of the eighteenth century, discovered a relationship between heat energy given off by an animal and energy in the ingested food (Hoar 1966). This was done by placing a guinea pig in a closed box surrounded by ice and recording the amount of ice that melted in a specified period of time. Knowing that about 80 calories of energy are required to melt one gram of ice, he then calculated the amount of heat energy released by the animal, and this was found to be related to the energy in the food.

The early direct methods of Lavoisier have been refined in calorimetry chambers. Professor Armsby built one early in the twentieth century at The Pennsylvania State University using the same basic principle as the crude icebox calorimeter, but circulating water removed the heat energy from the chamber. Calculations were based on the flow rate, temperature changes, and other thermal characteristics of the water and chamber.

INDIRECT METHODS.   Indirect measurements of heat production in use today are simpler and less expensive than the direct methods used by the earlier investigators. These indirect methods are used to measure the amount of oxygen consumed and the amount of carbon dioxide produced by the animal. Heat production is calculated from the measured oxygen and carbon dioxide volumes and their relationships to exothermic metabolic processes.

The use of a mask through which the oxygen intake can be monitored is one of the most inexpensive methods. The advantage of this is that the animal need not be confined in a small chamber; domestic animals usually become quite at ease, once accustomed to wearing the mask. Hammel (1962) and Hart et al. (1961) used this method in their studies of reindeer and caribou, respectively.

A respiration chamber designed for indirect calorimetry for deer is currently in use at the University of New Hampshire (Figure 7-2). The oxygen and carbon dioxide content of the air leaving the chamber is measured and is compared with that of air entering the chamber. The differences are attributed to the metabolic processes of the experimental animal. The chamber is equipped with temperature controls, so experiments on metabolic responses to changes in chamber temperature can be measured.

CALCULATION OF HEAT PRODUCTION.   The calculation of heat production using indirect calorimetry is illustrated by equation (7-1) from Brody (1945).

$$C_6H_{12}O_6 + 6O_2 = 6CO_2 + 6H_2O + 678 \text{ kcal} \qquad (7-1)$$

This equation shows that the oxidation of one mole of hexose ($C_6H_{12}O_6$: 180 g) requires 6 moles (134.4 liters) of oxygen (O_2). Six moles (134.4 liters) of carbon dioxide (CO_2) and 6 moles of water (H_2O) are produced, plus 678 kcal of heat. The heat production per liter of oxygen consumed is determined by dividing the heat production (678 kcal) by the liters of oxygen consumed (134.4 liters), resulting in 5.047 kcal per liter of oxygen consumed. The heat production per liter of carbon

FIGURE 7-2. The respiration chamber for deer in the Ritzman Laboratory at the University of New Hampshire, Durham. (Photograph courtesy of Helenette Silver.)

dioxide produced can be found in the same way; 5.047 kcal of heat are produced for each liter of CO_2 produced. This value is for carbohydrate oxidation only. The oxidation of mixed fat results in the release of 4.69 kcal per liter of oxygen consumed, and 6.6 kcal per liter of carbon dioxide produced. For the oxidation of mixed protein, 4.82 kcal are released per liter of oxygen consumed, and 5.88 kcal per liter of carbon dioxide produced.

The above figures show that the numbers of calories produced by oxidation vary with the carbohydrate, fat, and protein content of the food. The proportion of carbohydrate, fat, and protein in a food must be known or estimated before an accurate calculation can be made of the heat production from either oxygen consumption or carbon dioxide production.

The amount of protein oxidized is determined by dividing 100 by the percentage of nitrogen in the protein. Most animal proteins contain about 16% nitrogen, so the amount of protein can be estimated by multiplying the nitrogen content by 6.25, calculated from 100/16. Cereal proteins contain 17%–18% nitrogen, so the conversion factors are 100/17 = 5.9 and 100/18 = 5.6. In practice, however, the assumption that protein contains 16% nitrogen is sufficient at this time.

The relative amounts of fat and carbohydrate in a food that are oxidized can be determined from the nonprotein respiratory quotient (R.Q.). The R.Q. is the ratio of moles or volumes of CO_2 produced to moles or volumes of O_2 consumed. In equation (7-1) the R.Q. is equal to 1 ($6CO_2/6O_2 = 1.00$), indicating that carbohydrates had been consumed. If the nonprotein R.Q. were less than 1.00, it would indicate that some fats had been consumed. The R.Q. for mixed fats is 0.71. These are averages only, since fatty acids vary in their R.Q. (short-chain fatty acids have an R.Q. nearer to 0.8) and each protein and amino acid has its distinctive R.Q. (Brody 1945). The amount of heat produced by the oxidation of different mixtures of carbohydrates and fats can be computed from the R.Q. (Table 7-1).

The calories per liter of oxygen consumed range from 4.686 to 5.047. The mean value is about 4.86, and this occurs at an R.Q. of 0.85. The R.Q. of protein is about 0.82, so if an approximation is sufficient, the rate of heat production can be calculated by multiplying the liters of oxygen by 4.82 to 4.85 without correcting for protein metabolism. This method is often used by comparative physiologists (Hoar 1966).

The R.Q. is not always as simple as is implied in the preceding discussion. In ruminants, large quantities of CO_2 are produced by the rumen bacteria. This cannot be distinguished from the CO_2 originating in cellular respiration.

The calculation of heat production from oxygen consumption using an approximation of 0.85 for the R.Q. results in a good estimate of the energy expenditure of unrestrained animals. Measurement of oxygen consumption by unrestrained animals is difficult, however. Ventilation masks prohibit feeding, and the air must be sampled for gas analysis. If an implantable oxygen measurement transducer were available for placement in the trachea of an animal, the continuous measurement and wireless telemetry of this parameter would provide more precise measurements of the energy expenditures of free-ranging animals.

TABLE 7-1 THERMAL EQUIVALENTS OF O_2 AND CO_2 AND THE CORRESPONDING PERCENTAGES OF FAT AND CARBOHYDRATES OXIDIZED FOR DIFFERENT RESPIRATORY QUOTIENTS

| | | | | % O_2 Consumed by | | % Heat Produced by Oxidation of | |
| | O_2 | CO_2 | | Carbo- | | Carbo- | |
R.Q.	(kcal liter^{-1})	(kcal liter^{-1})	(kcal g^{-1})	hydrates	Fat	hydrates	Fat
0.70	4.686	6.694	3.408	0	100	0	100
0.75	4.729	6.319	3.217	14.7	85.3	15.6	84.4
0.80	4.801	6.001	3.055	31.7	68.3	33.4	66.6
0.85	4.863	5.721	2.912	48.8	51.2	50.7	49.3
0.90	4.924	5.471	2.785	65.9	34.1	67.5	32.5
0.95	4.985	5.247	2.671	82.9	17.1	84.0	16.0
1.00	5.047	5.047	2.569	100	0	100	0

SOURCE: Data from Brody 1945.

7-3 METABOLIC RATES OF RUMINANTS

RELATIONSHIPS TO BODY WEIGHT.   The relationship between heat production and body weight has been determined for a variety of species, and Benedict's "mouse to elephant" curve (Benedict 1938) is widely known among physiologists (Figure 7-3). The variation among species is quite low. The shape of the curve illustrates that the heat production per unit weight of a small animal is far greater than the heat production per unit weight of a large animal. Deviations from the mean may be attributed to species differences, to experimental error, or to variations in the proportions of metabolically active to metabolically inert tissue.

The body of an animal includes metabolically active tissue such as muscles, blood, fat, and other tissue that is being continually replaced or removed. Metabolically inert tissue is present also, including hair and antlers after both have reached maximum dimensions. Inert residues are also found in the gastrointestinal tract when an animal is on feed. Thus body weight does not truly represent the amount of metabolic tissue in an animal at the time of weighing, but if several measurements are made on different animals the error is reduced.

There has been considerable discussion of the mathematical relationship between heat production and body weight, which is expressed in equation (7-2).

$$Q_{mb} = c\, W_{kg}^b \tag{7-2}$$

where

Q_{mb} = energy expenditure for basal metabolism
c = constant
W_{kg} = weight of the animal in kg
b = an exponent that has been the focal point of most of the controversy $\cong 0.75$

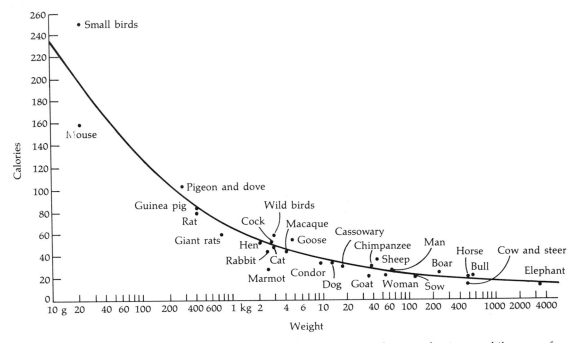

FIGURE 7-3.   Semilogarithmic chart showing the trend of the average heat production per kilogram of average body weight of each animal species. Weights range from 20 g to 4000 kg. (From Benedict 1938. Courtesy of Carnegie Institution.)

Physiologists have been searching for an exponent that will result in a constant value of c. If this were found, the basal heat production of any animal could be determined by simple multiplication after weighing the animal.

Reported values of the exponent b have ranged from 0.66 to 0.75. Kleiber (1961, p. 212) summarizes the results of metabolic rate measurements on a variety of mammals and concludes: "For all practical purposes, one may assume that the mean standard metabolic rate of mammals is seventy times the three-fourth power of their body weight (in kg) per day, or about three times the three-fourth power of their body weight (in kg) per hour." The results of several experiments may be found in "Nutrient Requirements of Farm Livestock," Number 2-Ruminants (1965), a publication sponsored by the Agricultural Research Council, and in Blaxter (1967). The general conclusion from these data is that the basal heat production approximates 70 $W_{kg}^{0.75}$ kcal per day. The National Research Council (1966) has adopted the exponent $b = 0.75$.

METABOLIC RATES OF WILD RUMINANTS.   Several measurements of the basal metabolic rates of wild ruminants have been made. Helenette Silver of the New Hampshire Fish and Game Department has made several measurements of white-tailed deer at different times of the year. Six series of measurements of four deer in winter coat resulted in an average of 75.7 kcal per $W_{kg}^{0.75}$ per 24 hours, and three series of measurements of three deer in summer coat yielded an average of 84.4 kcal per $W_{kg}^{0.75}$ per 24 hours (Table 7-2). The data indicate that there are consistent differences between individual deer since male no. 101 had a higher heat production than the other three deer measured in winter coat; in summer coat it was also higher than that of two others. The males generally had a higher heat production per kg and per $W_{kg}^{0.75}$, but with only two females and four males in the sample it is impossible to draw any definite conclusions.

The fasting metabolic rates of white-tailed deer have been measured by Silver et al. (1969) and are shown in Table 7-3. Again, there is considerable variation between individual deer. The females in winter coat vary from 64 to 128 kcal per $W_{kg}^{0.75}$ per day.

The resting heat production of young black-tailed deer weighing from 5 to 25 kg and ranging in age from 28 to 166 days is about two times the basal rate for adults (Table 7-4). The fawns being measured had been fed, however, so their heat production was higher owing to the effects of heat of fermentation and heat of nutrient metabolism.

A few measurements of metabolic rates of other wild ruminants have been reported in the literature. Krog and Monson (1954) measured a goat (*Oreamnos americanus*). The heat production was calculated to be 1027 kcal per 24 hours when the animal was confined in a chamber at temperatures from 20° to −20°C. The animal's heat production was 1.3 times higher at −30°C and 2.3 times higher at −50°C (Table 7-5).

TABLE 7-2 BASAL METABOLIC RATES OF WHITE-TAILED DEER

Deer I.D. No.	Sex	Age	Weight (kg)	Coat	Avg. Chamber Temperature (°C)	H.P. (kcal per W_{kg} per 24 hr)	H.P. (kcal per W_{kg} per 24 hr)	H.P. (kcal per $W_{kg}^{0.75}$ per 24 hr)	Avg. H.P. (kcal per $W_{kg}^{0.75}$ per 24 hr)
101	M	9 mo	32.9	W	19.2	1173	35.7	85.4	82.0
101	M	9 mo	31.3	W	16.0	1038	33.2	78.6	
102	M	9 mo	30.2	W	18.8	971	32.2	75.4	71.3
102	M	9 mo	27.9	W	16.0	814	29.2	67.1	
101	M	1 yr 4 mo	68.6	W	17.7	1988	29.0	83.4	85.3
101	M	1 yr 4 mo	68.2	W	18.4	2069	30.3	87.2	
101	M	1 yr 8 mo	49.4	W	18.2	1481	30.0	79.5	79.5
1	M	1 yr 8 mo	61.8	W	−0.4	1534	24.8	69.6	69.6
1	F	1 yr 8 mo	51.8	W	−3.9	1272	24.6	65.9	65.9
								Average for winter coat =	75.6
101	M	1 yr	35.7	S	18.7	1499	42.0	102.6	111.0
101	M	1 yr	36.5	S	19.7	1771	48.5	119.3	
2	M	2 yr	51.1	S	21.2	1327	26.0	69.4	69.4
2	F	2 yr	40.0	S	21.5	1160	29.0	72.9	72.9
								Average for summer coat =	84.4

SOURCE: Recalculated from Silver 1968.

TABLE 7-3 FASTING METABOLISM OF WHITE-TAILED DEER

Weight (kg)	Month of Measurement	Heat Production per 24 Hours		
		Total Kcal	Kcal per Kg	Kcal per $W_{kg}^{0.75}$
Adult males, winter coat				
82.0	Dec	2596	31.7	95.3
80.0	Apr	3192	39.9	119.3
80.0	Jan	2880	36.0	107.7
69.0	Jan	2318	33.6	96.8
68.6	Oct	2339	34.1	98.1
68.2	Oct	2434	35.7	102.6
66.5	Nov	2042	30.7	87.7
64.5	Jan	1737	26.9	76.3
63.8	Jan	2286	35.8	101.3
60.1	Mar	1750	29.1	81.1
58.4	Feb	1758	30.1	83.2
52.0	Jan	1620	31.2	83.7
49.4	Feb	1742	35.3	93.5
44.0	Mar	1908	43.4	111.7
		Averages =	33.8	95.6
Adult females, winter coat				
70.0	Dec	1548	22.1	64.0
69.5	Sept	3078	44.3	127.9
64.4	Sept	2290	35.6	100.7
		Averages =	34.0	97.5
Fawns, winter coat				
32.9	Feb	1380	41.9	100.5
31.3	Feb	1221	39.0	92.3
30.2	Feb	1142	37.8	88.6
27.9	Feb	958	34.3	78.9
		Averages =	38.3	90.1
Adult males, summer coat				
77.1	Aug	4150	53.8	159.5
54.6	July	2543	46.6	126.6
50.1	June	2789	55.7	148.1
47.9	May	2393	50.0	131.4
		Averages =	51.5	141.4
Adult females, summer coat				
66.0	June	2945	44.6	127.2
60.6	Aug	3268	53.9	150.5
58.8	July	3675	62.5	173.1
57.6	June	3134	54.4	149.9
54.6	June	2632	48.2	131.0
		Averages =	52.7	146.3
Yearlings, summer coat				
36.5	June	2084	57.1	140.3
35.7	May	1764	49.4	120.8
		Averages =	53.3	130.6

SOURCE: Recalculated from data in Silver et al. 1969.

TABLE 7-4 RESTING HEAT PRODUCTION OF BLACK-TAILED DEER FAWNS IN A THERMONEUTRAL ENVIRONMENT BUT NOT IN THE POST-ABSORPTIVE STATE

Body Weight (kg)*	$W_{kg}^{0.75}$	Age in Days*		Resting Heat Production* (kcal per day)		H.P. per $W_{kg}^{0.75}$ (kcal per day)		Multiple of BMR†	
		M	F	M	F	M	F	M	F
5	3.34	28	30	382	498	114.4	149.1	1.6	2.1
10	5.62	67	70	880	837	156.6	148.9	2.2	2.1
15	7.62	98	106	1180	1076	154.9	141.2	2.2	2.0
20	9.46	126	137	1460	1408	154.3	148.8	2.2	2.1
25	11.18	153	166	1722	1669	154.0	149.0	2.2	2.1

*Data from Nordan, Cowan, and Wood 1970.
†BMR = Q_{mb} = 70 $W_{kg}^{0.75}$.

Hammel (1962) computed the heat production of a reindeer (*Rangifer tarandus*) from the measured oxygen consumption both when the animal was standing quietly and when it was pulling a loaded sled (Table 7-6). At rest, reindeer heat production was similar to that reported for other species. It rose to about eight times the resting rate when exercising. The rectal temperature during rest ranged from 38.1° to 38.3°C, rising to a range of 38.8° to 39.2°C during exercise.

The metabolic rates of eight caribou calves were measured by Hart et al. (1961) (Table 7-7). These calves showed a response to the thermal conditions during the test as the metabolic rate rose to over ten times the basal metabolic rate for adult homeotherms. Several measurements of metabolic rates of female caribou during fasting and at maintenance have been reported by McEwan (1970) (Table 7-8). The results indicate that the metabolic rates are somewhat higher than 70 $W_{kg}^{0.75}$.

Brockway and Maloiy (1967) measured the metabolic rates of two red deer (*Cervus elaphus*) during fasting. The average of three measurements was 90 $W_{kg}^{0.75}$ kcal per day. The fasting metabolism of the wildebeest (*Connochaetes taurinus*) and eland (*Taurotragus oryx*) was 104.3 and 111.2 $W_{kg}^{0.73}$ kcal per day, respectively (Rogerson 1968).

TABLE 7-5 METABOLISM OF A MOUNTAIN GOAT (*Oreamnos americanus*)

Sex	Age	Weight (kg)	Dates Measured	Ambient Temperature (°C)	Total H.P. (kcal per 24 hr)	Kcal per $W_{kg}^{0.75}$	Multiple of BMR
M	1½ yr	32 ± 1.5	Feb 3–Mar 14	20 to −20	1027	76.3	1.09
				−30	1304	96.9	1.38
				−50	2362	175.6	2.51

SOURCE: Data from Krog and Monson 1954.

TABLE 7-6 METABOLISM OF A REINDEER (*Rangifer tarandus*)

					Heat Production		
Sex	Age	Weight (kg)	Date Measured	Ambient Temperature (°C)	Kcal per Hr	Kcal per 24 Hr	Kcal per $W_{kg}^{0.75}$
Metabolism while standing quietly:							
M	5–6 yr	100	Feb 26	−16	157	3768	119.2
			Feb 27	−12	177	4248	134.3
			Feb 27	−12	192	4608	145.7
			Mar 1	−16	179	4296	135.9
			Mar 2	− 7	173	4152	131.3
			Mar 7	+ 3	172	4128	130.5
				Averages:	175		132.8
						Multiple of BMR = 1.90	
Metabolism while pulling a loaded sled:							
			Feb 26	−16	739	17736	560.9
			Feb 27	−12	605	14520	459.2
			Feb 29	−21	699	16776	530.5
			Mar 7	+ 4	712	17088	540.4
			Mar 7	+ 4	752	18048	570.7
			Mar 8	(+)4	705	16920	535.1
			Mar 8	+ 4	605	14520	459.2
				Averages:	688		522.3
						Multiple of BMR = 7.46	

SOURCE: Data from Hammel 1962.

FASTING METABOLISM OF DOMESTIC RUMINANTS.   The number of metabolism experiments completed on domestic cattle and sheep is far greater than the number on wild ruminants.

The basal heat production of subadult sheep has been measured by Ritzman and Benedict (1930). The metabolic increment (I_m) over the basal metabolic rate varies from nearly 3 at one week of age to 1.4 at 16 weeks of age. The decline in heat production with age is essentially linear (Figure 7-4). The relationship between I_m and the weight of lambs is also linear (Figure 7-5), indicating that heat production decreases per unit metabolic weight as the lambs grow.

7-4 METABOLIC RATES OF OTHER ANIMALS

Several investigators have measured oxygen consumption or heat production of wild animals under laboratory conditions. The values are interesting for comparative purposes, but the analytical ecologist must recognize that the results under laboratory conditions are not directly applicable to the wild. The results discussed below are included mainly to call attention to the existence of data on several species. Additional data may be found in the biological handbook *Metabolism* (Altman and Dittmer 1968).

TABLE 7-7 METABOLISM OF INFANT CARIBOU

Calf No.	Weight (lb)	Weight (kg)	Wind (mi per hr)	Fur	Location	T_a (°C)	T_{br}* (°C)	Heat Production per $W_{kg}^{0.75}$ Kcal per Hr	Heat Production per $W_{kg}^{0.75}$ Kcal per 24 Hr	Multiple of BMR
1	11.7	5.3	0	dry	lab	20	40.0	4.11	98.64	1.4
2	11.3	5.1	0	dry	lab	22	40.2	9.04	216.96	3.1
3	11.7	5.3	0	dry	lab	19	39.0	6.42	154.08	2.2
3	12.8	5.8	0	dry	lab	20	40.0	9.64	231.36	3.3
5	12.0	5.5	0	dry	lab	14	39.2	12.07	289.68	4.1
7	13.4	6.1	0	dry	†	10	39.2	8.80	211.20	3.0
8	11.4	5.2	0	dry	†	10	38.8	6.60	158.40	2.3
1	11.6	5.3	0	dry	tent	4	40.5	12.23	293.52	4.2
2	11.0	5.0	0	dry	tent	4	39.2	12.86	308.64	4.4
4	11.4	5.2	0	dry	†	−3	39.2	15.02	360.48	5.1
6	12.0	5.5	0	dry	†	0	39.1	17.37	416.88	6.0
7	12.5	5.7	0	dry	†	0	38.8	12.20	292.80	4.2
5	14.5	6.6	0	dry	†	0	38.8	14.68	352.32	5.0
3	11.7	5.3	2	dry	†	0	39.0	16.01	384.24	5.5
1	11.8	5.4	15	dry	exposed	−1	38.1	19.20	460.80	6.6
1	11.6	5.3	14	dry	exposed	3	39.7	21.17	508.08	7.3
2	11.0	5.0	12	dry	exposed	6	39.0	14.20	340.80	4.9
3	12.8	5.8	14	dry	exposed	−5	41.0	21.61	518.64	7.4
2	11.6	5.3	14	dry	exposed	−5	41.0	22.94	550.56	7.9
7	19.7	9.0	21½	dry	exposed	3	39.0	20.69	496.56	7.1
6	12.0	5.5	18	damp	exposed	−1	38.5	28.65	687.60	9.8
7	10.6	4.8	15	damp	exposed	−0.5	39.2	20.88	501.12	7.2
5	12.0	5.5	15	wet	exposed	−1	38.5	26.28	630.72	9.0
6	13.4	6.1	17	wet	exposed	0	39.0	31.47	755.28	10.8
5	16.1	7.3	15	wet	exposed	−0.5	38.8	23.51	564.24	8.1
7	19.7	9.0	28	wet	exposed	3	38.3	29.69	712.56	10.2

SOURCE: Data from Hart et al. 1961.
* Rectal temperature.
† Location not given.

TABLE 7-8 ENERGY METABOLISM OF FEMALE BARREN-GROUND CARIBOU (*Rangifer tarandus*)

Age	Weight (kg)	Activity and Diet	Heat Production Total Kcal per Day	Heat Production Kcal per $W_{kg}^{0.75}$	Heat Production Kcal per $W_{kg}^{0.75}$*
21 mo	80.0	Resting at maintenance	3215	124.5 ± 1.94	120.2
21 mo	90.0	Resting at maintenance	3575	107.0 ± 3.04	122.3
9 mo	57.3	3rd day of fasting	2400	115.4	115.2
9 mo	56.0	6th day of fasting	2100	102.5	102.6
3 yr	94.0	21st day of fasting	2750	91.0	91.1

SOURCE: Data from McEwan 1970.
* Calculations made by the author.

Beck and Anthony (1971) calculated the mean standard metabolic rates of the long-tailed vole (*Microtus longicaudus*), concluding that the values were 60% to 70% higher than those predicted based on body weight. He cites other investigators who also found that microtine rodents had higher standard metabolic rates than those predicted based on body weight.

Packard (1968) found the standard metabolic rates of montane voles (*Microtus montanus*) to be 75% higher than those predicted from the empirical relationship of metabolism to body weight. Grodzinski and Gorecki (1967) summarize the data on daily energy budgets of small rodents in the temperate zone of central Europe. They state that "the well-known dependence of metabolism on body size (Kleiber 1961) definitely holds true among the discussed rodent species." The species studied ranged in weight from 8 to 37 g.

The metabolic rates of different species of birds have been expressed in a single equation by King and Farner (1961):

$$\log m = \log 74.3 + 0.744 \log W \pm 0.074 \tag{7-3}$$

where

m = metabolic rate (kcal per 24 hrs)
W = body weight in kg

This equation is apparently satisfactory for birds whose body weights range from 0.125 to 10 kg. It is nearly the same as an equation by Kleiber in 1947 [see King and Farner (1961)], indicating that the base line equation describing the metabo-

FIGURE 7-4.   The relationship between heat production and age of lambs. (Data from Ritzman and Benedict 1930.)

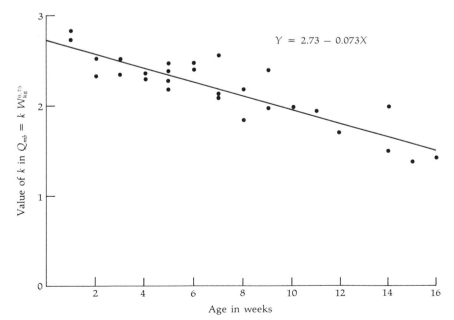

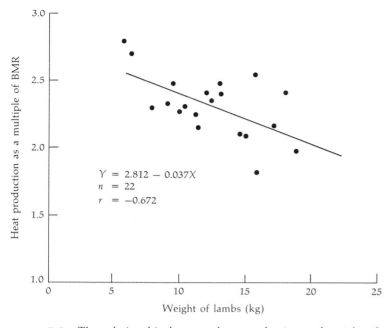

FIGURE 7-5.   The relationship between heat production and weight of lambs. (Data from Ritzman and Benedict 1930.)

lism of birds and mammals may, for practical purposes, be combined. Such equations are only base lines, of course; it is necessary to consider the metabolic rates of free-ranging birds participating in different activities and various productive functions in order to analyze the energetics of animals in their natural habitats. Mullen (1963) has pointed out that the relationship between metabolic rate and body weight varies, and that the effect of any parameter on metabolic rate cannot be fully evaluated until changes in body weight are considered. This makes it imperative that data on metabolic rates be accompanied by weight data, with changes in weight if possible. Many species exhibit cyclic weight changes, and these are important considerations when applying metabolic rates to the analysis of energetics of free-ranging animals in natural habitats. It appears that metabolic-rate variation due to natural functions may be considerably greater than variations in rates determined under "standard" conditions, and the analytical ecologist needs to relate these functions to the free-ranging animal if ecological considerations are to be made.

7-5 FACTORS INFLUENCING ENERGY METABOLISM AND HEAT PRODUCTION

The energy metabolism and heat production of free-ranging animals is highly variable, depending on the activity of the animal, its diet, thermoregulatory functions, sex, reproductive condition, time of day and year (daily and seasonal rhythms occur), hair or feather characteristics, weather factors, parasites and pathogens, and various social and psychological effects. A free-ranging animal has an *ecological metabolic rate* that is an expression of the energy "cost of living" for the conduction of daily activities and other life support processes. This ecological metabolism varies from one activity to another; the cost of actively escaping from a predator is different from the cost of resting. Further, the cost of escaping from one type of predator may be different than the cost of escaping from another type. To illustrate, a rabbit may run from a fox, but hide under a log to escape from an avian predator.

There is considerable difficulty in measuring the energy cost of the different activities in the life of a free-ranging animal. Tucker (1969) has published results of experiments on birds flying in a wind tunnel, showing rather marked differences in the metabolic cost in relation to ascending, descending, and level flight at different wind velocities (Figure 7-6). While flying, budgerigars (the common parakeet) and laughing gulls expend energy at eleven to twenty times the rate at which they metabolize while at rest. He has compared the cost of flying (birds and insects) with other transport costs, including walking and running (man and several domestic and wild mammals), swimming (a fish), and mechanical modes of transport (Figure 7-7). The cost of transport by swimming for a salmon is less than the cost of transport for any other animal or machine considered by Tucker.

FIGURE 7-6. Energetic cost of flight for gulls and budgerigars. When flying, these birds expend energy at eleven to twenty times the resting metabolic rate. (From "The Energetics of Bird Flight" by Vance A. Tucker. Copyright © 1969 by Scientific American Inc. All rights reserved.)

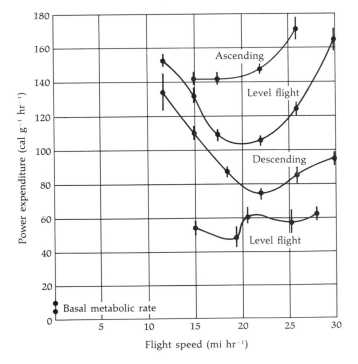

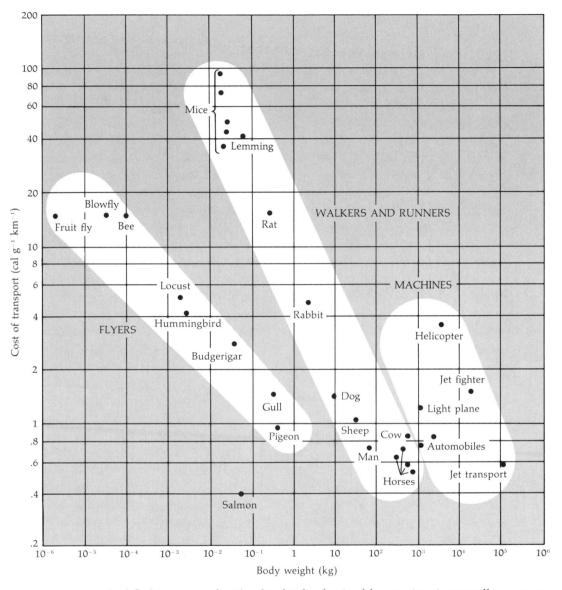

FIGURE 7-7. Bird flight, compared with other kinds of animal locomotion, is generally more economical than walking or running. Large flying birds travel farther for each calorie per unit of weight than a light plane or jet fighter. The young salmon's performance shows, however, that a fish can travel more economically than gulls, pigeons, horses, or any other kind of animal. (From "The Energetics of Bird Flight" by Vance A. Tucker. Copyright © 1969 by Scientific American, Inc. All rights reserved.)

However, the range in the cost of transport for the animals considered is wide. The cost for machines in comparison with that of some of the animals is low.

The efficiency of energy use has a direct effect on the physiology and behavior of an animal in the field. This results in constantly changing nutritional requirements, and differences between the energy required for daily life activities and the amount ingested are reflected in weight gains or losses as body tissue is used to ameliorate the differences.

The metabolism of blue-winged teal (*Anas discors*) carrying heart-rate transmitters was measured with birds in metabolism cages and in 1200-m² enclosures (Owen 1969). The relationship between heart rate and metabolism was determined in the chambers and applied to the birds in the enclosures. The results showed a 30% increase in heart rate when the birds were preening and a 60% increase when they swam rapidly. These increases may not be in direct proportion to metabolism, however, since stroke volume, oxygen transport, respiration efficiency, and so forth may also change.

Research currently in progress at the BioThermal Laboratory has provided us with many significant insights into physiological responses of white-tailed deer. These will be published upon completion of the current experiments. The extensive literature describing the effect of some important factors on domestic animals provides additional insight into mechanisms that may be important in the environment of free-ranging animals.

THE RELATIONSHIP BETWEEN HEAT PRODUCTION AND SURFACE AREA.   In the nineteenth century great interest developed in the relationship between the heat production of an animal and its surface area. The "surface area law" was formulated because heat loss from any object is proportional to surface area. In a homeothermic animal, heat production must be proportional to heat loss if homeothermy is to be maintained. Thus it was concluded that heat production must be proportional to surface area.

In considering the many modes of heat exchange (radiation, conduction, convection, and evaporation), it becomes obvious that surface area is only one of several parameters that must be considered in the calculation of heat loss. Thermal insulation values of hair coats, for example, are variable over different areas of an animal, between species, and seasonally for many species. This insulation represents a barrier to heat loss from the animal. If two animals with equal surface areas are exposed to identical thermal conditions, the one with the higher-quality insulation will lose less heat, and less heat will be necessary to maintain a given body temperature.

Another factor entering into the relationship between surface area and heat production is the variability in the surface area participating in heat exchange between an animal and its environment. A greater surface area is exposed when an animal is standing than when it is lying down. A living animal changes not only its surface area but also its heat production, evaporative heat loss, muscular activity, diet, hair insulation through piloerection, and other characteristics to maintain homeothermy.

If all of the characteristics of the thermal energy regime were known, it would be possible to calculate the critical surface area for an individual animal (Moen 1968). This critical surface area would be determined by the relationship of surface area to weight that is just sufficient to result in a net thermal exchange of zero. The dynamic characteristics of thermal exchange make the calculations virtually impossible, but the concept is easily understood. It is clear that the relationship observed between surface area and heat production is far more complex than it appeared to early investigators.

HEAT INCREMENTS DUE TO DIET. An animal on feed has a greater heat production than it does in a post-absorptive state. The difference between heat production on feed and on fast is called the heat increment. This heat increment results from the release of heat during fermentation in the gastrointestinal tract and from the heat of nutrient metabolism in the assimilation of body tissue. This heat energy is of value to the animal during colder weather since it helps maintain a balance between heat production and heat loss. When other metabolic processes are sufficient to maintain this balance, the heat increment represents a quantity of heat that must be dissipated to prevent a rise in body temperature.

How do variations in diet affect these heat increments? Dukes (1955) cites work by Benedict and Ritzman who found that steers on timothy hay produced heat at a rate 50% greater than the fasting metabolic rate, while steers on an alfalfa diet produced heat at a rate 60% greater than the fasting rate. Blaxter (1967) presents a table showing the differences in heat production between sheep and steers on fasting, maintenance, and full-feed diets. For sheep the ratio of the maintenance to the fasting diet was 1.47 : 1, and of the full-feed diet to the fasting diet, 1.88 : 1. Ritzman and Benedict (1931) determined a ratio of 1.26 : 1 for sheep 0–4 hours after feeding compared with 48–52 hours after feeding.

The effect of diet on metabolism may not be limited to the individual animal alone, but may also carry over into newly born offspring. Alexander (1962) suggests that a low summit metabolism in lambs may be associated with poor prenatal nutrition. This may cause mortality since the lamb must increase heat production if it is to survive cold weather shortly after birth. Adult sheep have been observed to be so underfed that their body temperatures fell 4°C when the surrounding temperature dropped from 35°C to 20°C (Graham 1964). This shows clearly that food can be important in the maintenance of homeothermy.

HEAT INCREMENTS DUE TO ACTIVITY. Heat production increases when an animal is active. Benedict and Ritzman (1923) stated that the activity of steers in the stall increased heat production by 15% over basal rates. Ritzman and Benedict (1931) consider the heat produced while standing compared with lying to be 15% greater for sheep and 17% greater for cattle. Dukes (1955) cites Hall and Brody who found an energy increment of standing over lying of 9% for cattle, with a 13% increment for one very fat steer.

Crampton and Harris (1969) suggest that the energy requirements for maintenance can be estimated by multiplying the basal metabolic rate by 1.33. This increment is applicable on a daily basis, with activities such as walking or running costing much more and standing or bedding somewhat less. Analyses of the energy cost of the daily activity of wild ruminants are given in Chapter 16.

NONSHIVERING THERMOGENESIS. Homeotherms have the physiological capability of increasing their heat production without overt activity. Summit metabolism is one example of this; newborn lambs can increase their heat production up to five times the basal rate without muscular activity. Postnatal tissue is the initial source of the energy. This is followed by the lamb's increasing dependence on milk and forage for energy.

SEX DIFFERENCES.   The few data on wild ruminants do not indicate that there are differences between male and female when both are in the same physiological and psychological condition. When the effects of gestation and lactation are considered, observed differences in energy requirements can be attributed to these reproductive processes.

REPRODUCTIVE CONDITION.   Each of the three stages in the reproductive cycle— breeding, gestation, and lactation—has a definite effect on energy metabolism. Male ruminants expend a large amount of energy during the breeding season, and this activity is accompanied by a reduction in forage intake and a marked weight loss. The increase in energy expenditure is most likely due to the increase in the overt activity of the animal rather than to changes in the rate of tissue metabolism per se.

The female must expend additional energy during gestation to maintain the uterus, for fetal growth, for the increased demands on the circulatory, respiratory, and excretory systems, and to handle the endocrine influences on her own metabolism (Brody 1945). Ritzman and Benedict (1931) reported a slight decrease in the metabolism of cows during the first three months of pregnancy, and it is generally accepted that the energy metabolism on a per weight basis is no higher for pregnant females until the last one-third of pregnancy when fetal growth is accelerated (Morrison 1948). Any small changes in the metabolic rate during the first two-thirds of pregnancy can be considered by expressing metabolism on a metabolic body-weight basis. During the stage of rapid fetal growth, elevated metabolism may result when the growth processes demand additional energy for synthesis of fetal tissue.

There is an obvious physiological demand on the female during lactation since milk must be produced. The heat production of cattle during peak lactation is approximately 100% above the nonlactating level and is associated with the higher food consumption and milk production rather than with muscular activity. The increase is directly proportional to milk production (Brody 1945). Lactating ewes consumed about two times as much feed but grew only 84% as much wool as those ewes whose lambs were removed at birth (Corbett 1964). When feed intake was held constant, wool growth during lactation ceased. Dolge (1963) reports that dairy cattle fed at a high grain level (challenge feeding) showed a marked increase in milk production but lost an average of 61 pounds of body weight during the 70-day test period. This indicates the high cost of lactation as body reserves were mobilized to meet the demands for energy and other nutrients. Calculations for white-tailed deer, discussed in Chapters 16 and 17, indicate a similar high cost for these wild ruminants.

Egg production by female birds has the same effect on the annual cycle of energy costs as gestation and lactation in mammals. West (1968) has studied the bioenergetics of willow ptarmigan (*Lagopus lagopus*) throughout the year, finding that egg-laying females had a higher energy requirement than nonlaying but molting females and males. The extensive literature on domestic poultry provides a good foundation for these kinds of considerations in wild birds.

RHYTHMIC CHANGES IN THE BASAL METABOLIC RATE.   Marked changes occur in the rate of metabolism during a 24-hour period. Nocturnal animals have a low BMR during the daytime (Benedict 1938). Muscular-activity rhythms and basal-metabolic rhythms follow similar time patterns according to Brody (1945). He points out that the diurnal variations in metabolism exceed the effect of the heat increment of food in rats. Thus the diurnal variation is an important consideration when interpreting variations in the metabolic rates, both within and between species, since the time of day at which the measurement was made is a source of variation.

Seasonal variations may occur also. Early studies on cattle indicate a seasonal lability (Ritzman and Benedict 1930; Benedict 1938), but additional research has shown that this conclusion was based on experimental artifacts. Helenette Silver and her associates at the University of New Hampshire, however, have an indication of depressed fasting metabolism in white-tailed deer in the winter (Silver et al. 1969). The number of deer studied, however, is not sufficient to warrant a definite conclusion about seasonal lability, but current research at the Bio-Thermal Laboratory indicates that it is a definite physiological rhythm. Additional studies by Patrick Karns of the Minnesota Department of Natural Resources also indicate seasonal changes in physiological parameters (personal communication).

INSULATION CHARACTERISTICS.   Warm-blooded animals maintain a fairly constant body temperature, so it is reasonable to assume that a decrease in the insulation value of the coat will cause an increase in the heat production of the animal if the heat loss is sufficient to cause the animal to have a negative thermal balance. The effectiveness of the coat on white-tailed deer has been demonstrated by Silver et al. (1969); they sheared a deer and then noted a marked increase in its heat production while it was in the cold respiration chamber. The same effect has been observed when sheared sheep are released in a cold environment; the lack of insulation can cause heat losses great enough to cause mortality. Shearing increases the heat tolerance of sheep exposed to high thermal energy environments (Wodzicka 1960).

Lambs with hairy coats have been observed to conserve heat more effectively than lambs with fine coats (Alexander 1962), and this results in a lower metabolic requirement for those with hairy coats. Brody (1945) attributed the ability of cattle to withstand temperatures of $-40°F$ to their coat insulation and to highly developed vascular control of peripheral tissues.

Mallard and black ducks (*Anas platyrhynchos* and *Anas rubripes*) with oil on their feathers show higher metabolic rates as the oil concentrations are increased (Hartung 1967). The metabolic rates return to normal levels several days after the oil has been removed. This can be attributed to a reduction of the insulating capability of the feathers, since oil causes derangement of the feather barbules.

WEATHER. Complete consideration of all the weather factors that affect the thermal balance of an animal is an enormous task. The concept of thermoneutrality has been discussed for many years. The complexity of the concept has not always been recognized, however. Identification of the many physiological, behavioral, and thermal factors that affect thermoneutrality helps to provide a broader understanding of the concept. Factors affecting the upper limit (*critical hyperthermal environment*) and lower limit (*critical hypothermal environment*) of the thermoneutral range are discussed in a paper by the author (Moen 1968) and are considered further in Chapters 6, 13, and 15.

PATHOGENS AND PARASITES. Whenever an organism must support the life processes of another organism living upon or within it, the energy requirements of the host must increase. The growth and survival of both species may be benefited, however. These beneficial interactions, called *mutualism* by Odum (1959), can be characterized by the relationship between a ruminant and the rumen flora and fauna. Pathogens and parasites, however, increase the cost of metabolism by the host organism, especially when their numbers exceed the normal complement of associated organisms that are either beneficial to or without effect on the host.

VanVolkenburg and Nicholson (1943) found eleven kinds of parasites in the nasal passage, skin, abomasum, and small intestine of dead deer on the Edwards Plateau, Texas. They concluded that infestations of parasites are apparently unimportant among deer on ranges with sufficient food. This is logical, perhaps, but it is important that variability between both the ability of individual deer to withstand the effect of parasites and the effect of different species of parasites on their host be recognized. Georgi and Whitlock (1967), for example, established a direct relationship between the exposure of sheep to infection by *Haemonchus contortus* and the onset of erythrocyte loss. They show clearly that the rate of iron loss in sheep infected with *H. contortus* was greater than that for noninfected sheep. The loss of iron promotes loss of erythrocytes, resulting in a reduction in the efficiency of the oxygen transport system, which in turn may depress the rate of tissue metabolism.

Evans and Whitlock (1964) conclude that ". . . other things being equal, an animal (sheep) with a low erythrocyte volume has a smaller chance of surviving a natural challenge with this parasite than an animal having a greater erythrocyte volume." The experimental evidence in sheep suggests that there is a need for basic research on the *reaction* of wild ruminants to pathogens and parasites rather than the mere identification and counting of them.

7-6 SOCIAL AND PSYCHOLOGICAL EFFECTS ON HEAT PRODUCTION

Numerous sociological and psychological interactions influence heat production. Pfander (1963) lists such things as confinement versus natural range, herd versus individual response, numbers per group, space per animal, noise level, and other disturbances as factors that affect the heat production of domestic animals. Voles (*Clethrionomys glareolus*), kept in groups of two to four animals, lowered their daily metabolism rate by 13.3% (Gorecki 1968). Group reductions by different species are expected to occur owing to thermal factors or social factors, depending on the thermal regime, population density, endocrine levels, and so forth.

Noise from cars, tractors, snowmobiles, and other sources definitely affects the heat production of deer at the BioThermal Laboratory, although not to the same extent at all times during the day or year. Newborn fawns are quite sensitive; even "well-trained" adults exhibit excitement from rather common and frequent noises.

The effect of confinement on the psychology of the animal is very difficult to quantify. Well-trained wild animals appear to be more calm, but the real cause of any variability between animals is hard to identify. Elevation of the metabolic rates of wild ruminants over the rates of domestic ones should not be considered an inherent difference in cell metabolism between these two groups, since elevated cellular metabolism can occur without visible signs of excitement in the animal. I have observed this in a male white-tailed deer that was rescued from icy water (Moen 1967). The deer showed a steady increase in rectal temperature during the warming process from a low of 26° to a high of 39.5°C, after which it stabilized at 38°C. My presence in the room caused the rectal temperature to rise within five minutes to 38.5°C, but the animal did not show any overt signs of fear. It decreased to 38°C when I left the room. The increase in the rectal temperature indicates that heat production was increased by the fear response. The rapidity of this change is surprising, considering the animal's size. Similar changes are being observed in the physiological telemetry experiments in progress on white-tailed deer at the BioThermal Laboratory.

LITERATURE CITED IN CHAPTER 7

Agricultural Research Council. 1965. *The nutrient requirements of farm livestock.* No. 2. *Ruminants.* London: Agricultural Research Council, 264 pp.

Alexander, G. 1962. Energy metabolism in the starved new-born lamb. *Australian J. Agr. Res.* **13**(1): 144–164.

Altman, P. L., and D. S. Dittmer, eds. 1968. *Metabolism.* Bethesda, Maryland: Federation of American Societies for Experimental Biology, 737 pp.

Beck, L. R., and R. G. Anthony. 1971. Metabolic and behavioral thermoregulation in the long-tailed vole, *Microtus longicaudus. J. Mammal.* **52**(2): 404–412.

Benedict, F. G. 1938. *Vital energetics; a study in comparative basal metabolism.* Publication No. 503. Washington, D.C.: Carnegie Institution, 215 pp.

Benedict, F. G., and E. G. Ritzman. 1923. *Undernutrition in steers.* Publication No. 324. Washington, D.C.: Carnegie Institution, 333 pp.

Blaxter, K. L. 1967. *The energy metabolism of ruminants.* London: Hutchinson, 332 pp.

Brockway, J. M., and G. M. O. Maloiy. 1967. Energy metabolism of the red deer. *J. Physiol.* **194**: 22p–24p.

Brody, S. 1945. *Bioenergetics and growth.* New York: Reinhold, 1023 pp.

Corbett, J. L. 1964. Effect of lactation on wool growth of merino sheep. *Proc. Australian Soc. Animal Prod.* **5**: 138–140.

Crampton, E. W., and L. E. Harris. 1969. *Applied animal nutrition*. 2d ed. San Francisco: W. H. Freeman and Company, 753 pp.

Dolge, K. L. 1963. Current nutritional problems and challenges in feeding dairy cattle. In *Bridging the gap in nutrition*, ed. R. H. Thayer. Midwest current nutritional problems clinic, 1st, Kansas City. Kansas City: Midwest Feed Manufacturers Association, pp. 134–156.

Dukes, H. H. 1955. *The physiology of domestic animals*. 7th ed. Ithaca, New York: Comstock, 1020 pp.

Evans, J. V., and J. H. Whitlock. 1964. Genetic relationships between maximum hematocrit values and hemoglobin types in sheep. *Science* **145**(3638): 1318.

Georgi, J. R., and J. H. Whitlock. 1967. Erythrocyte loss and restitution in ovine haemonchosis. Estimation of erythrocyte loss in lambs following natural exposure. *Cornell Vet.* **57**(1): 43–53.

Gorecki, A. 1968. Metabolic rate and energy budget in bank vole. *Acta theriologica* **13**(20): 341–365.

Graham, N. McC. 1964. Energy costs of feeding activities and energy expenditure of grazing sheep. *Australian J. Agr. Res.* **15**(6): 969–973.

Grodzinski, W., and A. Gorecki. 1967. Daily energy budgets of small rodents. In *Secondary productivity of terrestrial ecosystems*, ed. K. Petruservicz. Warsaw, pp. 295–314.

Hammel, H. T. 1962. *Thermal and metabolic measurements on a reindeer at rest and in exercise*. Tech. Doc. Rept. AAL-TDR 61-54 Artic Aeromedical Lab., USAF, Seattle, 34 pp.

Hart, J. S., O. Heroux, W. H. Cottle, and C. A. Mills. 1961. The influence of climate on metabolic and thermal responses of infant caribou. *Can. J. Zool.* **39**(4): 845–856.

Hartung, R. 1967. Energy metabolism in oil-covered ducks. *J. Wildlife Management* **31**(4): 798–804.

Hoar, W. S. 1966. *General and comparative physiology*. Englewood Cliffs, New Jersey: Prentice-Hall, 815 pp.

King, J. R., and D. S. Farner. 1961. Energy metabolism, thermoregulation and body temperature. In *Biology and comparative physiology of birds*, ed. A. J. Marshall. New York: Academic Press, pp. 215–288.

Kleiber, M. 1961. *The fire of life*. New York: Wiley, 453 pp.

Krog, H., and M. Monson. 1954. Notes on the metabolism of a mountain goat. *Am. J. Physiol.* **178**: 515–516.

Marston, H. R. 1948. Energy transactions in the sheep. I. The basal heat production and heat increment. *Australian J. Sci. Res.* **B1**: 93–129.

McEwan, E. H. 1970. Energy metabolism of barren ground caribou (*Rangifer tarandus*). *Can. J. Zool.* **48**: 391–392.

Moen, A. N. 1967. Hypothermia observed in water-chilled deer. *J. Mammal.* **48**(4): 655–656.

Moen, A. N. 1968. The critical thermal environment: a new look at an old concept. *BioScience* **18**(11): 1041–1043.

Morrison, F. B. 1948. *Feeds and feeding*. 21st ed. Ithaca, New York: Morrison, 1207 pp.

Mullen, W. J. 1963. Body size and metabolic rate in the fowl. *Agr. Sci. Rev.* **1**: 20–26, 49.

National Research Council. 1966. *Biological energy interrelationship and glossary of energy terms*. Publication No. 1411. Washington, D.C.: National Academy of Sciences, National Research Council, 35 pp.

Nordan, H. C., I. McT. Cowan, and A. J. Wood. 1970. The feed intake and heat production of the young black-tailed deer (*Odocoileus hemionus columbianus*). *Can. J. Zool.* **48**(2): 275–282.

Odum, E. P. 1959. *Fundamentals of ecology.* 2d ed. Philadelphia: Saunders, 546 pp.

Owen, R. B., Jr. 1969. Heart rate, a measure of metabolism in blue-winged teal. *Comp. Biochem. Physiol.* **31:** 431–436.

Packard, G. C. 1968. Oxygen consumption of *Microtus montanus* in relation to ambient temperature. *J. Mammal.* **49**(2): 215–220.

Pfander, W. H. 1963. Factors involved in determining nutritive requirements of beef cattle and sheep. In *Bridging the gap in nutrition,* ed. R. H. Thayer. Midwest current nutritional problems clinic, 1st, Kansas City. Kansas City: Midwest Feed Manufacturers Association, pp. 116–133.

Ritzman, E. G., and F. G. Benedict. 1930. *The energy metabolism of sheep.* Technical Bulletin No. 43. Durham: New Hampshire Agricultural Experiment Station, 23 pp.

Ritzman, E. G., and F. G. Benedict. 1931. *The heat production of sheep under varying conditions.* Technical Bulletin No. 45. Durham: New Hampshire Agricultural Experiment Station, 32 pp.

Rogerson, A. 1968. Energy utilization by the eland and wildebeest, In *Comparative nutrition of wild animals,* ed. M. A. Crawford. Symp. Zool. Soc. Vol. 21, pp. 153–161.

Silver, H. 1968. Deer nutrition studies. In *The white-tailed deer of New Hampshire,* ed. H. R. Siegler. Survey Report No. 10. Concord: New Hampshire Fish and Game Department, pp. 182–196.

Silver, H., N. F. Colovos, J. B. Holter, and H. H. Hayes. 1969. Fasting metabolism of white-tailed deer. *J. Wildlife Management* **33**(3): 490–498.

Tucker, V. A. 1969. The energetics of bird flight. *Sci. Am.* **220**(5): 70–78 (Offprint No. 1141).

Van Volkenburg, H. L., and A. J. Nicholson. 1943. Parasitism and malnutrition of deer in Texas. *J. Wildlife Management* **7**(2): 220–223.

West, G. C. 1968. Bioenergetics of captive willow ptaimigan under natural conditions. *Ecology* **49**(6): 1035–1045.

Wodzicka, M. 1960. Seasonal variations in wool growth heat tolerance in sheep. II. Heat tolerance. *Australian J. Agr. Res.* **11**(1): 85–96.

SELECTED REFERENCES

Bauley, E. D. 1965. Seasonal changes in metabolic activity of non-hibernating woodchucks. *Can. J. Zool.* **43:** 905–909.

Blaxter, K. L. 1962. The fasting metabolism of adult wether sheep. *Brit. J. Nutr.* **16:** 615–626.

Blaxter, K. L., ed. 1965. *Energy metabolism.* Proc. 3rd Symp. Troon. Scotland, May 1964. (European Assoc. Animal Prod., publication No. 11). London: Academic Press, 450 pp.

Chatonnet, J. 1963. Nervous control of metabolism. *Proc. Federation Am. Soc. Exptl. Biol.* **22:** 729–731.

Cloudsley-Thompson, J. L. 1961. *Rhythmic activity in animal physiology and behaviour.* New York: Academic Press, 236 pp.

Deighton, T., and J. C. D. Hutchinson. 1940. Studies on the metabolism of fowls. II. The effect of activity on metabolism. *J. Agr. Sci.* **30:** 141–157.

Helms, C. W. 1963. Tentative field estimates of metabolism in hunting. *Auk* **80:** 318–334.

Lasiewski, R. A., and W. Dawson. 1967. A re-examination of the relation between static metabolic weight and body weight in birds. *Condor* **69:** 13–23.

Ludwick, R. L., J. P. Fontenot, and H. S. Mosby. 1969. Energy metabolism of the eastern gray squirrel. *J. Wildlife Management* **33**(3): 569–575.

Morrison, P. R. 1948. Oxygen consumption in several mammals under basal conditions. *J. Cellular Comp. Physiol.* **31:** 281–291.

McNab, B. K. 1963. A model of the energy budget of a wild mouse. *Ecology* **44**(3): 521–532.

Owen, R. B., Jr. 1970. The bioenergetics of captive blue-winged teal under controlled and outdoor conditions. *Condor* **72:** 153–163.

Pearson, O. P. 1947. The rate of metabolism of some small mammals. *Ecology* **28:** 127–145.

Tucker, V. A., and K. Schmidt-Koenig. 1971. Flight speeds of birds in relation to energetics and wind directions. *Auk* **88**(1): 97–107.

Verbeek, N. A. M. 1964. A time and energy budget study of the brewer blackbird. *Condor* **66**(1): 70–74.

Wekstein, D. R., and J. F. Zolman. 1969. Ontogeny of heat production in chicks. *Proc. Federation Am. Soc. Exptl. Biol.* **28**(3): 1023–1028.

Wesley, D. E., K. L. Knox, and J. G. Nagy. 1970. Energy flux and water kinetics in young pronghorn antelope. *J. Wildlife Management* **34**(4): 908–912.

Wilson, T. A. 1965. Natural mortality and reproduction for a food supply at minimum metabolism. *Am. Naturalist* **99:** 373–376.

Wooden, G. R., K. L. Knox, and C. L. Wild. 1970. Energy metabolism in light horses. *J. Animal Sci.* **30**(4): 544–548.

DIGESTION

8-1 THE DEFINITION OF DIGESTION

Digestion results in the conversion of food into a form that can be absorbed by the body and assimilated into body tissue. Dukes (1955) defined digestion to include all activities of the alimentary canal and its glands in preparation of foods for absorption and the rejection of residues. The mechanical factors include chewing, swallowing, regurgitation, stomach and intestinal movements, and defecation. Secretory factors are associated with the activity of the digestive glands, and chemical factors include the reactions of the enzymes secreted by the digestive glands with the plant enzymes and other chemicals in the ingested food. Ruminants also have the additional chemical functions of the microorganisms in the gastrointestinal tract.

The transformation of food to metabolically useful nutrients is a biochemical process, and the whole process of digestion must be studied from that point of view if the functional relationships of foods to an animal are to be understood. Identification of different food species that are ingested is only an aid in approaching the study of biochemical aspects of nutrition. Food-habit lists are useful for only the most general analysis of animal-environment relationships. According to Crampton and Harris (1969, p. 5), "Feeds are merely the carriers of the nutrients and the potential energy . . . in a . . . diet."

8-2 A RESEARCH PHILOSOPHY

The utilization of plant species by wild and domestic ruminants on open range has been described in the literature in many different ways, including expressions of the number of twigs browsed, percentage of twigs browsed, volume of food in the rumen, frequency of occurrence of foods in the rumen, and animal-minutes spent ingesting each species. Statistical analyses of such data have been made, and biological inferences have been published.

Data like these may, however, be quite unrelated to the functional nutritive relationships between an animal and its range. For example, the chemical composition of two plant species may be very much alike, and an animal may be able to digest and assimilate each plant species to a similar degree. The animal makes no biochemical distinction between the two species, but the descriptive biologist separates them according to taxonomic differences. Such a separation of the two species is unnecessary from a nutritional point of view.

Some very pertinent questions can be asked at this point. How much do we know about the chemical interaction between wild ruminants and their environment? What does the environment contain chemically, and of what value is each chemical component to the animals? Do greater chemical differences exist between different plant species than between the same species on different soils? Do deer select different plant species for foraging or do they forage randomly on a variety of species just because they are dispersed throughout the habitat?

We know that there are chemical differences between plant species, and we know there are differences in the chemical characteristics of the soil. Field observations indicate that wild ruminants do have preferences for certain species, but preference lists are not similar for different geographical areas. Knowledge of differences in the food characteristics between habitats or of differences between animal species in their preferences for forage plants does not provide any significant insight into the chemical interaction between wild ruminants and their environment. Little is known about the requirements of wild ruminants and how well the environment supplies these requirements. Much more information is needed before the functional relationships between an organism and its complex and dynamic environment will be understood.

8-3   CHEMICAL COMPOSITION OF FOOD MATERIALS

Foods are complex structures that can be organized into groups having similar physical or chemical characteristics. Chemical analyses of foods have usually followed a proximate-analysis scheme devised by workers at the Weende Experiment Station in Germany over 100 years ago. This proximate analysis results in the grouping of chemically similar components of foods into six categories, including water, ether extract, crude fiber, nitrogen-free extract, crude protein, and ash. A chemical organization of foods is shown in the following outline, with the categories used in proximate analyses shown in parentheses.

I. Water
II. Dry matter
  A. Organic substances
    1. Nitrogenous compounds (crude protein)
      a. True protein
      b. Nonprotein nitrogenous materials

2. Non-nitrogenous substances
   a. Carbohydrates
      (1) Soluble carbohydrates (nitrogen-free extract)
      (2) Insoluble carbohydrates (crude fiber)
   b. Fats (ether extract)
B. Inorganic substances (ash)
   1. Salts
   2. Mineral matter

The Association of Official Agricultural Chemists periodically publishes descriptions of the standard analytical procedures for proximate analyses. Subsequent reference in this book to these procedures is indicated by the abbreviation "AOAC Handbook."

The chemically similar groups measured in the proximate analyses do not necessarily have similar nutritional significance and must be interpreted accordingly for different species of animals. Crampton and Harris (1969) devote an entire chapter to the discussion of the proximate analysis of feeds. The summary that follows is based mostly on their writings.

WATER. Water is a simple food substance chemically, but is difficult to measure quantitatively in different foods. The usual procedure is oven-drying at about 105°C until a constant sample weight is reached. This usually occurs in 24 to 48 hours. Heating at 105°C may also cause a loss of other volatile substances such as essential oils, as well as the decomposition of some sugars. These losses would then be considered a part of water loss. The importance of that error is obviously related to the composition of the plant material. Vacuum ovens are used to eliminate some of these errors, and distillation procedures are also used. Both are more time consuming and expensive than the straightforward drying in a forced-air oven.

NITROGENOUS SUBSTANCES—CRUDE PROTEIN. The total amount of nitrogen in the food is determined by methods of analysis described in the AOAC Handbook, and the crude protein is obtained by multiplying the amount of nitrogen in the food by 6.25. The numerical factor 6.25 is derived from the assumption that protein contains 16% nitrogen, that all of the nitrogen is in the protein, and that all urinary nitrogen excreted is derived from protein oxidation. Thus $100/16 = 6.25$, and urinary nitrogen excreted $\times 6.25 =$ the amount of protein oxidized (Brody 1945). This procedure is not completely valid, since the nitrogen content of the protein of different feeds ranges from 16% to 19%. When the latter percentage is correct for a particular feed, the conversion factor should be $100/19 = 5.26$. Crampton and Harris (1969) include two tables showing (1) the errors resulting from the use of a constant factor 6.25 in estimating the protein content of foods and (2) selected conversion factors for proteins. The greatest errors resulting from the use of 6.25 seem to be characteristic of the oil-seed and cereal proteins that have a higher percentage of nitrogen (greater than 16%) in the protein. Another error is dependent on the amount of nonprotein nitrogen (NPN) in the forage.

In view of the many other unknown factors in the total animal-range relationship, the N × 6.25 calculation seems to be a satisfactory approximation of the quantity of protein in a food. This calculation gives no indication of the amino acid composition of the food, but this is of little importance for ruminants since the metabolic processes of rumen microorganisms result in the synthesis of proteins from nonprotein sources.

CARBOHYDRATES—NITROGEN-FREE EXTRACT. The nitrogen-free extract (NFE) component of carbohydrates is made up of starches and sugars. These are highly digestible sources of energy that are converted through digestion from starches (starch, dextrin, and glycogen) and sugars (monosaccharides, disaccharides, and trisaccharides) to glucose. In a proximate analysis, the amount of NFE is determined by subtraction, with the water, ether extract, crude fiber, crude protein, and ash being subtracted from the original weight of the sample. Errors in the determination of any of these components are reflected in the figure obtained for the NFE.

CARBOHYDRATES—CRUDE FIBER. Crude fiber is the insoluble residue of a food after successive boiling with dilute acid (H_2SO_4) and dilute alkali (NaOH), according to procedures outlined in the AOAC Handbook. This material may not be insoluble when exposed to the digestion process of an animal, however. This is particularly true for ruminants whose microorganisms have the ability to break down cellulose for their own metabolic needs, producing volatile fatty acids (VFAs, including acetic, butyric, and propionic) that are absorbed from the rumen and supply energy to the host animal. Crampton and Harris (1969) present data showing that 50% to 90% of the crude fiber in plants may be digested in ruminants while nonruminants may digest as little as 3% or as much as 78%. The digestibility of crude fiber in mature, dormant, or dead plant material eaten by wild ruminants may be considerably less than 50%.

ETHER EXTRACT. The ether extract obtained from a food consists of glycerides of fatty acids, free fatty acids, cholesterol, lecithin, chlorophyl, alkali substances, volatile oils, and resins. The last four are not biological nutrients; they are extracted only incidentally. The ether extract of foods will depend, of course, on the chemical differences between them. Since an animal cannot use each of the extracted compounds equally well, the value of the ether extract in the diet is dependent on its utilization. Thus the usual practice of attributing 9.35 kcal of gross energy per gram of ether extract or about 9 kcal of metabolizable energy per gram is an oversimplification of the biological situation.

ASH. Ash is the inorganic residue left after food material has been burned at about 600°C. For ruminants, the use of quantitative determinations of ash has limited value because of the high variability in the amount and kind of ash in plant materials. Some of these minerals may depress digestion. For example, the presence of silica in plant tissue may result in an average decline of three units digestibility per unit of silica in the dry matter of grasses and legumes (Van Soest and Jones 1968).

SUMMARY OF PROXIMATE ANALYSIS. The proximate analysis of feeds is only an approximation of the chemical content of plant materials, which must be related to the digestion processes of living organisms with caution. Differences between the results of analyses completed in different laboratories indicate that they are subject to human error (Dietz and Cernow 1966).

One of the more subtle characteristics of the proximate analysis is that it results in groups that are related to the chemical characteristics of food materials, rather than to the nutrient content of foods. Indeed, the former is useful only because there is some correlation between the chemical characteristics of the isolates and the properties of feeds that have nutritional significance. Crampton and Harris (1969, p. 52) summarize it: "The Weende analysis does not define the nutrient content of feeds. It is an index of nutritive value only because the fractions that it isolates are correlated with some of the properties of feeds that have nutritional significance. Consequently, it is a useful descriptive device in establishing the characteristics of feeds. As with any other specialized tool, to use it correctly and to its fullest potential requires much other nutritional knowledge and judgement. An appreciation of its design, weaknesses, and limitations, though often stressed in destructive criticism, is more correctly an aid in making full legitimate use of a scheme of feed description which has broad and basic value."

8-4 THE NUTRITIVE EVALUATION OF FORAGES

The shortcomings of the proximate-analysis system of nutrient analysis have been recognized for many years. Short (1966a, p. 163) states: "The proximate analysis of important species of deer browse has many times been shown to have little value in predicting how a deer digests a particular forage item." Since the value of any food material depends on its chemical constituents and the ability of the animal to use the nutrients in metabolic processes, it is imperative that the real nutritive relationships be described in meaningful terms.

One new approach to the nutritive evaluation of forages is based on the anatomy of the plant cell in relation to the nutritive availability of the different chemical compounds in a plant cell. The distribution of these chemical compounds within a plant cell is shown in Figure 8-1.[1] The new system separates the highly digestible cell contents (98% to 100% digestible) from the differentially digestible cell-wall constituents. The cell contents are soluble in neutral detergent. The fiber-bound protein and hemicelluloses are soluble in acid detergent, but the cellulose, lignin, and liquified nitrogenous compounds are insoluble in acid detergent. The division of forage organic matter by a system of analysis using detergents is summarized in Table 8-1.

---

[1] Dr. Peter Van Soest, Department of Animal Science, Cornell University, is a leader in this work and has published several articles on the new system.

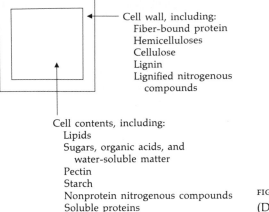

Cell wall, including:
  Fiber-bound protein
  Hemicelluloses
  Cellulose
  Lignin
  Lignified nitrogenous
    compounds

Cell contents, including:
  Lipids
  Sugars, organic acids, and
    water-soluble matter
  Pectin
  Starch
  Nonprotein nitrogenous compounds
  Soluble proteins

FIGURE 8-1. The composition of a plant cell.
(Data from Van Soest 1965.)

How does this approach compare with the older method of proximate analysis? First of all, the nitrogen-free extract in the older scheme includes carbon, hydrogen, and oxygen-rich compounds called carbohydrates. In proximate analysis, the amount of NFE is determined by subtracting all other chemically determined weights from the initial sample weight, resulting in an accumulative error that is dependent on the errors in the determination of other chemical groups. The NFE, however, contains both highly digestible starches and sugars and very indigestible lignin and poorly digestible xylan (Van Soest 1965). Thus the interpretation of the NFE value for a forage analyzed in the older scheme is dependent

TABLE 8-1 DIVISION OF FORAGE ORGANIC MATTER BY SYSTEM OF ANALYSIS
USING DETERGENTS

Fraction	Components
Cell contents (soluble in neutral detergent)	Lipids
	Sugars, organic acids, and water-soluble matter
	Pectin
	Starch
	Nonprotein nitrogenous compounds
	Soluble proteins
Cell-wall constituents (fiber insoluble in neutral detergent)	
(1) Soluble in acid detergent	Fiber-bound protein
	Hemicelluloses
(2) Acid-detergent fiber	Cellulose
	Lignin
	Lignified nitrogenous compounds

SOURCE: Van Soest 1965.

on the ratio between soluble carbohydrates and indigestible plant-cell materials. This ratio is not known after a complete analysis by the old proximate system, however. The new system of detergent analysis was devised to separate the cell components into chemical groups that had greater biological significance.

What, then, is the role of the proximate analysis data in analyzing the nutritional relationships between animal and range? The determinations of the nitrogen content and the ash in forages are still useful. The NFE and crude-fiber values for a forage may include considerable unknown variation in nutritive quality, and it will be necessary to reanalyze the forages using the detergent system.

8-5 THE ALIMENTARY CANAL

ANATOMY.   The alimentary canal in the ruminant includes the mouth, esophagus, four-part stomach, small intestine, large intestine, and rectum. The four-part stomach is unique to ruminants. It includes the rumen, reticulum, omasum, and abomasum. The first three develop as diverticula from the embryonic abomasum, or true stomach.

In the newborn ruminant, milk or water is diverted directly to the orifice entering the omasum through the esophageal groove (Dukes 1955) rather than going into the rumen and then on to the reticulum and omasum. This diversion is a reflex action in the young ruminant that gradually disappears as the animal matures.

The development of the rumen coincides with the increased ingestion of plant materials by the young ruminant. Plant materials that are eaten shortly after birth go to the rumen in which populations of microorganisms are building up as the rumen develops. Concomitant with rumen development is the change from the higher blood-glucose levels of the young ruminant to the lower levels typical of adult ruminants (McCandless and Dye 1950).

The alimentary canal of herbivorous animals has a relatively larger capacity than that of carnivorous animals. This increased capacity permits the extensive fermentation necessary for the breakdown by microorganisms of bulky, fibrous plant materials ingested by the herbivorous host. Carnivore diets, on the other hand, include mainly animal tissue with thin cell membranes, resulting in much more rapid digestion.

The relative sizes and positions of the stomach compartments are not constant throughout the life of a ruminant animal (Figure 8-2). In the newborn, the abomasum or true stomach is larger than the other three parts, and it is not until the age of one or two months that the volume relationship between the omasum plus abomasum and the rumen-reticulum is reversed. After that time, the rumen increases in size. At maturity the rumen comprises about 75% of the total stomach capacity and the reticulum about 10%. The volumetric relationship between rumen + reticulum and omasum + abomasum in white-tailed deer has been determined by Short (1964) and can be expressed as linear regression equations (8-1) and (8-2).

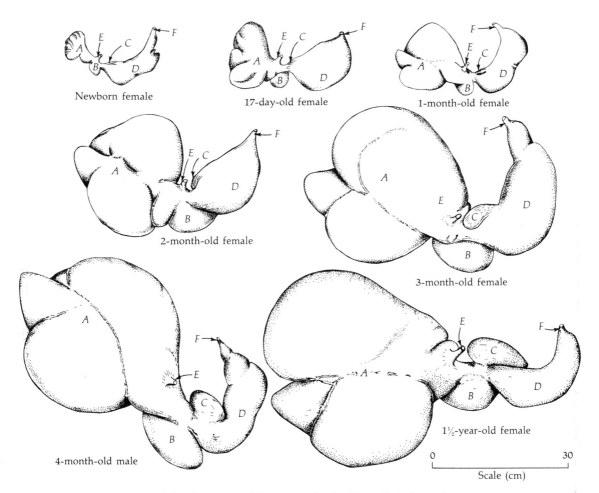

FIGURE 8-2. The sequential development of the stomach of white-tailed deer: A = rumen; B = reticulum; C = omasum; D = abomasum; E = esophagus; F = duodenum. (From Short 1964, *J. Wildlife Management*.)

Rumen + reticulum:

$$Y = 103.35 \, W_{\text{kg}} + 304.64 \tag{8-1}$$

Omasum + abomasum:

$$Y = 11.705 \, W_{\text{kg}} + 514.64 \tag{8-2}$$

where

$\quad Y$ = volume expressed as ml of water

$\quad W_{\text{kg}}$ = body weight in kg

The use of these equations and other data in Short (1964) permits an approximation of the relative volumes of rumen + reticulum and omasum + abomasum [equation (8-3) shown in Figure 8-3].

The ruminoreticular volume of mule deer, expressed as liters, is equal to about 10% of the body weight expressed in kilograms (Short, Medin, and Anderson 1965). This is considerably less than the percentage for cattle. The relationship between rumen volume and body weight is important because the rumen is a "holding tank" in which fermentation takes place. The metabolic products are a source of energy and protein for the animal. The capacity of the rumen represents a finite limit to the amount of nutrients that can be made available in a given length of time.

Stomach capacity in relation to body weight for age classes within a species are also worth considering. Short (1964) points out that the relatively small stomach capacity of deer fawns may be an important factor in winter mortality. The browse ingested by deer in the winter in forested areas may not be metabolized rapidly enough to supply the heat production necessary to maintain body temperature during periods of cold weather, thus causing a mobilization of the fat reserve.

One of the interesting deductions that can be made from the comparisons of growth between different stomach compartments is the anatomical and physiological basis for weaning. Tamate (1957) defines weaning as the period during which the capacity of the rumen equals that of the abomasum. He notes three levels of rumen capacities in goats: 20% in the preweaning stage, 44%–48% in the weaning stage, and 82%–85% in the postweaning stage. Adult proportions of the rumen and reticulum in lambs were reached at about 56 days of age (Church, Jessup, and Bogart 1962). *In vitro* fermentations in this study indicated that the rumen digestion was characteristic of that of adults by the time the lamb was three weeks of age, which is the time at which marked growth of the rumen occurs.

Using the time at which the animal achieves equal proportions between the rumen-reticulum and the omasum-abomasum as an indicator of physiological

FIGURE 8-3. The relative proportions of the different stomach compartments of white-tailed deer. The rumen plus reticulum occupies about 85% of the total volume for the remainder of the animal's life.

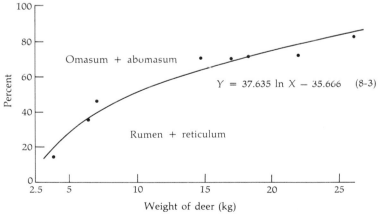

weaning, Short (1964) concludes that domestic goats and sheep are weaned at about 25 days of age, and white-tailed deer at about 35 days of age.

The weight at which the equal proportions are reached in white-tailed deer is 10 kg (see Figure 8-3). A growth equation in Murphy and Coates (1966) for suckling fawns is:

$$W_{lb} = 4.77 + 0.51t_d \qquad (8\text{-}4)$$

where

W_{lb} = weight in pounds

t_d = age in days

This equation is $2.17 + 0.23t_d$ if body weight is given in kg. A white-tailed deer can be predicted to reach a weight of 10 kg at age (t_d), using equation (8-5):

$$t_d = \frac{W_{kg} - 2.17}{0.23} \qquad (8\text{-}5)$$

The solution to equation (8-5) is 34 days, which is very close to the age of 35 days estimated by Short (1964).

White-tailed fawns can be weaned at an earlier age. One female fawn at the BioThermal Laboratory refused to take milk from the bottle at a weight of 15 lb. She was fed calf pellets and had access to grass in her pen, and she weighed as much in the fall as the other fawns on a feeding schedule that included milk.

Wood, Nordan, and Cowan (1961) fed their deer fawns evaporated milk and water in a 1:1 ratio six times per day and weaned them at 15 lb (7 kg) and 5 to 7 weeks of age. Seven-week-old orphaned pronghorn kids survived without milk (Bromley and O'Gara 1967). Thus it is obvious that young ruminants are capable of being fully functional ruminants at a much earlier age than the natural weaning process would indicate since wild fawns are usually not completely weaned until they are 3 months or older.

HISTOLOGY.   The four-part ruminant stomach has a tissue and cellular structure that is related to the functions of each of the parts. The rumen is a muscular organ with a middle layer of smooth muscle that is capable of rhythmic contraction. The smooth-muscle layer is enveloped between a highly vascular layer with many papillae and an outer serous layer that is continuous with membranes that support and suspend the stomach (Short 1964). The papillae lining the rumen (Figure 8-4) greatly increase the internal surface area of the organ, resulting in a more rapid absorption of the volatile fatty acids produced by the rumen microorganisms as by-products of their own metabolism.

The papillae are very rudimentary in the very young ruminants, and their development is dependent on the end products of rumen fermentation (Flatt, Warner, and Loosli 1958). The small and nonfunctional rumen at birth develops concomitantly with an increase in the dependence of the host on digestive action of the rumen microorganism. Short (1964) describes the anatomy of internal

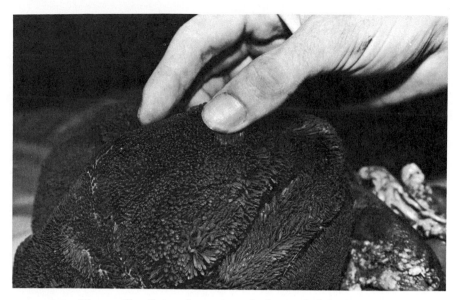

FIGURE 8-4. The papillae lining the rumen of white-tailed deer.

stomach characteristics for deer from birth through $5\frac{1}{2}$–years (Table 8-2). Note that the size of the papillae is nearly maximum by the age of 4 months.

The reticulum is not nearly as muscular as the rumen; the muscular middle layer is thin. It is connected to the rumen through the ruminoreticular fold. Its function is closely associated with the rumen because the ruminoreticular fold acts like a dam, retaining the more solid ingesta in the rumen. Contractions of the reticulum force liquid into the rumen, washing the material in the rumen. The finer particles of ingesta go into suspension. These are washed back into the reticulum by rumen contractions and then move on to the omasum. There are only occasional papillae and the epithelial lining is without glands.

The omasum has many leaves or folds covered by small granular papillae. The leaves increase the internal surface area of the omasum and play an important part in water absorption (Short 1964). The true stomach of the ruminant is the abomasum. It has a muscular layer composed of thin fibers in small bundles, an outer serous layer, and an interepithelial lining that includes secretory glands (Short 1964).

8-6 MECHANICAL AND SECRETORY PROCESSES IN DIGESTION

Mechanical factors function throughout the entire process of digestion to keep the food materials moving through the alimentary canal. The mechanical actions in the ruminant include the prehension and ingestion of food materials, mastication, swallowing, regurgitation, reswallowing, rumen contraction, intestinal contractions, and defecation.

During the mechanical movement of food through the alimentary canal, saliva and digestive juices are secreted and mixed with the food materials. The enzymes

TABLE 8-2 ANATOMICAL MEASUREMENTS (means of each age group) OF THE STOMACH OF WHITE-TAILED DEER

Age of Deer	Number of Deer	Rumen			Reticulum			Omasum				Abomasum Tissue Weight (g)	Stomach Contents				
		Tissue Weight (g)	Papillae Length (mm)	Papillae Width (mm)	Polygons (cm)	Papillae (mm)	Tissue Weight (g)	Length (cm)	Width of Leaves (cm)	Papillae (mm)	Tissue Weight (g)		Total Weight (g)	% Distribution		Nature of Contents	
														Omasum and Abomasum	Rumen and Reticulum	Omasum and Abomasum	Rumen and Reticulum
Newborn	1	7	R*	R*	0.1	R*	2	2.5	0.6	R*	2	25					
½ mo	3	26	0.5–1	0.2	0.4	0.5–1	7	4	0.7	1	5	47	195–300	65–96	35–4	Milk curds	Sand and grass
1 mo	2	47	2	0.5	0.5	1	7	4	1	1	3	32	262–348	33–45	67–55	Vegetation and milk curds	Vegetation
2 mo	2	183	4–5	1	0.6	1	27	4	1–1.5	1	17	70	1,321–1,373	16–22	84–78	80% vegetation, 20% milk curds	Vegetation
3 mo	2	189	5	1	1–1.2	1	21	5	2	1	11	45	1,612–1,720	5–14	95–86	Vegetation and trace of milk	Vegetation
4 mo	2	321	5–8	2	1.2	1	29	5	2.5–3.8	1	19	48	3,407–3,481	3–4	97–96	Vegetation and trace of milk	Vegetation
1½ yr	2	845	10	2	1.2	1	64	7.5	3.8	1	58	101	2,527–4,906	7–8	93–92	Similar contents	Similar contents
5½ yr	1	935	10	2	1.2	1	75	10	5	1	102	145	3,362	8	92	Similar contents	Similar contents

SOURCE: Short 1964, J. Wildlife Management **28**(3) p. 448.
* Rudimentary.

in these secretions assist the rumen microorganisms in the process of digestion. Since there are no glands in the linings of the rumen and reticulum, digestion is almost totally dependent on the activity of the microorganisms, with only a slight contribution from the saliva.

INGESTION.   Ingestion or prehension is the process of seizing and conveying food to the mouth (Dukes 1955). This may seem like a simple process but it is accomplished in different ways among wild ruminants. Deer, for example, remove browse by grinding it off with their molars. When feeding on low herbaceous vegetation, they use the lower incisors and upper gum to seize the plant material. Grains on the ground are picked up with the aid of the tongue.

SALIVATION AND MASTICATION.   Three salivary glands in the ruminant have been described by Dukes (1955). The parotid gland, located on the side of the cheek, secretes a thin, watery substance containing protein. The submaxillary gland, located below the lower mandible, secretes a thin watery substance containing the glycoprotein mucin. The sublingual gland, located under the tongue, also secretes both protein and mucin. The saliva functions as a lubricant during chewing, swallowing, and regurgitation. It is also important in maintaining the rumen fluid volume. Its high alkalinity makes it a good buffering agent for the maintenance of an appropriate pH in the rumen. The nitrogenous substances that make up the saliva also serve as a substrate for protein synthesis by rumen microorganisms, and this protein is useful to the ruminant host (Annison and Lewis 1959).

RUMINATION.   Four phases of rumination are listed by Dukes (1955): (1) regurgitation; (2) remastication; (3) reinsalivation; and (4) reswallowing. Regurgitation occurs with no contraction of the rumen since the skeletal muscles are used instead (Dukes 1955). Dzuik, Fashingbauer, and Idstrom (1963) observed that white-tailed deer would regurgitate ingesta, swallow without mastication, and regurgitate again in less than ten seconds. No possible explanations were given for the rapid reswallowing; it may be associated with some characteristic of the particular bolus.

The first signs of regurgitation appear in domestic goats at the age of 8–12 days (Tulbaev 1959). Goats on solid food assume a normal rumination rhythm during the next 5–8 weeks. At first, each regurgitation and remastication cycle lasts 15–25 seconds followed by a pause of 9–15 seconds. Ten to fifteen days after the introduction of solid food, remastication was prolonged to 20–30 seconds followed by a pause of 6–12 seconds.

The jaw movements during remastication are vertical and lateral, resulting in a circular motion that is centered on one side at a time. The innermost edge of the lower teeth and the outermost edge of the upper teeth are sharp. The teeth wear roughly, increasing the grinding efficiency until old age, when both the efficiency of grinding and the physical condition decline.

STOMACH AND INTESTINAL MOVEMENTS.   The rumen and reticulum are both physiologically and metabolically active parts of the ruminant digestive system. They move, churning the food that is being metabolized by the microorganisms, which results in a separation of the indigestible residue from the more digestible parts of the diet.

Ingested food is retained in the rumen and reticulum until it has a fine consistency. Contractions of the rumen and reticulum result in an interchange of food and liquids. The brisk regular contractions of the reticulum cause fluid to be washed backward into the rumen, flushing the contents with liquid. This is followed by contractions of the rumen, which return the fluid and small food particles to the reticulum and then to the omasum (Annison and Lewis 1959).

The contraction patterns of the rumen and the reticulum of white-tailed deer have been studied by Dzuik, Fashingbauer, and Idstrom (1963). Measurements indicated that contractions of the musculature of the rumen and reticulum are very closely associated with each other. The first or primary rumen contraction follows the "reticular doublet" or two successive contractions of the reticulum. The primary contraction involves all parts of the rumen. A secondary rumen contraction may occur, involving the dorsal sac, posterior dorsal blind sac, and the ventral sac of the rumen. The time required for the contractions that occur between one reticular doublet and the next is considered to be one complete ruminoreticular cycle.

The duration of the ruminoreticular cycle was found to be 20–30 seconds. The frequency of reticular and primary rumen contractions varied with the activity of the deer; they were less frequent when the deer were resting (2.2 contractions per minute). Secondary rumen contractions varied from 0.6 to 0.8 contractions per minute while the deer were standing to 0.2 to 0.5 contractions per minute while reclining. Thus the primary contractions were more frequent during periods of resting, but secondary contractions were more frequent during periods of standing. The significance of that reversal was not explained.

The secondary contractions of the rumen were almost always in a $1:1$ ratio with eructation, or belching. This is of interest because the eructation of gases represents an energy loss that is difficult to measure in wild ruminants.

Dzuik, Fashingbauer, and Idstrom (1963) concluded that there is a marked similarity in the ruminoreticular contractions of deer, cattle, and sheep. There were differences in frequencies but these were not stable for a single animal over a period of time. Their work yields insight into the mechanical processes to which food materials and rumen microorganisms are subjected. The similarity between deer and domestic ruminants indicates that domesticated animals are not vastly different from their wild relatives; much can be gained from comparative studies on domestic and wild ruminants.

In the omasum, contractions of the muscular walls compress and triturate or pulverize the food materials. At the same time, 60%–70% of the water content is absorbed. The food materials are then passed on to the abomasum where gastric juice is secreted and the fluid content is restored to approximately the original level in the omasum. The hydrochloric acid content of the gastric juice results in a pH of 1.5–3.0. Protozoa disintegrate there and most bacteria are killed. The food materials pass through the abomasum quite rapidly and move on to the small intestine (Annison and Lewis 1959).

DEFECATION. Undigested residues are eliminated by defecation. The fecal mass also includes degraded tissue that has been removed from the internal linings of the alimentary canal, as well as the remains of rumen microorganisms that have escaped digestion in the small intestine.

A considerable amount of emphasis has been placed on the defecation rate as a method of taking a census of wild populations. An assumption about the number of pellet groups released per animal per day (13 is a common estimate) has been used to estimate the number of deer in a given area. Defecation rates vary, however, depending on both the quantity and quality of the diet. Experiments have shown that known populations are generally underestimated by this method (Smith 1964).

8-7 CHEMICAL PROCESSES OF DIGESTION

The study of digestion or any other biological function is complex. There are many variables in digestion, including dietary components and the microorganism spectrum, but the end result of digestion is the same—raw material (food) is converted into a form that can be used by the metabolic machinery of the body for maintenance of basic life processes, support of activity, and tissue production.

One problem in studying an internal system is that the removal of any component may result in the disruption of the normal operation of the system. This is true in studies of rumen function. One technique that permits analyses of the chemical reactions in the rumen with little effect on the animal is the use of a rumen fistula. A fistula is a covered opening into the rumen through an incision on the posterior portion of the body wall (Figure 8-5). The cover is removable so that the contents of the rumen can be removed or altered for experimental

FIGURE 8-5. A rumen fistula in a white-tailed deer.

purposes. It is a useful technique for studying the internal characteristics of the rumen with a minimum of disturbance to the animal.

Successful installation of rumen fistulas has contributed to the development of another technique that has considerable promise in the field of ruminant nutrition. This is the *in vitro* method, or artificial rumen experiments. An artificial rumen system consists of a constant-temperature bath that maintains thermal conditions similar to those inside the animal and a container that simulates the rumen itself. This container is filled with the experimental forage and rumen fluid that has been extracted from an animal. The products of digestion are analyzed after an incubation period of about 48 hours. The equipment used for *in vitro* studies is shown in Figure 8-6.

An artificial rumen is not identical to a natural (*in vivo*) one. The dynamics of the living rumen musculature and the motion caused by gross animal activity are missing, as are the histological changes that occur continually in the rumen lining. Used properly, it is an excellent supplementary tool for research in ruminant nutrition.

Another recent technique for studying ruminant digestion is the "sack" technique. A small, indigestible, nylon-mesh sack is suspended in the rumen and the changes in its content of food material are noted. The restriction that the sack imposes on the food material is obviously unlike the natural flow in the rumen, but the results can be useful for interpreting certain aspects of ruminant nutrition.

RUMEN MICROORGANISMS.   Rumen microorganisms are absolutely essential for successful digestion. The microorganisms are absent at birth, and experiments

FIGURE 8-6.   The *in vitro* fermentation bath at the BioThermal Laboratory, Cornell University.

with axenic (germ-free) animals demonstrate clearly that the microorganisms are symbionts essential to ruminant animals (Pantelouris 1967). The necessity for these symbiotic organisms is due to the absence of the enzyme cellulase in the ruminant stomach. This enzyme is necessary for the breakdown of cellulose, a major component of the cell walls of plants. Both bacteria and protozoans are present in the anaerobic rumen environment. As many as 896 strains of bacteria have been isolated (Pantelouris 1967), and more than 100 species of protozoans (Annison and Lewis 1959).

No animal has the full range of microorganism populations, but the variety of microorganisms in the rumen is large enough so that gradual changes in the diet do not appreciably affect digestion. Further, different species of ruminants throughout the world have almost the same kinds of microorganisms. The number of each species or strains of microorganisms present in the rumen varies, however, even among animals of the same species on the same range. The ciliate *Eudiplodinium* was absent in two elk but was the only large ciliate found in others on the northern Yellowstone elk range (McBee 1964). Other genera were observed to vary among animals during the same season, and one genus, *Enoploplastron,* varied seasonally, disappearing when green grass became a major part of the diet.

The number of microorganisms in the rumen is extremely large. Microscopic counts of organisms in the rumen fluid of thirty-three elk showed from 2.9 to 74.2 billion bacteria per gram of rumen content, with an average of 35 billion per gram (McBee 1964). This is more than the average of 10 billion per milliliter reported by Annison and Lewis (1959) for domestic ruminants.

There are fewer protozoans in the rumen—10^6 or one million in each milliliter of rumen contents (Pantelouris 1967). Their bulk, however, may equal that of the bacteria since the protozoans are much larger (Annison and Lewis 1959).

The bacteria and protozoans in the rumen have their own distinct metabolic characteristics. Most of the important species are obligately anaerobic; they do not require oxygen for metabolism. They have nutrient requirements of their own and must have a certain quantity of protein available to supply their nitrogen requirements before the digestion of starch can occur. Starch is valuable as a substrate for rumen microorganisms, and its utilization is an important factor in maintaining flourishing populations.

Simple sugars are actively metabolized by both protozoans and bacteria in the rumen. Three distinct processes occur: (1) fermentation of sugars, which consists of energy-yielding catabolic reactions; (2) conversion of sugars to glycogenlike polysaccharides and the storage of this material; and (3) endogenous metabolism of the stored polysaccharide. The last two processes are more apparent in rumen protozoans but are undoubtedly part of bacterial metabolism as well. The stored polysaccharide in the protozoan is available for subsequent metabolism by the host animal. Its body protein is also of biological value, and the protozoans themselves are digested in the small intestine (Annison and Lewis 1959).

FERMENTATION.   The fermentation of food materials in the rumen is an essential step in the digestion process of ruminants. It takes place in a very particular type of environment—an anaerobic, highly reducing system at a slightly acid but buffered pH—in which a very specialized microorganism population develops (Annison and Lewis 1959). If the supply of food in the rumen is maintained by frequent intake, conditions in the rumen remain fairly constant inasmuch as the microorganisms have a regular supply of carbohydrates and proteins for the maintenance of their own metabolic activities. The soluble products of this activity are readily absorbed through the rumen wall, preventing their accumulation in the rumen.

The rumen functions as an open system that depends on the flow of food materials into it and the flow of microorganism metabolites and food residues out. If food remains undigested in the rumen, the animal's appetite is diminished and rumen movement stops. According to Nagy, Vidacs, and Ward (1967), experiments show that if the rumen is then filled with actively fermenting rumen fluid from other animals, both appetite and rumen movement begin again.

Rumen fermentation continues only if there is an adequate amount of carbohydrates and proteins to support microorganism metabolism. In addition to an adequate quantity, the balance between the quantity of carbohydrate and nitrogen is also important. A minimal level of protein that can be digested by rumen microorganisms is essential for supplying their own nitrogen requirements (Annison and Lewis 1959). Two main conclusions are usually drawn from feeding experiments on the energy and protein relationships in ruminants: (1) the rumen microorganisms attack the fibrous components of the ration more rapidly as the protein intake is increased and (2) they utilize the protein better in the presence of added carbohydrate (Annison and Lewis 1959).

Digestion by the microorganisms results in the breakdown of cellulose in the rumen. The efficiency of this process depends on the metabolism of simple carbohydrates because a large proportion of the energy requirements of the cellulose-splitting organisms is probably supplied by the fermentation of these materials (Annison and Lewis 1959). The complex interaction of the rumen microorganisms with their chemical environment does not result in an unlimited supply of energy to the host. Deer are not super-ruminants; Short (1963) found that *in vitro* digestion of cellulose was often greater for a steer than for white-tailed deer. The relative nutrient value of several browse species was related to the crude-fiber levels of the plant materials—as the crude-fiber content goes up the digestibility goes down. Short, Medin, and Anderson (1965) concluded that the cellulose content seems to limit the digestible energy available to deer in both natural forages and artificial rations.

Two criteria need to be considered in assessing the role of microorganisms in ruminant nutrition: one is that the organisms must be capable of carrying out a reaction known to take place in the rumen, and the other is that the organisms must be present in sufficient numbers to account for the extent of the reaction (Annison and Lewis 1959). Nagy, Vidacs, and Ward (1967) have studied both of these criteria. They determined that short-chain fatty acids were produced by the rumen fermentation of alfalfa hay in the same proportions normally found in the rumen contents of wild deer. This suggests that no major adjustments in the microbial spectrum are necessary for deer to digest different natural foods.

Changes in the diet of wild deer can cause mortality, however. Cases in which deer have died even though they had had a plentiful supply of hay have been reported in the literature. One possible explanation for this is that as the amount of substrate material decreases during starvation there is a loss of properly functioning microorganisms for the rumen. Food that is difficult for the microorganisms to digest could produce nutritional deficiencies for support of the rumen populations with a concomitant decrease in the rate of fermentation. When the point is reached at which the rumen contains only a small number of active microorganisms, the food residues, in normal passage through the intestinal tract, would tend to mechanically remove these microorganisms at a faster rate than the growth of the resident population (Nagy, Steinhoff, and Ward 1964). Appetite and rumen movement stopped completely when three 7-pound portions of sage brush were introduced into a steer through a rumen fistula. Thus the digestive efficiency of wild ruminants can be reduced when range conditions deteriorate to the point that animals are forced to eat certain plants that might otherwise be avoided.

pH. The pH of the rumen fluid varies seasonally. For elk, it was found to be generally below 6.0 during the summer, between 6.0 and 7.0 during autumn and most of the winter, and above 7.0 during late winter (McBee 1964). These values are very similar to those reported by Short (1963) for white-tailed deer on different diets. Aspen and white-cedar diets, typical winter forages of deer in the northern range, result in pH ranging from 6.42 to 7.09, but an alfalfa-concentrate diet results in a rumen pH of 5.23 and 5.48.

8-8 PRODUCTS OF FERMENTATION

HEAT ENERGY. Rumen microorganisms are metabolically active and consequently must produce certain end products of metabolism. These include heat energy, gases, volatile fatty acids, protein, and vitamins. The first three are waste products from the microorganisms themselves, but only the gases are waste products for the ruminant host. The heat energy is often called a waste, but that is an oversimplification of the role of heat energy in a homeotherm. When the thermal environment of an animal is such that heat production exceeds heat loss, the heat energy produced by the rumen microorganisms must be dissipated from the body surface to maintain a stable body temperature. But, if the thermal environment results in a heat loss greater than the heat produced by nutrient metabolism, then the heat of fermentation in the rumen is important for the maintenance of a stable body temperature.

GASES. The two gases produced by rumen fermentation are carbon dioxide (CO_2) and methane (CH_4). The actual volume of methane produced by cattle can approach 400 liters per 24 hours and by sheep up to 50 liters per day (Blaxter 1967). The amount of methane absorbed by the blood and exhaled through the lungs of cattle varied from 25% to 94% of the total measured CH_4 before feeding and 9% to 43% after feeding (Hoernicks et al. 1965). The remaining CH_4 was eliminated by eructation. The energy contained in the methane is about 8% of the total food energy. Blaxter (1967) points out that the rate of methane production varies with food consumption and activity. Maximum quantities are produced at the time the food is ingested; lesser amounts are produced when the animal is physically active (Figure 8-7).

VOLATILE FATTY ACIDS.   The volatile fatty acids (VFAs) are the most important product of ruminant fermentation, for they are readily absorbed from the rumen and serve as a major source of energy to the animal. Under normal conditions only small amounts of VFAs escape absorption in the rumen and pass on to the small intestine. The amount of energy absorbed from the rumen as a result of VFA production is of real biological significance. Annison and Lewis (1959) cite references indicating that at least 600 to 1200 kcal of energy are absorbed in the form of VFAs from the sheep rumen every 24 hours. In cattle, which have a much greater absolute rumen capacity, 6000 to 12000 kcal become available from the VFAs produced by rumen fermentation. In white-tailed deer, the energy contained in an average gram of rumen acid (with assumptions on molar percentages of the different acids) is estimated to be 4.15 kcal (Short 1963). The average amount

FIGURE 8-7.   Typical variation in the methane production of a sheep during a 24-hour period. Note the peaks in production associated with eating and with the physical activity of the animal. (Data from Blaxter 1967.)

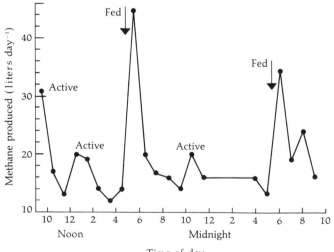

of VFA contained in 3000 ml of rumen contents of a deer on an alfalfa-concentrate diet was 25.3 g. This would provide about 105 kcal. If this diet were to provide 1000 kcal in a 24-hour period, the rumen contents would have to be absorbed 9.52 times. The total VFA content of 3000 ml of rumen contents of a deer on an aspen diet was 11.6 g of VFA, and this would provide 48.1 kcal. The VFA content of a deer on a white-cedar diet is 15.8 g, providing 65.6 kcal. The rumen contents due to these two diets would have to be turned over 20.79 times and 15.24 times, respectively, if they were to provide 1000 kcal of energy in a 24-hour period. These turnover rates appear to be unreasonable; further studies of the digestive physiology and the production and absorption of individual VFAs are needed.

The first accurate analyses of the rumen contents of domestic animals were completed in 1945 and 1946 (Annison and Lewis 1959). Acetic, butyric, and propionic acids were invariably found to be present. The total concentration of these VFAs in the rumen and the amount of each individual acid depends on both the composition of the ration and the feeding schedule. Seasonal differences were observed in mule deer (Short 1966b), and variations in the proportions of VFAs were observed between white-tailed deer killed at different times of the day (Short 1963).

The concentration of VFAs in the rumen is an indication of the rate of fermentation; higher concentrations of acids are present when easily fermented foods are eaten (Short 1963). High concentrations alone are not precise indicators of the actual rate of production, however. Annison and Lewis (1959) point out that the rumen concentration of VFAs at any given time depends not only on the rate of production of acids in the rumen, but also on the rate of absorption of acids from the rumen, the rate of passage from the rumen to the omasum, the dilution with saliva, the utilization of VFAs by the rumen microorganisms, and the conversion to other rumen metabolites. They also observed that the amount of butyric acid in the blood from the wall of the rumen was less than the amount expected based on the relative amount of butyric acid leaving the rumen. This indicates that the butyric acid was used by the rumen wall itself.

The VFAs in the rumen seem to come from the degradation of carbohydrates. In monogastric animals, most of the caloric energy is absorbed from the small intestine in the form of glucose (Annison and Lewis 1959). Glucose and VFA concentrations in a very young ruminant are very similar to those in a monogastric animal, but as the age of the ruminant increases, the rumen becomes more functional, the concentrations of VFAs increase, and the blood sugar falls to a level about one-half that of nonruminants.

The value of a particular diet should be determined on the basis of the rates at which the nutrients are made metabolically available to the animal through rumen fermentation. Short (1963) stresses the need for additional understanding of energy relationships before management of deer can be properly based on nutritional facts. The framework within which these energy considerations can be made is discussed in Part 6. The accuracy of any calculations related to the diet depends on the quality of the basic energy data and on the quality of the measurements of feeding data and other factors. It is unfortunate that so little research has been done on basic energy relationships of wild ruminants; the alternative is to make first approximations that will permit analyses of the total animal-environment relationship and an error analysis of the effect of approximations.

PROTEIN. It has been observed that the protein content of the rumen is often higher than the protein content of the food that is being ingested. The terminal two to three inches of growth on browse sampled by Bissel (1959) contained 6.9% crude protein, but the rumen contents of nine deer contained 17.6% crude protein. Fourteen deer were studied in January and February when plants were dormant and the range and rumen values for protein content were 7.1% and 17.2%, respectively, in January and 6.1% and 15.1% in February. Three deer that were fed alfalfa pellets with 15.7% protein content had rumen protein contents of 21.0%, 16.2%, and 14.7%.

The increase in the nitrogen content in the rumen over that of the ingested food materials can be accounted for in several ways. Ammonia that is formed by protein digestion is absorbed by the blood, converted into urea by the liver, and then either excreted by the kidney or returned to the rumen as a component of saliva (Dukes 1955). The urea secreted into the rumen is rapidly hydrolyzed by bacterial urease. On low-nitrogen diets this additional supply of ammonia helps to promote an active microbial population with a concomitant increase in the synthesis of microbial protein (Ullrey et al. 1967, citing Kay 1963).

The microbial protein synthesized is of direct benefit to the host animal. The rumen microorganisms synthesize protein from amino acids and nonprotein nitrogen sources, and the rapid turnover of the microbial population results in the passage of a large number of organisms to the omasum and abomasum where they are digested (Pantelouris 1967). A considerable portion of the protein requirement of ruminants is supplied in this way (Annison and Lewis 1959).

The greatest benefit derived from protein synthesis by rumen microorganisms is their conversion of nonprotein nitrogen to microbial protein that can be digested and absorbed by the host. Urea is such a converted compound. It is currently being used to economic advantage by cattle feeders who add it to the diet as a low-cost supplement to low-protein diets. Deer ingest urea and other nonprotein nitrogenous compounds by consuming water that has been accumulating in pools in which feces and urine have been deposited, by lapping urine from other deer, and by the cleaning of the anal region of the fawns by the doe. Captive white-tail fawns less than two weeks of age have shown varying degrees of interest in lapping the urine from their penmates, sometimes preferring urine to milk.

Other sources of protein are available to the ruminant, including respiratory secretions and the epithelial cells that are sloughed off. These must be replaced, however, so this source of protein cannot be considered of value over a long period of time.

VITAMINS AND MINERALS.   Vitamins and minerals are essential components of an animal's diet, with the former being used in nutrient metabolism and the latter deposited as a part of the structural components of different body tissue. It is not practical for the range manager to supply these as additives to feeds as the domestic feeder can. Free-ranging animals should have sufficient quantities of vitamins and minerals when the energy and protein needs of the body are met, except in local areas where the soil may be deficient or overly rich in some materials so a chemical imbalance results. This may develop on several western ranges where the calcium-phosphorus ratio is so high that the metabolism of phosphorus is affected. A deficiency of phosphorus or a wide calcium-phosphorus ratio could cause retarded growth, a high feed requirement, unthrifty appearance, weak young, decreased lactation, failure to conceive, stiffness of joints, and other abnormalities (Dietz 1965).

The vitamin requirements of adult ruminants are satisfied in part by the vitamin syntheses of rumen microflora. Water-soluble vitamins of the B complex are synthesized in the rumen if appropriate foods have been eaten (Annison and Lewis 1959). Vitamins A, D, and E are not synthesized and must be supplied in the diet, but fat-soluble vitamin K is synthesized in the rumen (Annison and Lewis 1959). The young animal that does not have a functional rumen requires the full vitamin complement as a component of the diet. In the wild, this is apparently met by the ingestion of milk and herbaceous plants.

The role of minerals and vitamins in the nutrition of wild ruminants needs to be analyzed in a dynamic way over the entire annual cycle. The methods used for energy and protein analyses in Part 6 are applicable to minerals and vitamins.

8-9 PASSAGE OF DIGESTA THROUGH THE GASTROINTESTINAL TRACT

The rate of passage of digesta through the gastrointestinal tract is an important factor in an evaluation of the nutritive relationships between an animal and its range. Although it is common practice to report the rumen contents of animals killed in the field, this may give a false picture of nutrient absorption and assimilation. The forages more resistant to the enzymatic activities of rumen microorganisms will remain in the rumen the longest. Hence the forages that may be least important nutritionally may be most abundant in the rumen.

An exercise illustrating the importance of turnover rate is shown in Table 8-3. Different foods (marbles) are placed in the rumen (a dish) at different rates, and they pass through the rumen at different rates. The absolute quantities of the different colored marbles at a given time are measured. These represent the "standing crop" or food in the rumen. The number of different marbles that enter the dish and pass on as digested food is then compared with the daily estimates. Over a 12-day period, a distinct difference between the percentage of observed abundance of marbles of each color through daily sampling and the actual abundance of marbles that have passed through is observed. It is a clear indication of the need to consider both the quantity of food in the rumen and the rate of its passage through the rumen.

TABLE 8-3 TURNOVER RATE IN THE "RUMEN" AND ITS RELATIONSHIP TO ACTUAL ABUNDANCE

Data

Black marbles	Enter 3 every third day, 3-day turnover time
Blue marbles	Enter 2 every other day, 1-day turnover time
White marbles	Enter 1 every other day, 2-day turnover time
Yellow marbles	Enter 1 every other day, 1-day turnover time

Abundance Through Time

						Days								
	1	2	3	4	5	6	7	8	9	10	11	12	Total	% of Total
Black marbles	3—————		3—————			3—————			3—————				12	29
Blue marbles	2	0	2	0	2	0	2	0	2	0	2	0	12	29
White marbles	1——		1——		1——		1——		1——		1——		6	14
Yellow marbles	1	1	1	1	1	1	1	1	1	1	1	1	12	29
													42	

Observed Abundance Through Daily Sampling

	Total Observed	No. of Days	Average	% Observed Abundance	% Actual Abundance
Black marbles	36	12	3	50	29
Blue marbles	12	12	1	17	29
White marbles	12	12	1	17	14
Yellow marbles	12	12	1	17	29

Several factors contribute to variation in the results of experiments on the rate at which digesta passes through the gastrointestinal tract of domestic ruminants. Church (1969) has concluded that the physical nature of the feed appears to be a most important factor controlling the rate of passage from the rumen. Other factors include the specific gravity of the ingesta, particle size, digestibility of the food, and the level of feed intake.

Some of the variation in reported results can be attributed to the methods used for marking the food. Dyes, polyethylene markers, and chemical markers in the forage have their own unique rate of passage and do not follow perfectly the movements of the ingesta. This error does not eliminate the usefulness of the data, but it must be considered when interpreting the observed results.

Some general relationships between the retention time for the forage and the characteristics of the forage are evident from the literature. Ingalls et al. (1966) calculated passage from the rumen by dividing the amount of food in the rumen by the daily intake. The retention time of alfalfa dry matter was 0.62 days, trefoil, 0.68 days, timothy, 0.84 days, and canary grass, 0.94 days. The more rapid digestion of legumes compared with grasses is reflected in the faster turnover rate. The retention time for browse species that make up major portions of the diets of some wild ruminants, especially in the winter, is probably longer.

Smaller particles are digested more quickly and pass through the gastrointestinal tract faster than larger ones. The surface area of a given weight of smaller particles is greater than the surface area of an equivalent weight of larger particles. This results in more efficient digestion. Thus the faster rate of passage does not necessarily mean a loss of nutrients. If the level of feeding is high enough so that the rate of passage results in incomplete digestion, the nutritive value of the forage may be reduced.

The many factors that affect turnover rates are difficult to study, but they appear to be very important in determining the nutritive characteristics in an animal-range relationship. A faster turnover rate may compensate for a lower digestibility up to a point, and conversely, the benefit of a high digestibility may be reduced by a rapid turnover rate. The relationship between turnover rate, digestibility, and nutritive value to the animal is considered in the calculations of carrying capacity described in Chapters 16 and 17.

8-10 DIGESTION AND ABSORPTION IN THE GASTROINTESTINAL TRACT

Most of the research on nutrition in ruminants has been directed toward the four-part stomach. This can be attributed to two things: (1) it is the unique characteristic of ruminants, and (2) it is considerably easier to study than are the remaining parts of the gastrointestinal tract.

There are no data available on the digestion and absorption characteristics of the entire gastrointestinal tract of wild ruminants. The pattern is probably similar to that of domestic ruminants in which most digestion takes place in the rumen, but some active digestion also takes place in the small and large intestines. Nutrients are absorbed all along the gastrointestinal tract. Research on lambs by Vidal et al. (1969) illustrates the importance of the lower gut in the ruminant, not only in the absorption of nutrients of dietary origin but also in both the production and reabsorption of endogenous material.

8-11 SUMMARY

The physiology of ruminant nutrition has been given considerable attention by researchers in the past 100 years, and the basic processes are generally understood. These have been described briefly in this chapter. Comparative work on the nutrition of wild ruminants indicates that the basic patterns are similar to those of domestic ruminants.

The relationships between the nutritive processes within the animal and between the animal and its range have not been studied as a complete system. The analysis of the relationships between nutrient use, animal requirements, and the range supply follows in the next chapter, and again in Chapters 16 and 17.

LITERATURE CITED IN CHAPTER 8

Annison, E. F., and D. Lewis. 1959. *Metabolism in the Rumen.* London: Metheun and New York: Wiley, 184 pp.

Association of Official Analytical Chemists. 1965. *Official Methods of Analysis of the AOAC,* ed. W. Horwitz. 10th ed. Washington, D.C.

Bissell, H. 1959. Interpreting chemical analyses of browse. *Calif. Fish Game* **45**(1): 57–58.

Blaxter, K. L. 1967. *The energy metabolism of ruminants.* London: Hutchinson, 332 pp.

Brody, S. 1945. *Bioenergetics and growth.* New York: Reinhold, 1023 pp.

Bromley, P. T., and B. W. O'Gara. 1967. Orphaned pronghorns survive. *J. Wildlife Management* **31**(4): 843.

Church, D. C. 1969. *Digestive physiology and nutrition of ruminants.* Vol. 1. Published by D. C. Church. Produced and distributed by the Oregon State University Bookstores, Inc., Corvallis, 316 pp.

Church, D. C., G. L. Jessup, Jr., and R. Bogart. 1962. Stomach development in the suckling lamb. *Am. J. Vet. Res.* **23**(93): 220–225.

Crampton, E. W., and L. E. Harris. 1969. *Applied animal nutrition.* 2d ed. San Francisco: W. H. Freeman and Company, 753 pp.

Dietz, D. R. 1965. Deer nutrition research in range management. *Trans. North Am. Wildlife Nat. Resources Conf.* **30**: 274–285.

Dietz, D. R., and R. D. Curnow. 1966. How reliable is a forage chemical analysis? *J. Range Management* **19**(6): 374–376.

Dukes, H. H. 1955. *The physiology of domestic animals.* 7th ed. Ithaca, New York: Comstock, 1020 pp.

Dzuik, H. E., G. A. Fashingbauer, and J. M. Idstrom. 1963. Ruminoreticular pressure patterns in fistulated white-tailed deer. *Am. J. Vet. Res.* **24**: 772–783.

Flatt, W. P., R. G. Warner, and J. K. Loosli. 1958. Influence of purified materials on the development of the ruminant stomach. *J. Dairy Sci.* **41**(11): 1593–1600.

Hoernicke, H., W. F. Williams, D. R. Waldo, and W. P. Flatt. 1965. Composition and absorption of rumen gases and their importance for the accuracy of respiration trials with tracheostomized ruminants. In *Energy metabolism,* proceedings of the 3rd symposium held at Troon, Scotland, May, 1964, ed. K. L. Blaxter, pp. 165–178.

Ingalls, J. R., J. W. Thomas, M. G. Tesar, and D. L. Carpenter. 1966. Relations between *ad libitum* intake of several forage species and gut fill. *J. Animal Sci.* **25**(2): 283–289.

Kay, R. N. B. 1963. Reviews of the progress of dairy science. Section A. Physiology. Part I. The physiology of the rumen. *J. Dairy Res.* **30**(2): 261–288.

McBee, R. H. 1964. *Rumen physiology and parasitology of the northern Yellowstone elk herd.* Dept. of Botany and Bacteriology, and the Veterinary Research Laboratories, Montana State College. 28 pp.

McCandless, E. L., and J. A. Dye. 1950. Physiological changes in intermediary metabolism of various species of ruminants incident to functional development of rumen. *Am. J. Phys.* **162**: 434–446.

Murphy, D. A., and J. A. Coates. 1966. Effects of dietary protein on deer. *Trans. North Am. Wildlife Nat. Resources Conf.* **31**: 129–139.

Nagy, J. G., G. Vidacs, and G. M. Ward. 1967. Previous diet of deer, cattle, and sheep and ability to digest alfalfa hay. *J. Wildlife Management* **31**(3): 443–447.

Nagy, J. G., H. W. Steinhoff, and G. M. Ward. 1964. Effects of essential oils of sagebrush on deer rumen microbial function. *J. Wildlife Management* **28**: 785–790.

Pantelouris, E. M. 1967. *Introduction to animal physiology and physiological genetics.* 1st ed. Oxford and New York: Pergamon, 497 pp.

Short, H. L. 1963. Rumen fermentations and energy relationships in white-tailed deer. *J. Wildlife Management* **27**(2): 184-195.

Short, H. L. 1964. Postnatal stomach development of white-tailed deer. *J. Wildlife Management* **28**(3): 445-458.

Short, H. L. 1966*a*. Effects of cellulose levels on the apparent digestibility of feeds eaten by mule deer. *J. Wildlife Management* **30**(1): 163-167.

Short, H. L. 1966*b*. Seasonal variations in volatile fatty acids in the rumen of mule deer. *J. Wildlife Management* **30**(3): 466-470.

Short, H. L., D. E. Medin, and A. E. Anderson. 1965. Ruminoreticular characteristics of mule deer. *J. Mammal.* **46**(2): 196-199.

Smith, A. D. 1964. Defecation rates of mule deer. *J. Wildlife Management* **28**(3): 435-444.

Tamate, H. 1957. The anatomical studies of the stomach of the goat. II. The post-natal changes in the capacities and the relative sizes of the fours division of the stomach. *Tohoku J. Agr. Res.* **8**(2): 65-77.

Tulbaev, P. O. 1959. Factors which bring about rumination during development. *Nutr. Abstr. Rev.* **29**: 511.

Ullrey, D. E., W. G. Youatt, H. E. Johnson, L. D. Fay, and B. L. Bradley. 1967. Protein requirements of white-tailed deer fawns. *J. Wildlife Management* **31**(4): 679-685.

Van Soest, P. J. 1965. Symposium on factors influencing the voluntary intake of herbage by ruminants: voluntary intake in relation to chemical composition and digestibility. *J. Animal Sci.* **24**(3): 834-843.

Van Soest, P. J., and L. H. P. Jones. 1968. Effect of silica in forages upon digestibility. *J. Dairy Sci.* **51**(10): 1644-1648.

Vidal, H. M., D. E. Hogue, J. M. Elliot, and E. F. Walker, Jr. 1969. Digesta of sheep fed different hay-grain ratios. *J. Animal Sci.* **29**(1): 62-68.

Wood, A. J., H. C. Nordan, I. McT. Cowan. 1961. The care and management of wild ungulates for experimental purposes. *J. Wildlife Management* **25**(3): 295-302.

SELECTED REFERENCES

Abrams, J. T., ed. 1966. *Recent advances in animal nutrition.* Boston: Little, Brown, 261 pp.

Barnett, A. J. G., and R. L. Reid. 1961. *Reactions in the rumen.* London: Edward Arnold, 252 pp.

Dougherty, R. W., ed. 1965. *Physiology of digestion in the ruminant.* Washington, D.C.: Butterworth, 480 pp.

Hungate, R. E. 1966. *The rumen and its microbes.* New York: Academic Press, 533 pp.

Hungate, R. E., G. D. Phillips, A. McGregor, D. P. Hungate, and H. K. Buechner. 1959. Microbial fermentation in certain mammals. *Science* **130**(3383): 1192-1194.

Karn, J. F., D. C. Clanton, and L. R. Rittenhouse. 1971. *In vitro* digestibility of native grass hay. *J. Range Management* **24**(2): 134-136.

Knox, K. L., J. G. Nagy, and R. D. Brown. 1969. Water turnover in mule deer. *J. Wildlife Management* **33**(2): 389-393.

Lewis, D., ed. 1961. *Digestive physiology and nutrition of the ruminant.* London: Butterworth, 297 pp.

Loe, W. C., O. T. Stallcup, and H. W. Colvin, Jr. 1959. Effect of various diets on the rumen development of dairy calves. *J. Dairy Sci.* **43**: 395.

Longhurst, W. M., N. F. Baker, G. E. Connolly, and R. A. Fisk. 1970. Total body water and water turnover in sheep and deer. *Am. J. Vet. Res.* **31**(4): 673–677.

Maloiy, G. M. O., R. N. B. Kay, E. D. Goodall, and J. H. Topps. 1970. Digestion and nitrogen metabolism in sheep and red deer given large or small amounts of water and protein. *Brit. J. Nutr.* **24**(3): 843–855.

Maynard, L. A., and J. K. Loosli, 1969. *Animal nutrition.* New York: McGraw-Hill, 613 pp.

McBee, R. H., J. L. Johnson, and M. P. Bryant. 1969. Ruminal microorganisms from elk. *J. Wildlife Management* **33**(1): 181–186.

Mitchell, H. H. 1963. *Comparative nutrition of man and domestic animals.* New York: Academic Press.

Nagy, J. G., and G. L. Williams. 1969. Rumino reticular VFA content of pronghorn antelope. *J. Wildlife Management* **33**(2): 437–439.

Oh, H. K., T. Sakai, M. B. Jones, and W. M. Longhurst. 1967. Effect of various essential oils isolated from Douglas fir needles upon sheep and deer rumen microbial activity. *Appl. Microbiol.* **15**(4): 777–784.

Oh, J. H., M. B. Jones, W. M. Longhurst, and G. E. Connolly. 1970. Deer browsing and rumen microbial fermentation of Douglas fir as affected by fertilization and growth stage. *Forest Sci.* **16**(1): 21–27.

Pearson, H. A. 1969. Rumen microbial ecology in mule deer (*Odocoileus hemionus*). *Appl. Microbiol.* **17**(6): 819–824.

Phillipson, A. T. 1970. *Physiology of digestion and metabolism in the ruminant.* Proc. 3rd Intern. Symp., Cambridge, England. Newcastle upon Tyne, England: Oriel Press.

Prins, R. A., and M. J. H. Geelen. 1971. Rumen characteristics of red deer, fallow deer, and roe deer. *J. Wildlife Management* **35**(4): 673–680.

Rittenhouse, L. R., C. L. Streeter, and D. C. Clanton. 1971. Estimating digestible energy from digestible dry and organic matter in diets of grazing cattle. *J. Range Management* **24**(1): 73–75.

Short, H. L. 1966. Methods for evaluating forages for wild ruminants. *Trans. North Am. Wildlife Nat. Resources Conf.* **31**: 122–128.

Short, H. L., C. A. Segelquist, P. D. Goodrum, and C. E. Boyd. 1969. Rumino-reticular characteristics of deer on food of two types. *J. Wildlife Management* **33**(2): 380–383.

Steen, E. 1968. Some aspects of the nutrition of semi-domestic reindeer. *Proc. Symp. Zool. Soc. London,* No. 21, ed. M. A. Crawford. 429 pp.

Symposia on ruminant nutrition. 1970. *Federation Proc.* (Jan., Feb.) **29**(1): 33–54.

Texter, C. E., Jr., C. C. Chou, H. C. Laureta, and G. R. Vantrappen. 1968. *Physiology of the gastrointestinal tract.* St. Louis: The C. V. Mosby Company, 262 pp.

SELECTED REFERENCES: NUTRITIVE ANALYSES

Billingsley, B. B., Jr., and D. H. Arner. 1970. The nutritive value and digestibility of some winter foods of the eastern wild turkey. *J. Wildlife Management* **34**(1): 176–182.

McEwan, E. H., and P. E. Whitehead. 1970. Seasonal changes in the energy and nitrogen intake in reindeer and caribou. *Can. J. Zool.* **48**(5): 905–913.

Oh, H. K., M. B. Jones, and W. M. Longurst. 1968. Comparison of rumen microbial inhibition resulting from various essential oils isolated from relatively unpalatable plant species. *Appl. Microbiol.* **16**(1): 39–44.

Segelquist, C. A., H. L. Short, F. D. Ward, and R. G. Leonard. 1972. Quality of some winter deer forages in the Arkansas Ozarks. *J. Wildlife Management* **36**(1): 174–177.

Short, H. L. 1966. Effects of cellulose levels on the apparent digestibility of feeds eaten by mule deer. *J. Wildlife Management* **30**: 163–167.

Short, H. L. 1971. Forage digestibility and diet of deer on southern upland range. *J. Wildlife Management* **35**(4): 698–706.

Short, H. L., and A. Harrell. 1969. Nutrient analysis of two browse species. *J. Range Management* **22**(1): 40–43.

Short, H. L., and J. C. Reagor. 1970. Cell wall digestibility affects forage value of woody twigs. *J. Wildlife Management* **34**(4): 964–967.

Short, H. L., D. R. Dietz, and E. E. Remmenga. 1966. Selected nutrients in mule deer browse plants. *Ecology* **47**(2): 222–229.

Tew, R. K. 1970. Seasonal variation in the nutrient content of aspen foliage. *J. Wildlife Management* **34**(2): 475–478.

Ullrey, D. E., W. G. Youatt, H. E. Johnson, L. D. Fay, D. B. Purser, B. L. Schoepke, and W. T. Magee. 1971. Limitations of winter aspen browse for the white-tailed deer. *J. Wildlife Management* **35**(4): 732–743.

Ward, A. L. 1971. *In vitro* digestibility of elk winter forage in southern Wyoming. *J. Wildlife Management* **35**(4): 681–688.

INGESTION AND
NUTRIENT UTILIZATION

9-1 VARIATIONS IN NUTRIENT INTAKE

There is a continuous flow of nutrients through the metabolic pathways in an animal-range relationship. There is variation in both the ingestion rate and the turnover rate of different forages as the food is processed by the animal. A time lag also occurs in the conversion of forage to metabolically useful energy for activity and tissue synthesis. An analysis of the characteristics associated with ingestion and the subsequent use of food material provides an understanding of the chemical communication between the animal and its range at a fundamental life-support level.

SEASONAL VARIATIONS. Wild ruminants exhibit seasonal differences in the rate of ingestion of different forages. The general pattern observed by many investigators shows a marked drop in consumption that begins in the fall and continues through the winter, with a reversal in the trend in the spring and summer. Males show a greater reduction than females. Silver (unpublished data) reports a 60% decline in intake from September through March for penned adult white-tailed deer on a pelleted diet of grains and alfalfa. Fowler, Newsom, and Short (1967) observed a decline in food consumption of white-tailed deer in Louisiana during the winter, accompanied by a 10% weight loss in bucks and a 3% weight loss in does. A decline in both feeding activity and cedar consumption was recorded for penned animals at the Cusino Wildlife Research Station in Upper Michigan during January and February (Ozoga and Verme 1970).

Nordan, Cowan, and Wood (1968) have studied the intake of black-tailed deer through several annual cycles, and they observed a very obvious decline in feed intake that begins when male deer exhibit rutting behavior. Males on a high plane of nutrition have lost up to 35% of their body weight during the rutting season, refusing food completely for nearly sixty days. Female deer showed a similar pattern of feed intake and weight loss during the breeding season, but not to the extent exhibited by the males.

One interesting and significant observation reported by Nordan, Cowan, and Wood is that a male deer on a low plane of nutrition did not increase his intake when put on *ad libitum* feeding during the rutting season, but exhibited a feed intake that was characteristic of the rut. This indicates that the physiological circumstances of the rut prevailed rather than the available food conditions.

Sexually immature animals do not always show a marked decrease in either food consumption or weight during the breeding season. Two white-tailed deer fawns used for experimental purposes at the University of Minnesota maintained their consumption of a pelleted horse ration, gaining 15 pounds, which was equal to about 30% of their weight in December. The small size of the two fawns when acquired in December (42 and 44 lb) was an indication of malnutrition, but compensatory growth occurred even though they were penned outside during a cold Minnesota winter (Moen unpublished data).

Fawns killed in New Hampshire during the eight-week hunting season in November and December showed a dressed weight loss of 0.6 lb week^{-1} for males and 0.25 lb week^{-1} for females. Yearlings of both sexes showed a weight loss of over 2 lb week^{-1}, and adult males lost as much as 5.75 lb week^{-1} (White 1968). These differences in weight indicate that physical and sexual maturity may be important; Fowler, Newsom, and Short (1967) concluded that weight loss of white-tailed deer is related to the sexual cycle.

The indications that food intake and weight loss vary seasonally and with the age and sexual maturity of an animal suggest that they are subject to some physiological controls that almost certainly involve the endocrine system. The relationship between food intake and weight loss is more than a correlation of intake or weight with "winter stress." There is a real need for experimental work in order to understand the physiological mechanisms employed in the regulation of intake and weight loss in wild ruminants.

INDIVIDUAL VARIATIONS.   The selection of particular forage species and even individual plants within a species has been observed by many investigators. Shade-grown sprouts were much preferred to the coarse sprouts produced after clear-cutting procedures (Cook 1939), and marked selectivity in summer feeding was also observed, with the new hardwood sprouts being much preferred to small seedling trees (Cook 1946).

Marked differences in selectivity of forage species has been observed between different geographical areas. Red pine (*Pinus resinosa*) in northern Minnesota was scarcely touched (Burcalow and Marshall 1958), but small red pines planted in western Minnesota were browsed heavily although other food was abundant in the surrounding fields and woodlots (Moen 1966). Soil fertility appears to be an important factor; Swift (1948) presents data showing that deer grazed preferentially in wheat and clover fields that had been fertilized, often traveling through an unfertilized field to get the preferred plants.

Deer seem to make their initial selection of forage through olfaction. Observations of deer in the field and in pens at the Hopland Field Station, University of California, indicate that they smell the plant first, then if they like the smell they will taste it, and if they like the taste they will continue feeding on it (Longhurst et al. 1968). Healy (1967), using a deer trained to a lead rope, also noted the importance of smell while the deer was browsing in a hardwood forest; it frequently located food items that were hidden from view. He also observed a preference not only for particular forage species, but for individual plants within a species. Individual leaves and twigs on a single plant seemed to be selected rather than chosen at random. The deer also exhibited a greater selectivity in areas where food was more abundant.

It has often been assumed that deer exhibit some kind of beneficial selection when feeding. Some plants produce aromatic and volatile compounds that have an inhibitory effect on rumen function. Longhurst et al. (1968) question whether deer can detect the presence of either nutrients or odors; the relationship between the nutrient content of plants and their palatability was not always a positive one.

The amount of inhibitory activity of plant chemicals on rumen digestion is dependent in part on the concentration of the inhibitory substance. A mixed diet that includes forages that vary from one extreme to the other in nutritive value can be tolerated, but a diet of forages with low nutritive value or with a high percentage of plants inhibitory to rumen function may result in weight loss or mortality. This is another example of a dynamic animal-range relationship that is dependent on the characteristics of both the animal and the range.

9-2 REGULATION OF NUTRIENT INTAKE

RELATIONSHIP TO WEIGHT. The nutritive requirements of an animal are related to weight in a nonlinear fashion that is expressed for energy in the "mouse-to-elephant" curve (see Figure 7-3), or by the mathematical expression $(c)W_{kg}^{0.75}$. An approximation for basal metabolism can be expressed for mammals using $c = 70$, and a gross approximation of the energy requirement for normal activity can be made with an activity increment $(I_{ma}) = 1.33$. Thus an active animal may have an energy expenditure equal to $1.33 \times 70 \times W_{kg}^{0.75}$. Further analyses of this approximation are made in Chapter 16.

A large animal is relatively more efficient at using energy than a small one, requiring less energy for basal metabolism per unit weight. The absolute energy requirement of a large animal is greater, however, so fewer large animals can be supported on any given range. The relationship between animal size and energy use is important to the range manager who is responsible for managing a finite quantity of resources to support an animal population.

RELATIONSHIP TO ENERGY EXPENDITURE.   The energy and protein requirements of an animal are related to both its behavioral and its physiological characteristics since everything an animal does "costs" something in terms of energy and protein. Life itself costs something—expressed as a basal metabolic rate—and activity beyond that which is stipulated as standard conditions in a BMR test results in an additional energy and protein requirement. Walking costs more than lying down, walking uphill costs more than walking on the level, and running costs more than walking. Crampton [Crampton and Harris (1969) and personal communication] believes that the several nutrients required by animals are related to the energy expenditure of the animal, so energy metabolism is the base for determining the nutrient requirements.

Some biological functions require more energy than others. Breeding activity appears to be quite costly for male white-tailed deer inasmuch as they lose 5–6 pounds of body weight per week during the breeding season (White 1968). The reduced food intake that accompanies sexual activity is not sufficient to sustain the activity level of the animal, especially in males.

Pregnancy demands additional energy and protein, especially during the last third of the gestation period. Two living units—the mother and the fetus or feti—are supported by the food ingested by the mother alone. As rapid growth of the fetus occurs in the late stages of pregnancy, the nutrients required are supplied by increased ingestion and by mobilization of body reserves in the female.

Lactation is a costly process. Dairymen know that high feeding levels of concentrated nutrients are necessary to sustain high milk production in the dairy cow, and it is no doubt a costly process for the wild ruminant as well. There are, however, many natural controls operating in a free-ranging population, so lactation is very likely a more efficient process for wild ruminants who are regulated by natural biological constraints rather than economics. Lactation ceases, for example, when the young are weaned naturally. On the other hand, dairy farmers try to sustain a high milk production for as long as possible, and this requires a high feeding level.

CONTROL OF INTAKE.   The availability of nutrients in any forage is a function of the quantity of the forage eaten and the efficiency with which the forage is digested and used metabolically. Extensive work has been done on the digestibility of forages, particularly for domestic animals, but little has been done on the factors affecting intake and the turnover rate in the gastrointestinal tract.

Three classes of effects of forage composition upon nutritive value may be distinguished according to how chemical constitution affects intake, digestibility, and the relationship between them (Van Soest 1965): (1) chemical constitution affects intake but has no direct or reliable effect on digestibility; (2) a positive

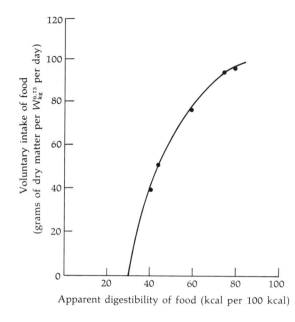

FIGURE 9-1. The relationship between voluntary intake and apparent digestibility by sheep. (Data from Blaxter 1967.)

relationship between intake and digestibility is promoted; and (3) a negative relationship between intake and digestibility is promoted.

The first class includes forages that give off aromatic compounds that may or may not relate to the nutritive value of the forage, such as those reported by Longhurst et al. (1968). The second class includes forage species that can be shown to have a high digestibility and are preferred by the animal. This relationship is illustrated in Figure 9-1. The third class is represented by feeds with a small fiber fraction that has little or no effect on intake. Very young, highly digestible herbage is an example of this class (Compling 1964), as are concentrates and grains. When the feed consumed has a high nutritive value, the intake of digestible energy may then be limited by the requirements of the animal (Van Soest 1965) rather than by some characteristic of the feed.

The three classes presented above are real biological relationships that have been shown to exist by researchers in animal science. It is necessary to consider the chemical and physical characteristics of a particular forage and the efficiency with which that forage can be used by an animal before intake can be considered to be characteristic of any one of these three possible relationships.

The first class—including those forages that give off aromatic compounds without a predictable relationship between intake and digestibility—cannot be useful in predicting the nutritive value of the diet because there is no known mathematical relationship between the odor associated with the forage and its nutritive value. The third class—including very young, highly digestible herbage as well as grains and concentrates—is probably quite unimportant for free-ranging animals because their diets contain a natural mixture of forages and seeds.

The positive relationship between intake and digestibility may be the most important one to consider for free-ranging ruminants, especially when they are on a browse diet. The basic idea behind this relationship is that the amount of food in the digestive tract limits the amount that can be ingested. It is expressed in the "rumen load" or "fill" theory; the animal does not ingest more food after the rumen is filled to capacity. The more digestible the food mass is, the sooner the rumen will be emptied and the animal will be able to eat again.

What components of the forage material are residual in the rumen for any length of time? The work of Van Soest at Beltsville, Maryland and later at Cornell University indicates that the cell wall is the most indigestible part of the plant cell. The intracellular material, that is, the material inside the cell wall, is highly digestible—on the order of 98% to 100%. The rumen contents include primarily the cell walls of the forage material that is being attacked by the rumen microorganisms. As the volume of cell-wall material increases in comparison with the volume of intracellular material, the voluntary intake declines because the turn-over rate of the cell-wall material is slow.

The rumen capacity must be considered together with turnover time. The simple illustration in Table 8-3 of the effect of turnover time on available nutrients is about as much as is known about turnover rates in wild ruminants. The alternative to this lack of knowledge of turnover rates is an analysis of the importance of variation in this parameter.

Voluntary intake by sheep in relation to the cell-wall content and digestible dry matter for 83 forages is shown in Figure 9-2 (Van Soest 1965). The voluntary intake declines rapidly when the percentage of cell-wall constituents is greater than 60. This relationship was observed for the forages from Maryland, Michigan,

FIGURE 9-2. Relationship between voluntary intake by sheep and cell-wall constituents of 83 forages from West Virginia. Reression equation: $Y = 110.4 - 1716/(100 - X)$. (Data from Van Soest 1965, p. 837.)

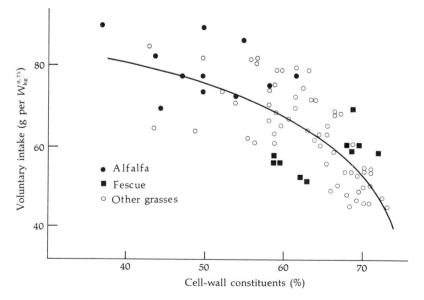

TABLE 9-1   VOLUNTARY INTAKE BY SHEEP OF SOME FORAGES IN RELATION TO THEIR CELL-WALL CONTENT AND THEIR DIGESTIBLE DRY MATTER

Forage	Voluntary Intake (g per $W_{kg}^{0.75}$)	Rank*	Cell Wall (% of forage DM)	Rank*	Digestible Dry Matter (%)	Rank*
Trefoil	89	1	44	1	60	4.5
Good alfalfa	79	2	47	2	60	4.5
Mature alfalfa	67	4	59	3.5	56	6
Early orchard grass	78	3	59	3.5	72	1
Mature orchard grass	29	7	77	7	50	7
Good brome grass	63	5	60	5	65	2
Mature brome grass	51	6	70	6	64	3

SOURCE: Data from Van Soest 1968a.
* Rank added by author.

and Utah; it was the only consistent relationship among many chemical factors that were studied in these forage experiments.

There is a nearly perfect inverse relationship between the rankings of voluntary intake with the percentage of cell wall (Table 9-1). This is biologically reasonable since the cell-wall fraction of the forage, including hemicellulose, cellulose, and lignin, is the fraction that ferments slowly, accumulates in the digestive tract, and limits further feed consumption.

Grinding of domestic animal feeds has been shown to increase intake. This supports the hypothesized relationship between cell-wall constituents and the intracellular components since the mechanical grinding of feed exposes more cell-wall surfaces to be attacked by the microorganisms in the rumen. As might be expected, the grinding effect on plants with a low percentage of cell-wall composition, like alfalfa, is small (Van Soest 1968b).

A considerable amount of time and effort has been spent on the study of browse plants and their relationship to population levels of deer and other wild ruminants. The research has often been started with a vague hope that a correlation may be found between an easily observed parameter, such as plant abundance, and the density of the animal population. It is more useful to consider the basic, biological relationships that exist, however, and the application of basic information is often very straightforward. As Van Soest points out (1968b) "the problem [nutrition] will not be solved by superficial regressions that fail to disclose causative relationships between nutritive value and composition. It is the disclosure and understanding that is required."

Several other theories on the regulatory factors affecting intake have been proposed. These are mentioned here because some of them may have an effect that is superimposed on the basic relationship between the chemical and structural characteristics of the food and intake.

A thermostatic theory [Blaxter (1967) citing Brobeck] states that eating is a response to a drop in heat production. This may be generally true when applied to gross changes in the thermal regime. In a hot environment, in which the relationship between heat production and heat loss may result in an increase in the body temperature, an animal will reduce the food intake and, subsequently, its heat production. In a cold environment, in which heat loss may exceed heat production, an increase in the amount of food consumed and fermented will help maintain or restore the thermal balance.

A theory suggesting that the differences in glucose concentration between arterial and venous blood has been advanced by Mayer [cited by Blaxter (1967)] for man and simple-stomached animals, but this has been shown to be untenable for adult ruminants. Ruminants normally have a low blood-sugar concentration because of the direct absorption of glucose by rumen microorganisms; experimental infusion of glucose does not change the daily intake of food (Blaxter 1967). Glucose concentration may be a factor in the feeding schedule and milk intake of suckling ruminants since they are essentially simple-stomached animals for the first few days after birth.

A lipostatic or fat-balance theory has been advanced by Kennedy [cited by Blaxter (1967)], suggesting that, in the long run, the amount of body fat present regulates the intake of food. This cannot apply to male deer during the breeding season, however, since breeding bucks lose body fat but do not eat an amount of food sufficient to replace it until the following summer. Weight losses are common in adult deer of both sexes during the winter even if sufficient food is available.

In summary, the effect of aromatic compounds in the forage on the olfactory responses of deer and the effect of food quality—in terms of the cell-wall constituency of the forage—seem to be the two most important factors regulating the food intake at the moment at which the animal is ready to eat. There are seasonal changes in the animal's willingness to eat that are related to the maturity of the animal and the hormone balance at the time, especially in relation to sexual characteristics such as breeding, gestation, and lactation. The hormone balance of wild ruminants has received little attention, although it is apparent from the behavior of the animals that it is extremely important as a physiological base for overt behavior.

9-3 ENERGY UTILIZATION

The flow of energy through any living organism follows the Law of Conservation of Energy, which is that "energy can neither be created nor destroyed, but only changed in form." There is a time lag between ingestion and the subsequent

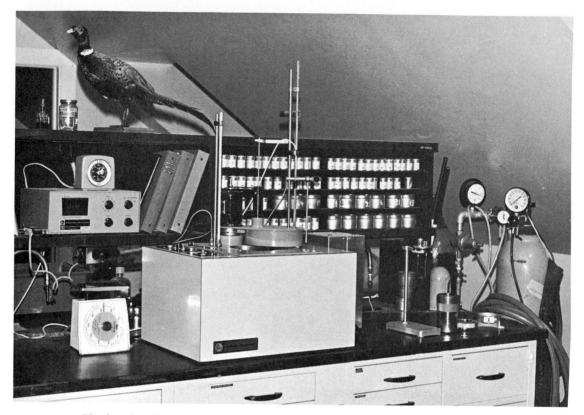

FIGURE 9-3. The bomb calorimeter at the BioThermal Laboratory, Cornell University.

release of energy. This time lag is a necessary characteristic of life for it permits an animal to ingest food whose usefulness will persist for a few hours after eating. Further, those nutrients stored as body tissue will be useful for days or weeks afterwards, increasing the animal's ability to survive periods of food shortages.

PATHWAYS OF ENERGY UTILIZATION. The flow of energy through an animal follows some very logical steps proceeding from gross energy within the food to net energy that is useful to the animal for maintenance and production. The gross energy per unit weight of food is measured in a bomb calorimeter (Figure 9-3) and is usually expressed in kilocalories per gram. Not all of this gross energy is available to the animal because the efficiency of an animal is less than 100%.

Elimination of undigested residue from the gastrointestinal tract is one indication of waste due to a biological efficiency of less than 100%. The gross energy in the food less the heat of combustion of the feces is called the *apparently digestible energy,* expressed as:

$$\left\{\begin{matrix} \text{Apparently} \\ \text{digestible energy} \end{matrix}\right\} = \left\{\begin{matrix} \text{Gross energy} \\ \text{in food} \end{matrix}\right\} - \left\{\begin{matrix} \text{Fecal} \\ \text{energy} \end{matrix}\right\}$$

Two other significant ways in which energy is lost in ruminants are in the gaseous products of digestion and in the heat of fermentation. Methane (CH_4) is the principal gaseous product of digestion and represents a loss of approximately 8% of the gross energy of the food. The amount of energy lost to the animal in methane can be estimated for roughage by the regression equation (9-1) from Blaxter (1967):

$$CH_4 = 4.28 + 0.059 \, DE \qquad (9\text{-}1)$$

where

CH_4 = energy lost in methane, expressed in kcal per 100 kcal gross energy
DE = apparent digestible energy of the food expressed as a whole number (50% = 50)

Thus if $DE = 50\%$, the CH_4 produced is 4.28 + (0.059) (50) = 7.23 kcal. At 80% apparent digestibility, 4.28 + (0.059) (80) = 9.0 kcal. Note that these bracket the mean of 8% given above.

The methane produced by the wildebeest (*Connochaetes taurinus*) is 7.0% to 7.6% of the gross energy in the feed, and for the eland (*Taurotragus oryx*), 6.8% to 7.4% (Rogerson 1968). The averages of 7.3 kcal/100 kcal for eland are similar to the amount produced by domestic ruminants.

The level of feeding influences the methane production, ranging from more than 10% on feeding levels lower than maintenance to 6% at three times the maintenance level. These extremes bracket the average of 8%, and the mean value falls between feeding levels of 1 to 2 times maintenance (Figure 9-4).

FIGURE 9-4.   Methane production in relation to level of feeding. (From "The Utilization of Foods by Sheep and Cattle" by K. L. Blaxter and F. W. Wainman. *J. Agr. Sci.* **57**:419–425, 1961. Cambridge University Press.)

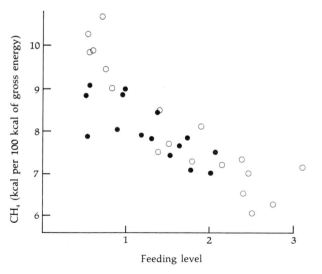

Heat is liberated during digestion owing to the exothermic nature of the chemical reactions in the rumen. It is sometimes called a waste product, but this is not accurate inasmuch as this heat helps the animal to maintain homeothermy when thermal conditions in the environment could cause a decline in body temperature. The energy pathways considered thus far can be expressed as:

$$
\begin{Bmatrix} \text{True} \\ \text{digestible} \\ \text{energy} \end{Bmatrix} = \begin{Bmatrix} \text{Gross} \\ \text{energy} \\ \text{of food} \end{Bmatrix} - \begin{Bmatrix} \text{Heat of combustion} \\ \text{of food residues} \\ \text{in feces} \end{Bmatrix}
$$

$$
- \begin{Bmatrix} \text{Heat of combustion} \\ \text{of fermentation} \\ \text{gases (methane)} \end{Bmatrix} - \begin{Bmatrix} \text{Heat of} \\ \text{fermentation} \\ \text{of feed} \end{Bmatrix}
$$

An additional way in which energy is lost that has not been measured is the heat of combustion of metabolic products secreted into the intestine. The true digestible energy heretofore illustrated can be called "absorbed energy" since it is the energy that passes through the wall of the intestine and is absorbed by the blood for distribution throughout the body.

Digestible energy that is absorbed by the blood is not completely useful. Some portions of the absorbed nutrients are diverted to urine, and the energy in the remaining food materials available for assimilation is called the *true metabolizable energy*. It can be expressed as follows:

$$
\begin{Bmatrix} \text{True} \\ \text{metabolizable} \\ \text{energy} \end{Bmatrix} = \begin{Bmatrix} \text{Gross} \\ \text{energy} \\ \text{of food} \end{Bmatrix} - \begin{Bmatrix} \text{Heat of combustion} \\ \text{of food residues} \\ \text{in feces} \end{Bmatrix}
$$

$$
- \begin{Bmatrix} \text{Heat of combustion} \\ \text{of fermentation} \\ \text{gases (methane)} \end{Bmatrix} - \begin{Bmatrix} \text{Heat of} \\ \text{fermentation} \\ \text{of feed} \end{Bmatrix} - \begin{Bmatrix} \text{Heat of} \\ \text{combustion} \\ \text{of urine} \end{Bmatrix}
$$

Metabolizable energy is converted to body tissue and is used for basal metabolic processes, activity, production, and other processes basic to life. These conversions require an expenditure of energy, called the heat of nutrient metabolism. Metabolizable energy less the heat of nutrient metabolism is called the *true net energy*, and it is used for the maintenance and productive purposes listed above.

The true net energy available to the animal contributes to two separate energy requirements—maintenance energy and production energy. The entire energy pathway can be expressed as:

$$
\begin{Bmatrix} \text{Gross} \\ \text{energy} \end{Bmatrix} - \begin{Bmatrix} \text{Fecal} \\ \text{energy} \end{Bmatrix} - \begin{Bmatrix} \text{Methane} \\ \text{losses} \end{Bmatrix} - \begin{Bmatrix} \text{Heat of} \\ \text{fermentation} \end{Bmatrix} - \begin{Bmatrix} \text{Urinary} \\ \text{energy losses} \end{Bmatrix}
$$

$$
- \begin{Bmatrix} \text{Heat of} \\ \text{nutrient metabolism} \end{Bmatrix} = \begin{Bmatrix} \text{Net energy available for} \\ \text{maintenance and production} \end{Bmatrix}
$$

These diversions are illustrated in Figure 9-5.

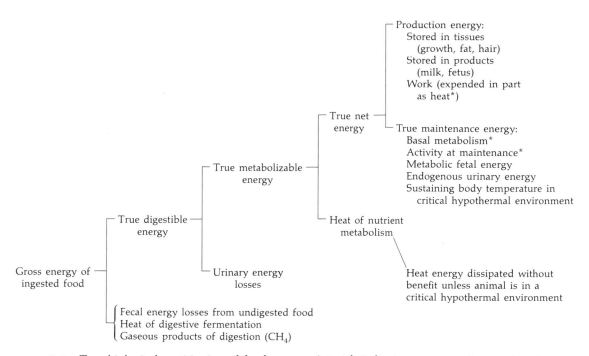

FIGURE 9-5.   True biological partitioning of food energy. Asterisk indicates processes that produce heat. (Modified from Crampton and Harris 1969.)

NET ENERGY FOR MAINTENANCE.   Maintenance energy includes the basal requirements for cellular metabolism without an increase in body weight and for voluntary activity and other life-support processes in which an animal participates as a part of daily life. This includes feeding behavior, reproductive behavior, flight behavior to escape predators, the maintenance of homeothermy when in a critical thermal environment, the support of a "normal" parasite and pathogen load, and any other requirement that is a part of the animal's life.

Some of the energy in the feces and urine comes from tissue that was produced earlier. Metabolic fecal energy and metabolic urinary energy is of body origin and is derived from the breakdown or catabolism of living tissue. The energy in the cell residues in the feces and urine must be separated from the indigestible food residues that are a major part of the feces and from the excretory products in the urine to allow computation of the net energy derived from the feed. To illustrate, some of the energy in the feces is due to the presence of epithelial cells that have been removed from the lining of the intestinal tract. The energy contained in these cells cannot be attributed to indigestible food material since these cells are the product of metabolic processes, and their loss will be replaced by new cells that are formed at a metabolic cost to the animal. If metabolic fecal and urinary energy is not subtracted from the total fecal and urinary energy, the efficiency of an animal in converting food to metabolic products would be underestimated.

If the thermal conditions in the environment result in a heat loss greater than the heat production by the animal from the heat of fermentation of foods and nutrient metabolism in the cells, the animal is in a critical hypothermal environment (Moen 1968). The animal may increase its heat production in order to maintain body temperature, and this energy demand must be considered a part of the maintenance requirements of the animal under these conditions. When the heat production due to basal metabolism and voluntary activity rises above heat loss, the excess heat energy must be dissipated to prevent a rise in body temperature.

NET ENERGY FOR PRODUCTION.   The energy required for production is used for tissue synthesis and storage. The growth of all body tissue, including not only the internal tissue but also the external covering of hair, requires energy, as does the production of ova and growth of the fetus in the pregnant female and the production of semen by the male. Further, lactation requires the production of milk that is ingested by the suckling animal with an energy cycle of its own. Fat is a readily available energy reserve. These production processes are programmed into the computations of the total energy requirement of free-ranging animals in Chapters 16 and 17.

SUMMARY OF ENERGY UTILIZATION.   As an animal completes the cycle of conception, birth, growth, and death, all of the energy that went into the animal can be accounted for, with a certain portion of it held in reserve by the body at death, to be dissipated as decomposers use it as a substrate for meeting their own energy requirements. The events that go into the transformation of energy form a very basic foundation on which the survival of an individual and a population, as well as the carrying capacity of the range, can be based. The energy pathways described in this chapter are expressed quantitatively in Chapter 16, which deals with animal requirements, and these equations are used in the calculation of carrying capacity in Chapter 17.

9-4 PROTEIN UTILIZATION

Proteins are complex chemical substances composed mostly of carbon, hydrogen, oxygen, and nitrogen. These elements are arranged into large units called amino acids, which are synthesized by rumen microorganisms from simpler nitrogenous compounds.

Not all of the protein ingested is available to the animal because the efficiency of an animal is less than 100%. Thus there is a difference between the total protein content of the food and the amount of protein actually available to the animal for tissue production. The pathways for nitrogen diversion and use are shown in Figure 9-6. The synthesis of body tissue is a part of the growth process of young animals. Even in adults tissue replacement is necessary because of catabolic processes that result from the maintenance of normal activity and life processes from day to day. Some tissue deposition, such as the production of hair, fetal

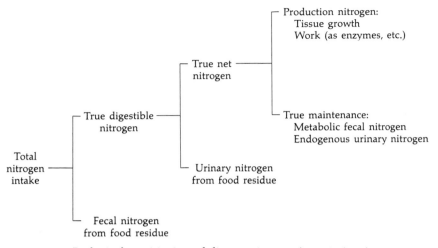

FIGURE 9-6. Biological partitioning of dietary nitrogen (protein/6.25).

tissue, and milk, varies seasonally. Protein requirements for maintenance of body tissue vary much less over time. The protein costs for maintenance and production throughout the annual cycle are expressed quantitatively in Chapter 16.

9-5 EFFICIENCY OF NUTRIENT UTILIZATION

DIGESTION COEFFICIENTS. Food ingested by a ruminant is attacked by microorganisms in the rumen. These microorganisms are engaged in their own struggle for survival, and the end products of their metabolism are of direct value to the ruminant host in this symbiotic relationship. Because of a lack of information on digestibility coefficients for different forages consumed by wild ruminants, it is logical to use instead digestibility coefficients for forage eaten by domestic ruminants. This raises a question about the similarity of different species of ruminants and the validity of comparisons between species.

A review of the literature on both domestic and wild ruminants suggests that any differences observed between different species is not so much due to differences between species per se but is due to the specific history of the different individuals. If two deer were on entirely different diets and rumen fluids from each were used to innoculate *in vitro* fermentation vials containing equivalent forage samples, the results would probably be quite different. If a deer and a sheep were on the same diet and their rumen fluids were used for innoculation, the results would probably be quite similar.

Three mule deer showed differences in cellulose digestion between the animals themselves and between feeding-trial periods (Short 1966). Protein digestion was similar in all feeding trials. It is clear that there is no single value or mathematical relationship available at the present time with which to express these relationships.

NET ENERGY AND PROTEIN COEFFICIENTS. The expression of net energy and net protein in forage is the best way to describe its nutritive value. This is so because the net values represent the value of the forage actually available to the animal for maintenance and production. Although the expression of net protein and net energy is easy to accept conceptually, there is a loss of accuracy in the predictions because of the number of biological variables and the chance for error in measurement associated with each variable. The calculation of apparently digestible energy by subtracting fecal energy from the gross energy in the food is fairly simple, with a possibility of error only in the measurement of the quantity of food ingested, the caloric value of the ingested food, the quantity defecated, and the caloric value of the feces. If the true digestible energy is to be determined, the caloric energy of the metabolic residues in the feces (separate from the indigestible residues) must be determined, and the measurement of both the quantity and the caloric value of these metabolic products introduces additional errors. Thus it is obvious that as greater detail is sought there will be less precision.

An analysis of basic relationships within any system will ultimately produce more progress in the management of the system than will the continued use of simplified approaches with hoped-for useful results. The use of net energy and net protein coefficients is conceptually sound and is evaluated further within an ecological framework in Chapter 17.

METABOLIC EFFICIENCY.   The net energy and net protein coefficients discussed above express the potential value of the forage when it is absorbed and utilized at the cellular level. The efficiency of cellular metabolism is dependent on the presence of cellular enzymes and a cellular environment favorable for particular biochemical reactions.

Vitamins are important constituents of the cellular enzymes. They are small but often essential parts of complex protein molecules and are usually needed in very small amounts. The water-soluble vitamins of the B group are synthesized by the microorganisms in the rumen, and the adult ruminant needs only Vitamins A, D, and E in its diet (Annison and Lewis 1959). The synthesis of some of these vitamins depends on the presence of certain trace elements. Cobalt, for example, is necessary for the synthesis of B_{12}. The young ruminant relies on the ingestion of the full complement of vitamins during the period before the rumen is functional.

Some minerals, such as phosphorus and calcium, are required in larger amounts because they are important components of bone tissue. The young, growing animal has a relatively higher requirement than the mature animal, and this requirement must be met by the diet. Some minerals, such as selenium, are necessary in trace amounts but are detrimental to normal biological processes if present in larger amounts. The mechanism of selenium toxicity is not understood (Crampton and Harris 1969), but symptoms indicate that the nervous system is impaired and normal growth is interrupted.

The requirements and tolerances of wild ruminants for vitamins and minerals are probably quite similar to those of domestic animals. After the computations of carrying capacity described in Chapter 17 are completed on an energy and protein base, the vitamin and mineral requirements can be added as further considerations if the vitamin and mineral composition of the diet is known.

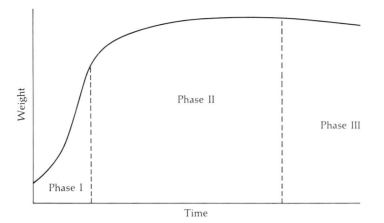

FIGURE 9-7. A theoretical weight change curve showing the
three phases of growth during the life span of an individual.

9-6 BODY GROWTH

The growth of animals is a continuous process from conception to the time of
death. The rate at which an animal grows is not constant, however. A theoretical
weight-change curve (Figure 9-7) can be divided into three phases: (1) an initial
phase during which there is accelerating growth; (2) a phase during which the
rate of weight change slows considerably; and (3) a phase during which the weight
change may be slightly negative. The weight fluctuates within each phase, of
course, including rather marked reductions in the weight of many adult wild
ruminants during the winter.

At the beginning of Phase I there is a period of adjustment to a vastly different
environment immediately after birth. The infant animal goes from a warm
constant-temperature uterine environment with an arterial nutrient supply and
the mother's excretory system for removal of metabolic waste to an independent
life with its own nutrient ingestion. The young animal must metabolize at a rate
compatible with its needs as a homeotherm, and this requires energy.

The mother and the young animal begin Phase I as a single nutritional unit,
gradually becoming separate units. The young animal grows rapidly on a milk
and forage diet, with an increasing dependence on forage. At weaning, the separation is complete nutritionally, but doe and fawn are still united socially to a
certain extent.

The transition from Phase I to Phase II of the weight curve is accompanied
by the weaning process and, in some of the ruminants, by sexual maturity. Female
white-tailed deer, for example, frequently breed as fawns, and male fawns have
been reported fertile (Silver 1965). Cowan and Wood (1955) reported the fertile
mating of black-tailed deer fawns (*Odocoileus hemionus columbianus*) during their
first winter.

The Phase II growth period may continue for several years. The weight of the
animal increases, but the rate of increase is slow. There are seasonal increases
and decreases that seem to be associated primarily with the breeding cycle and

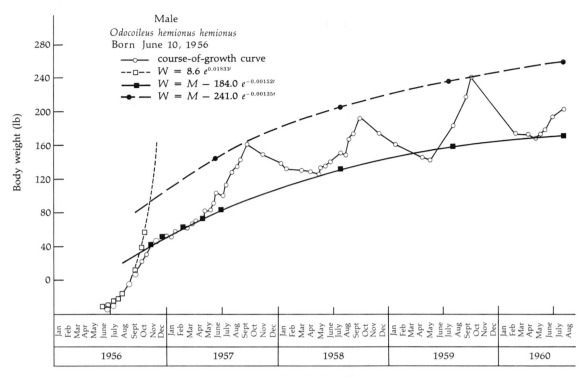

FIGURE 9-8.   Growth curves for a representative male of *Odocoileus hemionus hemionus*: W = weight in pounds; t = age in days. [Data from Wood, Cowan, and Nordan. Reproduced by permission of the National Research Council of Canada from the Canadian Journal of Zoology, 40, pp. 593–603 (1962).]

secondarily with the winter range conditions. The reproductive condition of the animal is at its peak during this phase; the animal is in its prime.

The termination of Phase II and the beginning of Phase III, or old age, is a gradual process. Animals lose their biological efficiency owing to a gradual loss of elasticity of body tissue. Further, teeth are becoming worn and the animal is apparently not quite as efficient at utilizing forage as it once was. Concomitant with a loss in weight is a loss in productivity; old white-tailed does produce fewer young than do young does (Cheatum and Severinghaus 1950).

Weight changes of a male mule deer are reported by Wood, Cowan, and Nordan (1962) and are shown in Figure 9-8. Weight during the first summer shows an increasing rate of gain until October when it begins to slow down. If the animal has been on an adequate diet, reproductive behavior appears. The weight increase is greater during the following summer when the animal is a yearling, but a decline follows during the animal's second winter. Note that the weight each winter is higher than the previous winter, and summer weights are progressively higher too. These increases from year to year are a part of the second growth phase, which continues until the maximum weight at maturity is reached. Similar growth equations for other black-tailed deer are reported by Wood, Cowen, and Nordan

TABLE 9-2 AGE-WEIGHT RELATIONSHIPS FOR DIFFERENT SUBSPECIES OF *Odocoileus hemionus*

Species	Sex	Growth Phase I		Growth Phase II
O. h. sitkensis	M	$W_{\text{lb}} = 7.8 \ e^{0.01324t*}$	a†	$W_{\text{lb}} = M - 325 \ e^{-0.00283t}; M = 240$
			b	$W_{\text{lb}} = M - 132 \ e^{-0.00174t}; M = 165$
O. h. columbianus (Vancouver Island genotype)	M	$W_{\text{lb}} = 9.5 \ e^{0.00996t}$	a	$W_{\text{lb}} = M - 131.5 \ e^{-0.00161t}; M = 170$
			b	$W_{\text{lb}} = M - 110.5 \ e^{-0.00123t}; M = 140$
O. h. columbianus (California genotype)	M	$W_{\text{lb}} = 8.6 \ e^{0.01134t}$	a	$W_{\text{lb}} = M - 166.9 \ e^{-0.00145t}; M = 210$
			b	$W_{\text{lb}} = M - 136.5 \ e^{-0.00237t}; M = 150$

SOURCE: Data from Wood, Cowan, and Nordan 1962.
*t is time in days.
†a = summer weights; b = winter weights.

(1962) and are listed in Table 9-2, and for caribou (*Rangifer tarandus groenlandicus*) by McEwan (1968), in Table 9-3.

The number of points in the graphs shown in Figures 9-7 and 9-8 indicate the amount of effort needed to construct a complete weight profile over a period of years. First approximations have been made for white-tailed deer, elk, and moose (Table 9-4). The equations are based on data reported in the literature and are thought to be realistic for first approximations. The effect of error in the age-weight relationship can be analyzed so that decisions can be made about the level of accuracy required in this parameter.

TABLE 9-3 AGE-WEIGHT RELATIONSHIPS FOR *Rangifer tarandus groenlandicus*

Animal	Sex	Growth Phase I	Growth Phase II
U8	M	$W = 7.2 \ e^{0.020174t}$	$W = M - 130 \ e^{-0.001429t}; M = 150$
U3	M	$W = 6.2 \ e^{0.020712t}$	$W = M - 120 \ e^{-0.001347t}; M = 150$
V17	M	$W = 6.0 \ e^{0.019917t}$	$W = M - 116 \ e^{-0.001311t}; M = 150$
V21	M	$W = 4.1 \ e^{0.022904t}$	$W = M - 120 \ e^{-0.001172t}; M = 150$
Wild	M		$W = M - 130 \ e^{-0.001429t}; M = 125$
Wild	M		$W = M - 120 \ e^{-0.0004755t}; M = 150$
U4	F	$W = 7.2 \ e^{0.01983t}$	$W = 110 - 80 \ e^{-0.001597t}$
U7	F	$W = 6.6 \ e^{0.020542t}$	$W = 90 - 80 \ e^{-0.003398t}$
V27	F	$W = 6.9 \ e^{0.018829t}$	
V28	F	$W = 6.4 \ e^{0.018819t}$	
Wild	F		$W = 90 - 60 \ e^{-0.001230t}$
Wild	F		$W = 90 - 55 \ e^{-0.0005852t}$

SOURCE: Data from McEwan [reproduced by permission of the National Research Council of Canada from the *Canadian Journal of Zoology*, **46**, pp. 1023–1029 (1968)].

TABLE 9-4 AGE-WEIGHT RELATIONSHIPS FOR DEER, MOOSE, AND ELK

Species	Sex	Growth Phase I	Ref.*	Growth Phase II	Maximum Weight	Ref.*
White-tailed deer	M & F	$W_{kg} = 3.0 + 0.229\, t_d$	1			
	M			$W_{kg} = e^{(1.230 + 0.440\, \ln t_d)}$	111 kg	1
	F			$W_{kg} = e^{(1.617 + 0.357\, \ln t_d)}$	71 kg	1
Moose	M & F	$W_{kg} = 13 + 1.0\, t_d$	2	$W_{kg} = e^{(2.198 + 0.549\, \ln t_d)}$		3
Elk	M			$W_{kg} = e^{(2.409 + 0.435\, \ln t_d)}$		4
	F			$W_{kg} = e^{(2.529 + 0.385\, \ln t_d)}$		4

*References include either the equation or the data from which the above equations were calculated.
1. Moen, unpublished data.
2. Verme 1970.
3. Blood, McGillis, and Lovaas 1967.
4. Murie 1951.

9-7 SUMMARY

The ingestion and utilization of food is a complex biological phenomenon. It is a chemical communication between animal and environment, absolutely essential for survival and growth. Wildlife biologists traditionally list the three requirements of an animal as food, cover, and space, but often too little attention is given to food requirements. Food habits and range conditions are often the only things studied in an analysis of the habitat requirements of a wild species. The complex nature of nutritional relationships and the fundamental importance of nutrition for survival and growth must be recognized if wildlife scientists are to develop more reliable knowledge about the habitat requirements of any species.

Other apsects of the total organism-environment relationship are discussed in Parts 4 and 5, with a return to nutritive considerations in relation to carrying capacity in Part 6.

LITERATURE CITED IN CHAPTER 9

Annison, E. F., and D. Lewis. 1959. *Metabolism in the rumen.* London: Metheun and New York: Wiley, 184 pp.

Blaxter, K. L. 1967. *Energy metabolism of ruminants.* London: Hutchinson, 332 pp.

Blaxter, K. L., and F. W. Wainman. 1961. The utilization of food by sheep and cattle. *J. Agr. Sci.* **57**: 419–425.

Blood, D. A., J. R. McGillis, and A. L. Lovaas. 1967. Weights and measurements of moose in Elk Island National Park, Alberta. *Can. Field-Nat.* **81**(4): 263–269.

Burcalow, D. W., and W. H. Marshall, 1958. Deer numbers, kill, and recreational use on an intensively managed forest. *J. Wildlife Management* **22**(1): 141–148.

Cheatum, E. L., and C. W. Severinghaus. 1950. Variations in fertility of white-tailed deer related to range conditions. *Trans. North Am. Wildlife Conf.* **15**: 170–190.

Compling, R. C. 1964. Factors affecting the voluntary intake of grass. *Pract. Nutr. Soc.* **23**: 80–88.

Cook, D. B. 1939. Thinning for browse. *J. Wildlife Management* **3**(3): 201–202.

Cook, D. B. 1946. Summer browsing by deer on cutover hardwood lands. *J. Wildlife Management* **10**(17): 60–63.

Cowan, I. McT., and A. J. Wood. 1955. The growth rate of the black-tailed deer. *J. Wildlife Management* **19**(3): 331–336.

Crampton, E. W., and L. E. Harris. 1969. *Applied animal nutrition.* 2d ed. San Francisco: W. H. Freeman and Company, 753 pp.

Fowler, J. F., J. D. Newsom, and H. L. Short. 1967. Seasonal variation in food consumption and weight gain in male and female white-tailed deer. *Proc. 21st Ann. Conf. SE Assoc. Game Fish Comms.* pp. 24–32.

Healy, W. 1967. Forage preferences of captive deer while free-ranging in the Allegheny National Forest. Master's thesis, Pennsylvania State University. 93 pp.

Longhurst, W. H., H. K. Oh, M. B. Jones, and R. E. Kepner. 1968. A basis for the palatibility of deer forage plants. *Trans. North Am. Wildlife Nat. Resources Conf.* **33**: 181–192.

McEwan, E. H. 1968. Growth and development of the barren-ground caribou. II. Postnatal growth rates. *Can. J. Zool.* **46**: 1023–1029.

Moen, A. N. 1966. Factors affecting the energy exchange of white-tailed deer, western Minnesota. Ph.D. dissertation, University of Minnesota. 121 pp.

Moen, A. N. 1968. The critical thermal environment: a new look at an old concept. *BioScience* **18**(11): 1041–1043.

Murie, O. J. 1951. *The elk of North America.* 1st ed. Harrisburg, Pennsylvania: Stackpole and Washington, D.C.: Wildlife Management Institute, 376 pp.

Nordan, H. C., I. McT. Cowan, and A. J. Wood. 1968. Nutritional requirements and growth of black-tailed deer (*Odocoileus hemionus columbianus*) in captivity. In *Comparative nutrition of wild animals,* ed. M. A. Crawford. Symp. Zool. Soc. London, No. 21. pp. 89–96.

Ozoga, J. J., and L. J. Verme. 1970. Winter feeding patterns of penned white-tailed deer. *J. Wildlife Management* **34**(2): 431–439.

Rogerson, A. 1968. Energy utilization by the eland and wildebeest. In *Comparative nutrition of wild animals,* ed. M. A. Crawford. Symp. Zool. Soc. London, No. 21. pp. 153–161.

Short, H. L. 1966. Effects of cellulose levels on the apparent digestibility of foods eaten by mule deer. *J. Wildlife Management* **30**(1): 163–167.

Silver, H. 1965. An instance of fertility in a white-tailed buck fawn. *J. Wildlife Management* **29**(3): 634–636.

Swift, R. W. 1948. Deer select most nutritious forages. *J. Wildlife Mangement* **12**(1): 109–110.

Van Soest, P. J. 1965. Symposium on factors influencing the voluntary intake of herbage by ruminants: voluntary intake in relation to chemical composition and digestibility. *J. Animal Sci.* **23**(3): 834–843.

Van Soest, P. J. 1968a. Chemical estimates of the nutritional value of feeds. In *Proceedings: Cornell Nutrition Conference for Feed Manufacturers,* pp. 38–46.

Van Soest, P. J. 1968b. *Structural and chemical characteristics which limit the nutritive value of forages. Forage: economics/quality* ASA Special Publication No. 13.

Verme, L. J. 1970. Some characteristics of captive Michigan moose. *J. Mammal.* **51**(2): 403–405.

White, D. L. 1968. Condition and productivity of New Hampshire deer. In *The white-tailed deer of New Hampshire,* ed. H. R. Siegler. Survey Report No. 10. Concord: New Hampshire Game and Fish Dept., pp. 69–113.

Wood, A. J., I. McT. Cowan, and H. C. Nordan. 1962. Periodicity of growth in ungulates as shown by deer of the genus *Odocoileus. Can. J. Zool.* **40**: 593–603.

IDEAS FOR CONSIDERATION

The nutritive pathways discussed in this part of the book are related more specifically to ruminants than to other mammals and birds. Pathways specific to these other animals should be described by students having different interests, utilizing the gross-to-net concept in identifying the pathways and quantifying the nutritive processes.

SELECTED REFERENCES

Bailey, E. D., and D. E. Davis. 1965. The utilization of body fat during hibernation in woodchucks. *Can. J. Zool.* **43**: 701–707.

Barrett, M. W., and E. D. Bailey. 1972. Influence of metabolizable energy on condition and reproduction of pheasants. *J. Wildlife Management* **36**(1): 12–23.

Blaxter, K. L., J. C. Clapperton, and F. W. Wainman. 1966. The extent of differences between six British breeds of sheep in their metabolism, feed intake and utilization, and resistance to climatic stress. *Brit. J. Nutr.* **20**: 283–294.

Byerly, T. C. 1967. Efficiency of feed conversion. *Science* **157**(3791): 890–895.

Elder, J. B. 1954. Notes on summer water consumption by desert mule deer. *J. Wildlife Management* **18**(4): 540–541.

Goldstein, R. A., and J. W. Elwood. 1971. A two-compartment, three-parameter model for the absorption and retention of ingested elements by animals. *Ecology* **52**(5): 935–939.

Graham, N. McC. 1967. Effects of feeding frequency on energy and nitrogen balance in sheep given a ground and pelleted forage. *Australian J. Agr. Res.* **18**: 467–483.

Hill, D. C., E. V. Evans, and H. G. Lumsden. 1968. Metabolizable energy of aspen flower buds for captive ruffed grouse. *J. Wildlife Management* **32**(4): 854–858.

Jarrett, C. J., and O. T. Fosgate. 1964. Sources and percentages of net energy of dairy rations as related to milk production. *J. Dairy Sci.* **47**(3): 299–301.

Jennings, J. B. 1965. *Feeding, digestion, and assimilation in animals.* New York: Pergamon, 228 pp.

McGowan, J. D. 1969. Starvation of Alaskan ruffed and sharp-tailed grouse caused by icing. *Auk* **86**(1): 142–143.

Richardson, A. J. 1970. The role of the crop in the feeding behaviour of the domestic chicken. *Animal Behaviour* **18**(14): 633–639.

Schmidt, W. D. 1965. Energy uptake in the mourning dove. *Science* **150**(3700): 1171–1172.

Ullrey, D. E., W. G. Yougatt, H. E. Johnson, L. D. Fay, B. L. Schoepke, and W. T. Magee. 1969. Digestible energy requirements for winter maintenance of Michigan white-tailed does. *J. Wildlife Management* **33**(3): 482–490.

Ullrey, D. E., W. G. Youatt, H. E. Johnson, L. D. Fay, B. L. Schoepke, and W. T. Magee. 1970. Digestible and metabolizable energy requirements for winter maintenance of Michigan white-tailed does. *J. Wildlife Management* **34**(4): 863–869.

Ullrey, D. E., H. E. Johnson, W. G. Youatt, L. D. Fay, B. L. Schoepke, and W. T. Magee. 1971. A basal diet diet for deer nutrition research. *J. Wildlife Management* **35**(1): 57–62.

West, G. C., and M. S. Meng. 1966. Nutrition of willow ptarmigan in Northern Alaska. *Auk* **83**: 603–615.

Wiegert, R. G. 1968. Thermodynamic considerations in animal nutrition. *Am. Zool.* **8**: 71–81.

Williams, V. J., T. R. Hutchings, and K. A. Archer. 1968. Absorption of volatile fatty acids from the reticulo-rumen and abomasum of sheep. *Australian J. Biol. Sci.* **21**(1): 89–96.

Young, V. R., and N. S. Scrimshaw. 1971. The physiology of starvation. *Sci. Am.* **225**(4): 14–21 (Offprint No. 1232).

Courtesy of Paul M. Kelsey
New York State Department of Environmental Conservation

PART

BEHAVIORAL FACTORS
IN RELATION TO
PRODUCTIVITY

Wildlife management has been directed primarily at the management of the habitat of game species and the use of the habitat by humans. Certain behavior patterns have been regulated by law, such as the length of the hunting season, the number of animals that can be taken, and so forth. This direction is mostly by necessity since the animals themselves are not under the direct control of man.

The success of management practices depends on the degree to which the relationships between the animal and its habitat are optimized. The successful management of the habitat is dependent on a knowledge of what the animal requires behaviorally as well as physiologically.

Behavior characteristics change with time. Animals that are gregarious during one period of the year (e.g., birds in the wintering grounds) are spaced widely on territories during other periods, such as the breeding season. Predation is a function of population densities, animal requirements for food, habitat characteristics, and a host of other interrelated factors.

This part of the book calls attention to the chronology of the events in the life of an animal, with a general review of some behavioral traits and intraspecific and interspecific relationships. The important point of the entire part relates to the kinds of interactions that occur between organism and environment in time. There is order to these interactions, and the recognition of this order, especially within a format utilizing the latest analytical methods, will result in significant progress toward an understanding of the biological interrelations in plant and animal communities.

10

THE ORGANISM AS A
FUNDAMENTAL UNIT
IN A POPULATION

The individual organism is of fundamental significance in ecological analyses because every organism lives and dies as an individual. It has many relationships with other individuals in the community, of course, and is influenced by them in many ways. These other organisms are an integral part of the environment of each individual whenever there are functional relationships between them, just as physical factors that have functional relationships with an organism are a part of its environment. Each individual, however, relates to both the biological and the physical factors in its own unique way.

There is a tendency to consider groups of individuals as the smallest unit that can be analyzed ecologically. The key point in this chapter is that even in a group the individual's productivity is its own, and the variation between individuals in a community is an exciting part of ecology! The "averaging" of several individuals in a group, describing the group with a single number or by a mean with a standard deviation, masks the drama within the community as each organism meets the ecological forces in its day to day existence.

Analysis of the effect of variation on different individuals reveals something about what they must do to cope successfully with ecological forces. Thus a parasite-laden white-tailed deer may need a larger fat reserve going into winter to cope with the extra demands that the parasites place on it. A pregnant deer may need a larger fat reserve to survive the winter, or it may benefit more from an early spring than a nonpregnant deer would. Both the pregnant and the nonpregnant deer may live in the same area, but the effect of winter on each of them is different because *they* are different. To be sure, they are both *Odocoileus virginianus*, but that is a similarity based on several gross features that man has decided upon as standards for identifying the species. Ecologically, they can be described in terms of their own energy and matter characteristics in relation to the energy and matter characteristics of their environment.

Analysis of the individual, however, is not an analysis of a single animal per se, but of individual characteristics. The purpose of such analyses is the determination of the importance of individual characteristics for the survival and productivity of the individual first, and then of the population. Pregnancy, for example, may have an effect on the ability of the pregnant individual to survive. If the individual survives and the pregnancy is successful, there is an addition to the population.

10-1   ENERGY, MATTER, AND TIME

One of the significant features of the existence of an individual organism is its energy and matter configuration through time. The energy requirements of individuals are constantly changing. The requirements are generally greatest per unit body weight during the early part of life. Adults have lower requirements for energy while resting, but higher requirements for productivity, especially for reproduction. Energy and matter is being reorganized at a rapid rate during gestation (especially the last two-thirds of the gestation period) and during lactation or egg-laying. These processes result in a synthesis of new tissue uniquely organized to become self-sufficient as a separate individual. The ability of this individual to survive depends on its success in coping with the forces of nature.

How can the differences between individuals in a population be analyzed? First of all, individuals can be represented by "points" on gradients between maximum and minimum values for different biological functions. For example, white-tailed deer may weigh between 2 kg and 120 kg. We all know that small, newborn deer are different from larger adult deer in many ways. They have spots rather than a solid coat color and are therefore a different part of a predator's visual environment than a solid-colored deer would be. They are lower to the ground, living in a different wind environment from that of the larger deer. A newborn fawn does not have as distinctive an odor as a larger, older deer. The younger deer are growing at a faster rate than the older deer, requiring more energy and protein on a relative or per-unit-weight basis. The older and larger deer have a higher absolute requirement though, since they need a considerable amount of energy just for maintenance. The pregnant adult deer need energy and protein for pregnancy, too. Thus the younger, smaller deer have different ecological roles from those of the older, larger deer, and each makes different contributions to the population.

Individual characteristics can be analyzed by studying the biological characteristics of deer ranging in weight from one extreme to the other, say, at 5-kg intervals up to 120 kg. This procedure demands a bit of knowledge about deer biology, of course, since the changing characteristics of deer of different weights must be known. Examples of such knowledge, expressed mathematically for use in computer analyses, are included in Parts 5 and 6.

10-2 BIOLOGICAL CHRONOLOGY

The chronological format for displaying the biological functions through an annual cycle are shown in Figure 10-1. Note that time is expressed in years (t_y) and days (t_d). The sum of those two, with the age in years converted to days ($t_y \times 365$), gives the total age in days. For deer, the chronology begins with parturition, with

FIGURE 10-1. Format for displaying biological functions through the annual cycle for (A) white-tailed deer and (B) moose: t_y = time in years; t_d = time in days.

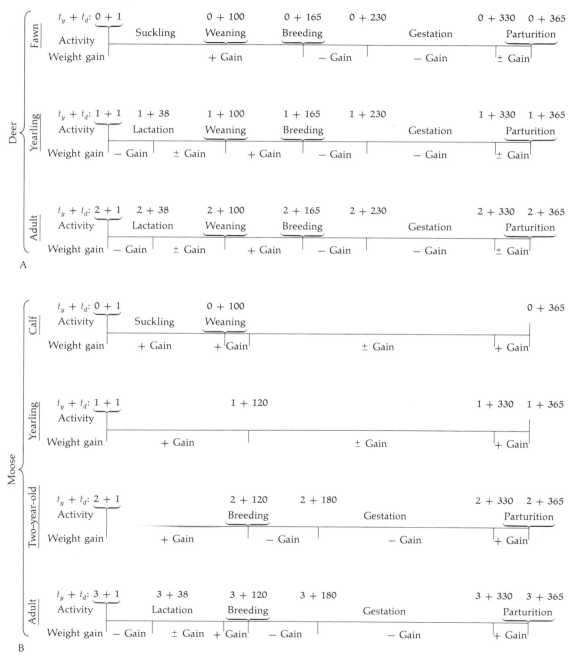

the fawn weaned 100 days later, bred at the age of 165 days, and giving birth to its own young on its first birthday. This is an idealized chronology, of course. It is not meant to be precise for all deer since the primary function of such a chronology is to provide a time base in a computer program. A similar example of chronology for moose is also shown in Figure 10-1.

Given such a time base, the computer program can be written with equations for calculating the energy or protein costs of pregnancy during the periods of 165 to 365 days. The program can be written to include decision-making capabilities so that any analysis outside of that time period will not include the pregnancy cost calculations. The effect of variation in the timing of biological functions can be analyzed with this approach. Suppose female fawns were not capable of reproduction until the age of 195 days (30 days later than that shown in Figure 10-1). This would delay conception by one month, with the fawn being dropped one month later in the spring (about July 1). Suppose that fawn was bred on day 225, delaying the birth of its fawn by yet another month (to August 1). Continuing that pattern for the late-born, it would be bred in February—a time when the bucks may not be capable of servicing a female. If she was not bred then, she would breed for the first time as a long yearling. At this older age, she probably would come into heat early in the breeding period, resulting in an early fawn the following summer. Her fawn might very well breed that fall, starting the cycle all over again.

The foregoing example illustrates the use of the chronology as a base, with a shift in the breeding time. This procedure can be used to test a hypothesis about the time of fawn or calf drop and the age of reproductive maturity, comparing the theoretical results with observed field data. This background knowledge about the significance of the time of breeding may help in interpreting the field data. Many other relationships can be tested in a similar way. The time base is used to test the effect of time variation on nutritive relationships in Part 6.

The biological or life-history characteristics of waterfowl can be expressed on a time base that includes significant reference points in their annual cycle. Let us illustrate the annual cycle of several species of waterfowl by first identifying their main biological activities (Figure 10-2). Note that both males and females are present on the wintering grounds and both participate in breeding activity during the flight northward. After arriving at the marsh, the male participates in territorial defense, "loafing," and molting prior to southward migration. The female builds a nest, lays eggs, incubates, rears the brood, molts, and then flies south. If the first nest is destroyed, the reproductive activities may be repeated.

The next step is to identify the period of time in which each of these biological activities occurs. Data for northward flight, arrival at Delta Marsh, and the beginning of nesting for five species of ducks described in Sowls (1955) have been positioned on chronology scales (Figure 10-3). The average date of arrival is shown along with earliest and latest recorded dates of arrival. The peak for beginning the building of nests is shown, along with earliest and latest days on which nests were started. Note that the distribution is skewed to the left because of renesting attempts. Other significant days in the annual cycle can be determined

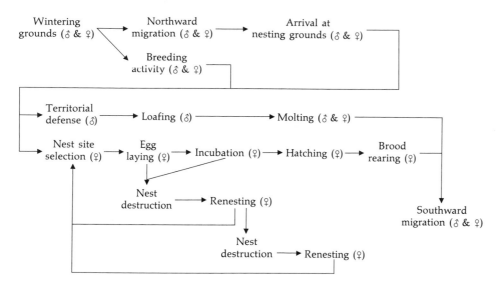

FIGURE 10-2. Significant reference points in the annual cycle of waterfowl.

from the data in Figure 10-3. One egg is laid per day, for example, so the number of days of egg-laying is equal to the number of eggs in the clutch. Incubation is from 21 to 23 days for ducks, which establishes the date of hatching. Thus, there are certain constraints in the chronology; nesting begins only after arrival at the marsh, egg-laying takes at least as many days as there are eggs in the clutch, brood-rearing cannot begin before the end of the 21-day incubation period, and so forth.

These characteristics of the biological cycle through time can be expressed mathematically for use as a time base for computer calculations of the nutrient requirements, behavioral interactions, and other characteristics of the animal throughout the year. The symbols used to designate biological events are listed in Table 10-1. The values for different characteristics of the five species are shown

TABLE 10-1 SYMBOLS USED IN THE CHRONOLOGY OF WATERFOWL

d_{nf} = day northward flight begins
d_{am} = day of arrival at marsh
d_{lb} = day laying begins
d_{nd} = day nest destroyed
d_{le} = day laying ends
d_{ib} = day incubation begins
d_{ie} = day incubation ends
d_{ln} = day leave nest = $d_{ie} + 2$
d_{fb} = day flight begins = $d_{ie} + 42$
d_{sf} = day southward flight begins

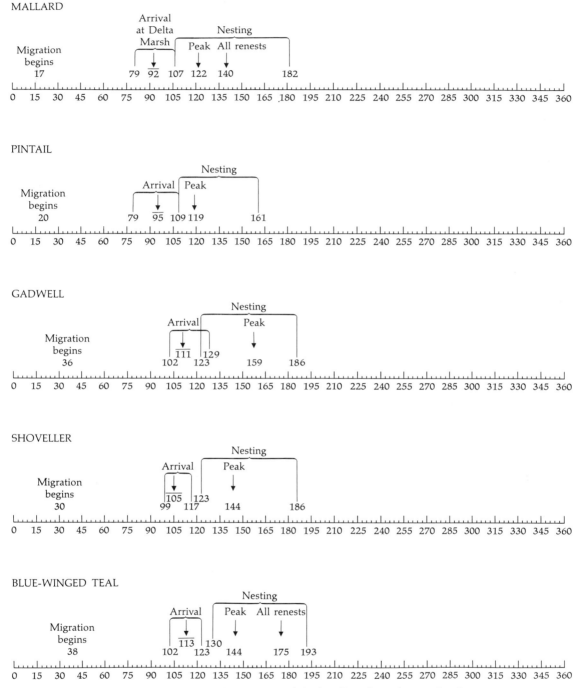

FIGURE 10-3. Chronology of events for five species of ducks. (Based on data in Sowls 1955.)

TABLE 10-2 VALUES (days) USED TO CALCULATE THE EXPECTED DATE OF OCCURRENCE OF THE BIOLOGICAL EVENTS DESCRIBED IN THE TEXT

Event	Mallard	Pintail	Gadwell	Shoveller	Blue-winged Teal
$d_{nf}{}^* = d_{am} -$	75	75	75	75	75
$d_{lb(1)} = d_{am} +$	30	24	48	39	31
$d_{le(1)} = d_{am} +$	40	33	58	51	42
$d_{ib(1)} = d_{am} +$	41	34	59	52	43
$d_{ie(1)} = d_{am} +$	62	55	80	73	64
$d_{ln(1)} = d_{am} +$	64	57	82	75	66

*A constant flight time is shown here for all species. Variations can be added for different weather conditions and other factors affecting migration time.

in Table 10-2 for the first nest. All of these are related to d_{am}, the day of arrival at the marsh, with the rationale shown below using blue-winged teal as an example.

$d_{nf} = d_{am} - 75$

d_{am} is considered the biologically significant day in the program. The northward flight from the wintering grounds to Delta Marsh is assumed to take 75 days.

$d_{lb(1)*} = d_{am} + 30$

Laying is assumed to begin 30 days after arrival at Delta. It cannot begin before arrival, and Sowls uses arbitrary dates for the latest date on which the first nests were begun by each of the five species.

$d_{le(1)} = d_{am} + 42$

The number added here is 30, taken from the preceding event, plus the number of eggs laid in the first clutch. Sowls states that one egg is laid each day until the clutch is complete.

$d_{ib(1)} = d_{am} + 43$

Incubation is assumed to begin the day after laying ends.

$d_{ie(1)} = d_{am} + 64$

For ducks, incubation ends 21 days after it begins. Published incubation periods vary from 21 to 23 days. The extra two days are considered in part in the next entry since not all ducks spend two full days at the nest after hatching.

*The 1 indicates the first nest.

$$d_{ln(1)} = d_{am} + 66$$

The ducklings leave the next two days after hatching, hence 2 days are added to d_{ie}.

$$d_{fb(1)} = d_{am} + 108$$

Forty-two days are allowed for growth after the ducklings leave the nest and begin to fly.

A renesting attempt can be added to the chronology for the blue-winged teal by using the following information:

$$d_{lb(2)*} = d_{am} + 30 + n \text{ eggs laid}$$
$$+ 5 \text{ if } d_{nd} < d_{am} + 41$$

Five days are added to the number of eggs laid before nest destruction. This is a recovery period during which the hen locates a new nest site and builds the nest.

$$d_{lb(2)} = d_{am} + (d_{nd} - d_{am}) + 5$$
$$+ \{[d_{nd} - (d_{am} + 42)] \, 0.644\}$$
$$\text{if } d_{am} + 42 > d_{nd(1)} < d_{am} + 66$$

Five days are allowed first, then a 0.644-day delay in the beginning of a renest is allowed for each day the duck was into the incubation period. This is based on a linear regression equation in Sowls.

$$d_{le(2)} = d_{nd(1)} + 5$$
$$+ \{[d_{nd} - (d_{am} + 42)] \, 0.644\} + 9$$

Nine days are added to the day on which laying begins, based on data from Sowls indicating that blue-winged teal lay 9 eggs in a renest.

$$d_{ib(2)} = d_{le} + 1$$

See $d_{ib(1)}$.

$$d_{ie(2)} + d_{le(2)} + 22$$

Twenty-two days are added to the end of egg laying, including 1 day for the interval between the end of laying and the beginning of incubation and 21 days for incubation.

$$d_{ln(2)} + d_{ie(2)} + 2$$

See $d_{ln(1)}$.

$$d_{fb(2)} + d_{ln(2)} + 42$$

See $d_{fb(1)}$.

---

*The 2 indicates the second nest.

$$d_{sf} = 244 \text{ if } d_{fb(1)} \text{ or } d_{fb(2)} < 244$$

$$d_{sf} = d_{fb(2)} \text{ if } d_{fb(2)} > 244$$

This assumes that they begin southward migration on the day that they learn to fly. This is the earliest possible date; it is likely that migration would begin after a few days of "practice."

The other four species included in the chronology of the first nesting attempt can be described for the second nesting attempt if the number of eggs in the second clutch is known. These are [based on data in Sowls (1955)]: mallard (9), pintail (9), gadwell (7), and shoveller (9).

This rather complex chronology is useful if a computing system is used to calculate biological characteristics over time. Thus to obtain information on the protein requirement of blue-winged teal on May 30, an input of day 150 (May 30, Julian calendar) into a computer program with the chronology stored in it, along with equations for calculating protein requirements for maintenance, activity, and production, will result in an output including egg-laying as a protein cost. In other words, the computer is programmed to make decisions resulting in the acceptance of equations for calculating cost items present at the time of year being analyzed and rejecting the irrelevant ones.

An interesting variation to test would be to vary the day of arrival at the marsh by intervals on either side of the average date, with concomitant variations in the span for the beginning of nesting, after which egg-laying, incubation, and brood-rearing follows at set intervals. For example, suppose nesting was delayed 30 days for blue-winged teal because of very cold, stormy weather. Other functions following the onset of nesting would also be delayed, since only 1 egg a day is laid, 21 days are required for incubation, and the growth rate of the ducklings has an upper limit. These delays may result in higher mortality should the marshes dry up while the birds are still young or should there be a large proportion of immature birds later in the fall, rendering them less capable of completing southbound migration. Thus the effect of changes in the phenology of the bird can be related to the effects of weather conditions, water levels, and other natural phenomena to develop predictive capabilities when different combinations of events occur.

The information necessary to develop chronological models of different species is available in research reports, in life-history books, and through observation in the field. The behavior of turkeys in Texas has been described by Watts and Stokes (1971), with a chronological display (Figure 10-4) similar to that shown earlier in this chapter for ducks (Figure 10-2). These events for turkeys can be quantified if additional data on clutch size, energy, protein and mineral contents of eggs, incubation period, requirements for brood-rearing, and others are determined. Some of this information is available in books such as Hewitt's *The Wild*

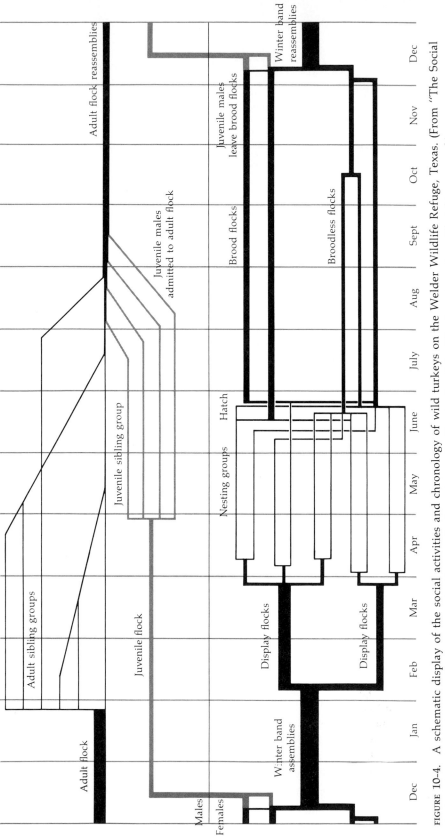

FIGURE 10-4. A schematic display of the social activities and chronology of wild turkeys on the Welder Wildlife Refuge, Texas. (From "The Social Order of Turkeys" by C. Robert Watts and Allen W. Stokes. Copyright © 1971 by Scientific American, Inc. All rights reserved.)

TABLE 10-3 YEARLY AVERAGE HATCHING DATES IN WISCONSIN—STATEWIDE AND GREEN AND MILWAUKEE COUNTIES

Year	Statewide June Avg.	No. Broods	Std. Err. (days)	Green Co. June Avg.	No. Broods	Std. Err. (days)	Milwaukee Co. June Avg.	No. Broods	Std. Err. (days)	Weighted Avg. June Avg.	No. Broods
1947	21	43	3.2	—	—	—	23	53	4.0	22	96
1948	18	60	2.4	17	56	1.6	18	78	1.6	18	194
1949	16	218	1.2	16	148	1.4	18	147	1.1	17	513
1950	21	166	1.1	24	147	1.3	24	148	1.2	23	461
1951	16	157	1.4	23	102	1.3	22	17	2.0	18	276
1952	15	238	1.0	13	112	1.3	17	60	1.8	15	410
1953	15	378	0.8	13	94	1.5	14	71	1.9	15	543
1954	20	306	0.9	21	77	1.8	21	48	1.1	20	431
1955	12	338	0.5	12	52	1.7	21	23	2.5	13	413
1956	21	261	1.0	18	29	3.1	26	8	6.2	21	298
Un-weighted mean	June 18			June 17			June 20			June 18	

SOURCE: Wagner, Besadny, and Kabat 1965.

Turkey (1967), with additional information in basic books on avian physiology [e.g., Sturkie (1965)].

Chronologies for pheasants, with variation within the earliest and latest hatching dates reported, can be established from data in Table 10-3 for Wisconsin (Wagner, Besadny, and Kabat 1965). The authors provide additional data on the observed "week of peak hatch," so variation of the data in Table 10-3 could realistically simulate previously observed conditions. The mortality of chicks in a brood is shown in Figure 10-5; note the decline in the average brood size from birth through 16 weeks. The data in Figure 10-6 show the relationship between average brood size and week of hatch; broods hatched later are smaller. Further, changes in brood size vary according to the time of hatching (Figure 10-7). Broods hatched early are larger, but they also suffer a higher mortality rate through the age of 16 weeks.

The pheasant data shown in Table 10-3 and Figures 10-5—10-7 are nonlinear over time. The technique of using first approximations that are mathematically simple can be usefully employed here. For example, the effect of hatching dates in Table 10-3 on the ecology of pheasant populations should be analyzed from June 12 through June 24, the earliest and latest yearly averages reported. The data in Figure 10-5 can be expressed in either one or three linear regression equations, calculating the change in average brood size with age. The data in Figure 10-6 can be expressed with a single linear regression equation. Figure 10-7 can be expressed with three or more simple equations approximating the data.

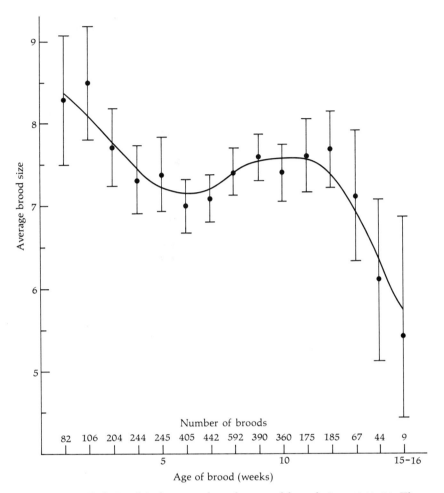

FIGURE 10-5. Relationship between brood age and brood size, 1946–56. The limits on each side of the mean represent twice the standard error of the mean. The line was drawn with three-point moving averages. (From Wagner, Besadny, and Kabat 1965.)

Once these first approximations are determined, they can be synthesized into a working model that begins with the onset of egg-laying, continuing through incubation and brood-rearing, with an analysis of brood mortality throughout the summer. After the techniques for relating these biological parameters in an ecological analysis have been worked out, the first approximations can be improved where necessary.

Many books containing life-history data of other species are available, including Taylor's *The Deer of North America* (1956), Allen's *Pheasants in North America* (1956), Mech's *The Wolves of Isle Royal* (1966), Errington's *Muskrats and Marsh Management* (1961), Peterson's *North American Moose* (1955), Jackson's *Mammals of Wisconsin* (1961), Palmer's *Fieldbook of Natural History* (1944), Leopold's *Game Management*

(1933), Jackson's *The Clever Coyote* (1951), Burrows's *Wild Fox* (1968), Stoddard's *Bobwhite Quail* (1931), Young's *The Bobcat of North America* (1958), Young's and Goldman's *The Puma, Mysterious American Cat* (1946), and many others. All of these are valuable in providing life-history information that can be expressed mathematically, including such things as reproductive phenology, growth rates, ingestion rates of different foods, activity regimes, and others.

FIGURE 10-6. Relationship between time of hatch and size of broods 4–10 weeks of age, 1946–56. The limits on each side of the mean represent twice the standard error of the mean. The line was drawn with three-point moving averages. (From Wagner, Besadny, and Kabat 1965.)

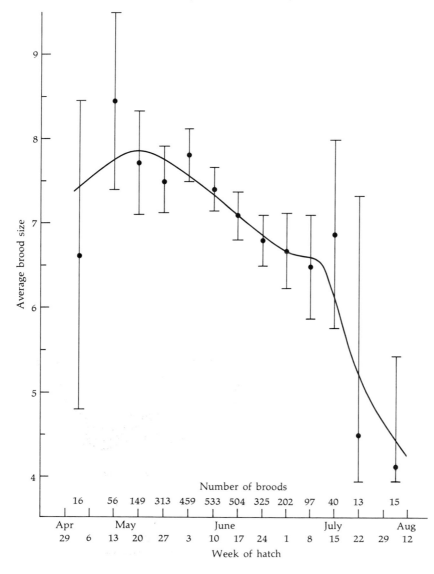

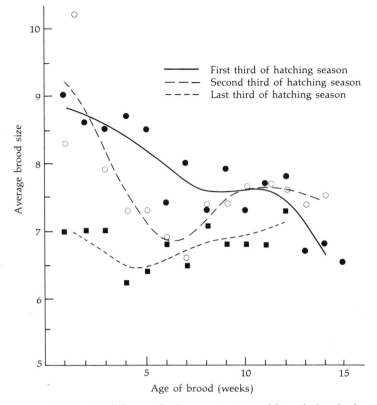

FIGURE 10-7. Week-by-week changes in sizes of broods hatched in the first, second, and last third of the hatching season. The graphs include data from all sources, 1946–56. The lines were fitted by the use of three-point moving averages. (From Wagner, Besadny, and Kabat 1965.)

LITERATURE CITED IN CHAPTER 10

Allen, D. W., ed. 1956. *Pheasants in North America.* Harrisburg, Pennsylvania: The Stackpole Co., 490 pp.

Burrows, R. 1968. *Wild fox.* Newton Abbot: David & Charles, 202 pp.

Errington, P. 1961. *Muskrats and marsh management.* Harrisburg, Pennsylvania: The Stackpole Co., 183 pp.

Hewitt, O. H., ed. 1967. *The wild turkey.* Harrisburg, Pennsylvania: The Stackpole Co., 589 pp.

Jackson, H. H. T. 1951. *The clever coyote.* Harrisburg, Pennsylvania: The Stackpole Co., 411 pp.

Jackson, H. H. T. 1961. *Mammals of Wisconsin.* Madison: University of Wisconsin Press, 504 pp.

Leopold, A. 1933. *Game Management.* New York: Scribner, 481 pp.

Mech, L. D. 1966. *The wolves of Isle Royal.* U.S. National Park Service Fauna Series 7, 210 pp.

Palmer, E. L. 1949. *Fieldbook of natural history.* New York: McGraw-Hill, 664 pp.

Peterson, R. L. 1955. *North American moose.* Toronto: Toronto University Press, 280 pp.

Sowls, L. K. 1955. *Prairie ducks.* Harrisburg, Pennsylvania: The Stackpole Co., 193 pp.

Stoddard, H. L. 1931. *The bobwhite quail; its habits, preservation, and increase.* New York: Scribner, 559 pp.

Sturkie, P. D. 1965. *Avian physiology.* Ithaca, New York: Cornell University Press, 766 pp.

Taylor, W. P., ed. 1956. *The deer of North America.* Harrisburg, Pennsylvania: The Stackpole Co., 668 pp.

Wagner, F. H., C. D. Besadny, and C. K. Kabat. 1965. *Population ecology and management of Wisconsin pheasants.* Technical Bulletin No. 34 Madison: Wisconsin Conservation Dept., 168 pp.

Watts, C. R., and A. W. Stokes. 1971. The social order of turkeys. *Sci. Am.* **224**(6): 112–118 (Offprint No. 1224).

Young, S. P., and E. A. Goldman. 1946. *The puma, mysterious American cat.* Washington, D.C.: The American Wildlife Institute, 358 pp.

Young, S. P. 1958. *The bobcat of North America.* Harrisburg, Pennsylvania: The Stackpole Co., 193 pp.

IDEAS FOR CONSIDERATION

Tabulate the life-history information about free-ranging animals in a chronological format, relating these biological functions to each other through time. Which factors are controlled internally, which are controlled externally, and how does variation in either or both of these sets affect the ecological productivity of the individual?

Prepare equations describing the cost of various activities and productive functions (see Chapter 16, for example), and calculate the specific "cost of living" throughout the animal cycle.

SELECTED REFERENCES

Baskett, T. S. 1947. Nesting and production of the ring-necked pheasant in north central Iowa. *Ecol. Monographs* **17**: 1–30.

Blood, D. A., D. R. Flook, and W. D. Wishart. 1970. Weights and growth of rocky mountain bighorn sheep in Western Alberta. *J. Wildlife Management* **34**(2): 451–455.

Cheatum, E. L., and G. H. Morton. 1946. Breeding season of white-tailed deer in New York. *J. Wildlife Management* **10**(3): 249–263.

Cowan, I. McT., and A. J. Wood. 1955. The growth rate of the black-tailed deer. *J. Wildlife Management* **19**(3): 331–336.

Cringan, A. T. 1970. Reproductive biology of ruffed grouse in southern Ontario. *J. Wildlife Management* **34**(4): 756–761.

Dodds, D. G. 1959. Feeding and growth of a captive moose calf. *J. Wildlife Management* **23**(2): 231–232.

Doutt, J. K. 1970. Weights and measurements of moose, *Alces alces shirasi. J. Mammal.* **51**(4): 808.

Dzieciolowski, R. 1969. Growth and development of red deer calves in captivity. *Acta Theriol.* **14**(10): 141–151.

Forester, D. J., and R. S. Hoffman. 1963. Growth and behavior of a captive bighorn lamb. *J. Mammal.* **44**(1): 116–118.

Gates, J. M., and E. E. Woehler. 1968. Winter weight loss related to subsequent weights and reproduction in penned pheasant hens. *J. Wildlife Management* **32**(2): 234–247.

Geist, V. 1968. On delayed social and physical maturation in mountain sheep. *Can. J. Zool.* **46**(5): 899–904.

Golley, F. B. 1957. Gestation period, breeding and fawning behavior of Columbian black-tailed deer. *J. Mammal.* **38**(1): 116–120.

Goodrum, P. D. 1972. Adult fox squirrel weights in eastern Texas. *J. Wildlife Management* **36**(1): 159–161.

Greer, K. R., and R. E. Howe. 1964. Winter weights of northern yellow-stone elk, 1961–62. *Trans. North Am. Wildlife Nat. Resources Conf.* **29**: 237–248.

Haugen, A. O. 1959. Breeding records of captive white-tailed deer in Alabama. *J. Mammal.* **40**(1): 108–113.

Illige, D. 1951. An analysis of the reproductive pattern of white tail deer in south Texas. *J. Mammal.* **23**(4): 411–421.

Johnson, D. E. 1951. Biology of the elk calf. *Cervus canadensis nelsoni. J. Wildlife Management* **15**(4): 396–410.

Johnston, D. W., and R. W. McFarlane. 1967. Migration and bioenergetics of flight in the Pacific Golen Plover. *Condor* **69**(2): 156–168.

Kirkpatrick, C. M. 1944. Body weights and organ measurements in relation to age and season in ring-necked pheasants. *Anat. Record* **89**(2): 175–194.

Klein, D. R., and H. Strandgaard. 1972. Factors affecting growth and body size of roe deer. *J. Wildlife Management* **36**(1): 64–79.

Knight, R. R. 1970. The Sun River elk herd. *Wildlife Monographs* **23**: 1–66.

Krebs, C. J., and I. McT. Cowan. 1962. Growth studies of reindeer fawns. *Can. J. Zool.* **40**(5): 863–869.

Lentfer, J. W. 1955. A two-year study of the rocky mountain goat in the crazy mountains, Montana. *J. Wildlife Management* **19**(4): 417–429.

McEwan, E. H., and A. J. Wood. 1966. Growth and development of the barren-ground caribou. I. Heart girth, hind foot length and body weight relationships. *Can. J. Zool.* **44**: 401–411.

Menaker, M. 1969. Biological clocks. *BioScience* **19**(8): 681–692.

Mitchell, G. J. 1971. Measurements, weights and carcass yields of pronghorns in Alberta. *J. Wildlife Management* **35**(1): 76–85.

O'Gara, B. W. 1970. Derivation of whole weights for the pronghorn. *J. Wildlife Management* **34**(2): 470–472.

Quay, W. B., and D. Müller-Schwarze. 1971. Relations of age and sex to integumentary glandular regions in Rocky Mountain Mule Deer (*Odocoileus hemionus hemionus*). *J. Mammal.* **52**(4): 670–685.

Ransom, A. B. 1966. Breeding seasons of white-tailed deer in Manitoba. *Can. J. Zool.* **44**(1): 59–62.

Ransom, A. B. 1967. Reproductive biology of white-tailed deer in Manitoba. *J. Wildlife Management* **31**(1): 114–123.

Raveling, D. G. 1968. Weights of *Branta canadensis interior* during winter. *J. Wildlife Management* **32**(2): 412–414.

Robinette, W. L., and J. S. Gashwiler, 1950. Breeding season, productivity, and fawning period of the mule deer in Utah. *J. Wildlife Management* **14**(4): 457–469.

Sadleir, R. M. F. S. 1969. *The ecology of reproduction in wild and domestic mammals.* London: Methuen, 321 pp.

Short, H. L., and W. B. Duke. 1971. Seasonal food consumption and body weight of captive tree squirrels. *J. Wildlife Management* **35**(3): 435–439.

Stephensen, S. K. 1962. Growth measurements and their biological interpretation of mammalian growth. *Nature* **196:** 1070–1074.

Thompson, D. R., and R. D. Taber. 1948. Reference tables for dating events in nesting of ring-necked pheasants, Bobwhite Quail, and Hungarian Partridge by aging of broods. *J. Wildlife Management* **12**(1): 14–19.

Thompson, D. R., and C. Kabat. 1949. Hatching dates of quail in Wisconsin. *J. Wildlife Management* **13**(2): 231–233.

Verme, L. J. 1970. Some characteristics of captive Michigan moose. *J. Mammal.* **51**(2): 403–405.

Wilson, P. N., and D. F. Osbourn. 1960. Compensatory growth after undernutrition in mammals and birds. *Biol. Rev.* **35**(3): 324–363.

Wood, A. J., I. McT. Cowan, and H. Nordan. 1962. Periodicity of growth in ungulates as shown by deer of the genus *Odocoileus*. *Can. J. Zool.* **40:** 593–603.

INTRASPECIES INTERACTION

A social structure within a population cannot exist without some form of communication between organisms. The ability to communicate depends on both the signal that is sent by one organism and the receiving ability of another organism. This has been discussed in Chapter 2, with the term "operational environment" applied to the stimulus-response combination.

11-1 SENSORY PERCEPTION

The sensory perceptions of animals include sight, hearing, smell, taste, touch, and thermal perceptions. The neurological perception levels are very difficult to quantify, especially for taste and smell. There really is no way to quantify the strength of these associations directly, although the responses of animals to such things as chemicals can be tabulated. Sight and hearing are more easily quantified since light and sound energy can be described in terms of wavelengths, frequencies, loudness, and other units.

There is general agreement among natural-history writers that most wild animals have good vision. Most appear to exhibit some difficulty in detecting the presence of motionless objects, but moving objects are usually detected very easily. The lack of ability to detect motionless objects is partly credited to the apparent lack of color vision. In the absence of color there are fewer stimuli that help to distinguish an object from its surroundings. In analyzing sight capabilities, careful distinction must be made between what an animal can see and what the animal sees and responds to since visual detection may occur without an observed behavioral response.

Many wild animals are active at night, and it appears that their night vision is considerably better than that of a human. Deer tend to feed in the early morning and late evening, indicating that activity periods transcend changing light conditions. Their readiness to bound off a road into a stand of trees at night indicates some ability to see in very dim light; however, inasmuch as humans cannot see the animals as they move through the forest at night, it is difficult to determine the number of collisions with branches and other objects.

Some wild ruminants can apparently see objects at greater distances than others. Pronghorns, bighorn sheep, and mountain goats seem to possess keen vision for distance. Their habitat is generally more open, and vision for long distances is useful for increasing their ability to detect danger. Their visual capabilities may be an indicator of evolutionary development that relates visual ability to the selection of certain habitat characteristics.

Caribou spend much of their time in rather open country, but their eyesight is considered inferior to that of most game animals (Kelsall 1957). They have good ability to detect moving objects, however, and Kelsall reports instances in which caribou bands have fled when a single person was walking at distances of more than one-half mile from the band.

Deer have a keen sense of hearing. Many accounts are recorded in the literature that attribute an almost unbelievable hearing ability to deer. There are strong indications that deer distinguish between noises that are a regular part of their habitat and those that are a bit different from the usual. My personal observations have indicated that the chattering of squirrels, the noise of the wind, the creaking of branches, and the like are accepted by deer. Distant gun shots do not cause concern. A sharp snap of a twig, however, makes them instantly alert.

The mobility of the ears is a striking characteristic of deer. The direction in which either ear is pointed can be controlled independently of the other. When both ears are directed toward a sound, their ability to hear is enhanced by the directional effect of ear orientation.[1]

The bugling of elk is a form of sound communication that is a fairly common practice among elk, even by cows at about the time of parturition (Murie 1951). Bugling by male elk is usually thought of as a challenge to a rival bull. Murie suggests that the rivalry may be exhibited in more meaningful ways than bugling. Male elk will bugle early in the rutting season with no effect on other male companions. This illustrates how the importance of a biological characteristic is dependent not only on the stimulus but also on the condition of the receiver.

Another means of communication between individual animals is touch. The muzzle and tongue are used to make contact with different parts of another animal, which is part of the behavior pattern of both young and mature animals. The mouth is also sensitive to the mechanical characteristics of forages. The hair coat is sensitive to pressure; the skin twitches to rid itself of flies. Peterson (1955) suggests that touch might enhance the ability of moose to locate submerged

---

[1]Notice the orientation of the ears of the deer in the photograph facing page 1. The ears are oriented forward in some of the deer and backward in the others.

vegetation: ". . . moose were observed to swim out to deep water, then suddenly dive for a few seconds, perhaps become completely submerged, and come up with a mouthful of pond weed (*Potamogeten* sp.)." He speculates that moose can feel the plants with their legs.

The sense of touch may be particularly important for animals that are born blind. Members of the cat family, for example, live in a world of sounds, odors, and tactile reflexes before their eyes open. The cuddling of human infants in the first few days of life is important for their development.

Observation of the sensory capabilities of moose have led to a generalization that may apply to most big-game animals. The ears often serve to alert an animal, the eyes are used to investigate the disturbance, and the sense of smell is the most important means of detection (Peterson 1955). Severinghaus and Cheatum (1956) state it another way: "Deer depend first on scent, second on hearing, and last on sight."

The importance of smell is indicated by many writers, and it is a consideration of every big-game hunter. The olfactory capabilities of wild ruminants may seem relatively greater, to an observer, than they actually are, because the sense of smell is so poorly developed in man. Nevertheless, the ability of ruminants, as well as that of most other wild mammals, to detect odors is particularly well developed.

Kelsall (1957) has observed deer, moose, bear, and caribou as they were catching a human scent. The first three always reacted as if the scent of man meant danger, but the caribou seem to doubt the evidence of scent. Although deer have been reported to move downwind from the source of a scent in order to confirm it, Kelsall has observed caribou moving downwind and toward the source as if to confirm it by sight. He further states that large herds seem less concerned than small herds when the scent of man is present.

One indication of the importance of scent in the activity of an animal is the relative flight distance when a stimulus is received by different senses. Peterson (1955) relates the observation of Sheldon that "After being frightened by scent, moose go much farther without stopping than when frightened by sight." The flight behavior of big game is discussed in Chapter 12.

A recent study of pheromones in black-tailed deer (*Odocoileus hemionus columbianus*) by Müller-Schwarze (1971) has revealed a number of distinct functions related to the various glands. The pathways taken are shown in Figure 11-1. Note how the pheromones are transferred from one part to another of a single deer (hindleg rubbed on forehead), from a deer to another object such as a branch, and from one deer to another through the air.

Four scents are considered important by Müller-Schwarze in social communication of the black-tailed deer. The *tarsal scent* serves to identify sex, age, or type of an individual at a close distance. The *metatarsal scent* acts like an alarm pheromone over moderately large distances. When a deer rubs its head on branches, the scent of the forehead glands left on them marks home range. The interdigital gland leaves scent on the ground, although infrequent responses were observed in the black-tailed deer.

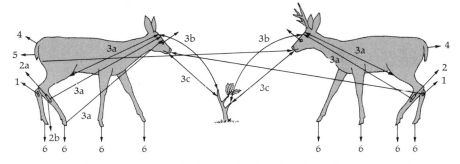

FIGURE 11-1. Pathways of social odors in black-tailed deer. Scents of tarsal organ (1), metatarsal gland (2a), tail (4) and urine (5) are transmitted through air. When the deer is reclining, the metatarsal gland touches ground (2b). The deer rubs hindleg over forehead (3a). Marked twigs are sniffed and liked (3c). Interdigital glands leave scent on ground (6). (From Müller-Schwarze 1971.)

Urine also provides olfactory information. Fawns urinate in their beds, and Müller-Schwarze observed that they did not use the same bed site again if the urine had been disturbed. Black-tailed fawns just two days old urinate on their hocks while rubbing their legs together; this habit continues through adulthood. It has also been observed in white-tailed deer. Black-tailed orphan fawns in captivity rub-urinated four times in 100 hours of observation, though none were observed by Müller-Schwarze in 300 hours of observation in the wild. Rub-urinating by a captive fawn living with its mother attracted the mother, indicating that it may be related to the separation of the doe and fawn. In adults, it appeared to serve to increase the distance between individuals.

Aggressiveness between males of many species of both birds and mammals can be observed. Pheasant (*Phasianus* sp.) males defend a "territory" against intrusion by other cocks. The females cross territorial lines and are courted by the resident male. No attempt seems to be made to herd the females or to retain a harem.

Ducks and geese establish territories during the breeding season too. Sowls (1955) discusses the home-range and territorial characteristics of several species. Home range is described simply as familiar area used by the birds. The defended territory is used exclusively by a single pair of one species. Hochbaum (1944) describes the territory as a place in which the paired drake and hen may be found day after day. It contains water, a loafing spot, nesting cover (adjacent or nearby), and food, and it is defended by the drake from intrusion by other sexually active birds of his own species. These territories are not strict geographical entities, however, inasmuch as they vary in size, they overlap, and there are differences in the degree of territoriality among birds.

Some species exhibit a social hierarchy within a group rather than a territorial separation of its members. The social order of Rio Grande turkeys (*Meleagris gallopavo*) on the Welder Wildlife Refuge near Corpus Christi, Texas, has been described by Watts and Stokes (1971). In the winter the males form flocks, in which the social status of the male is determined. Each male fights to establish

FIGURE 11-2.   Wrestling match between juvenile wild turkeys, who are members of the same sibling group, is one of several forms of combat that eventually determine which male will dominate the other members of the group. The birds usually fight until exhausted. Dominant males at the Welder Wildlife Refuge in Texas act as sires in the great majority of annual matings among the resident turkeys. (From "The Social Order of Turkeys" by C. Robert Watts and Allen W. Stokes. Copyright © 1971 by Scientific American, Inc. All rights reserved.)

his position among his siblings first (Figure 11-2); then the entire group fights other sibling groups to determine its position among them. The winter flocks of males begin breaking up in February, with groups of sibling males courting the females. The dominant group moves about within the ranks of the females, and the subordinate groups follow along on the periphery of the female group. The dominant male in the dominant sibling group does nearly all of the mating with the hens! A complete copulation takes about four minutes, and subdominant males do not have time to fulfill the mating attempt before they are detected by the dominant male, who drives them off and completes the mating with the hen himself.

Male ruminants are generally aggressive during the breeding season. The presence of horns or antlers makes them potentially dangerous not only to man or to some other animal, but also to both male and female members of his own species. Rutting behavior includes two kinds of associations: (1) the aggressive association between males and (2) the attraction between male and female during ovulation. The strength of these associations is marked, resulting in stronger interactions between individuals during the mating season than at any other time of the year.

One of the secondary sexual characteristics of members of the family *Cervidae* is the seasonal growth and shedding of antlers. Growth begins in later winter or early spring and accelerates during the summer. By August or September the velvet-covered antlers harden and the first signs of aggressive behavior begin to appear. Much has been written over the years about the "rubbing off of velvet," and the usual implication is that this is done in order to shed the velvet. Actually, the velvet on the antlers of white-tailed deer is shed in a matter of hours (Figure 11-3), and may not be accompanied by rubbing. The mutilation of small trees is an indication of an increase in aggressiveness that begins as the reproductive condition develops. Severinghaus and Cheatum (1956) reported that captive deer at the Delmar Laboratory exhibited aggressive behavior toward fences, trees, or any other resistant object as the neck muscles harden. This also serves to mark the territory, which Müller-Schwarze has described for black-tailed deer.

During the rut, male deer also paw and dig up a circular patch of ground a few feet in diameter, which results in the formation of a "scrape." These scrapes may also have some value as markers for a buck's territory, or at least for informing the other members of the population, both male and female, that an

FIGURE 11-3.   Male white-tailed deer with velvet coming off.

area is being used by the buck. Bronson (cited by Severinghaus and Cheatum 1956) reports that white-tailed bucks urinate in the pawed-up ground and roll or wallow in it, leaving scent indications as well.

The scrapes or wallows of white-tailed bucks have their counterparts in the wallows made by moose and elk bulls. Peterson (1955) records accounts of natural historians who have observed moose bulls dig a wallow with their front feet, urinate in it, and roll in it. Bull elk also dig wallows, using their antlers as well as their front feet to tear up the sod. Bugling may occur while the bull is lying in a wallow (Murie 1951).

Male ruminants are generally polygamous, servicing as many females as possible during the breeding season. The sexual readiness of the male coincides with that of the female, although the latter has an estrous cycle that results in a readiness to stand at intervals of several days until fertilization occurs. Limits imposed on the number that can be serviced include both the number of receptive females and the energy of the male for repeated copulation. Observations of penned deer at Delmar, New York, and at the BioThermal Laboratory at Cornell University show that the female will stand for more than one buck. As a male becomes spent from repeated servicing, a second male will copulate successfully. These observations may not be strictly applicable to wild conditions, but they do indicate the potential of deer for repeated copulation.

Martinka (1969) describes the social relationships of elk from calving through the end of the breeding season, including calving in the first part of June, aggregation for two months following calving, dispersal prior to the breeding season, breeding in September and October, and reaggregation following breeding and on through the winter. Bull elk maintain control of a harem of cows during the breeding season. This social organization may not result in highly organized reproductive behavior, however, since subdominant bulls may breed with the cows when the harem bull is occupied with defensive behavior.

The approach of parturition does not seem to trigger special preparations by the white-tailed doe. Severinghaus and Cheatum (1956) describe several types of parturition sites that have been observed, including open fields, snow patches, and nicely sheltered insect-free sites. It would be easy to become anthropomorphistic about the care with which a doe selects a place for giving birth, but it is apparently a matter of convenience, occurring wherever the mother happens to be at the time.

Barren-ground caribou cows seem to move with a sense of purpose prior to calving (Kelsall 1957). As calving time approaches they assume the leadership of the spring migration. When they reach the calving grounds, they tend to scatter widely and move slowly. The geographic location of calving grounds seems to vary, but it is evident that certain calving grounds are favored year after year even by widely separated herds.

The parental care of a fawn or calf begins immediately after birth. The mother licks the newborn, and although this is often said to be a cleaning process that may have nutritional benefits for the mother, the tactile stimuli very likely have an influence on the imprinting process. Not only does the newborn see its mother, but to feel the mother's gentle licking reinforces the relationship.

Barren-ground caribou calves are considered more precocious than the young of other ungulates, inasmuch as they may walk and run for several miles within an hour and one-half after birth (Kelsall 1957). Only two calves, each less than 4 hours old, were captured out of many hundreds seen, according to Kelsall. He also reports that calves less than 12 hours old will jump off an ice shelf and swim easily.

An interesting observation reported by Kelsall indicates the strength of the imprinting instinct in the caribou. The birth of a calf was observed and the calf was ear-tagged a short time after birth. The cow fled as the observers approached the calf. After handling the calf, the observers could not get away without the calf's following them. It was necessary to roll it on its side behind a boulder and dash away. The cow returned later and located the calf, apparently by accident. This account is similar to the explanation sometimes given by persons who have picked up white-tailed deer fawns that "the fawn followed us because it was lost." It is very likely a case of successful imprinting shortly after birth.

Imprinting has been observed in birds, too. It was recognized by Lorenz (1937) when he demonstrated that young geese and ducks become attached to whatever living creature they are first exposed to. Since that time, experiments have shown that ducks can be successfully imprinted to other species, man, and even mechanical objects (Figure 11-4).

11-2 PROTECTIVE BEHAVIOR AND FAMILY TIES

The protective instincts of the ruminant mother have been observed by many naturalists. In general, their accounts indicate that the female will actively defend her young against aggressors, whether of the same species or of another species. Peterson (1955), for example, relates an account of a bull moose coming between a cow and her calf, only to be driven off by the cow. Moose have been known to charge humans that were disturbing their young, but deer, elk, and caribou tend to step aside when human intruders appear.

The forefeet are often used for defense, and a blow from the hooves of deer or elk can cause serious injury. Several accounts of dogs having been killed by the hooves of elk are recorded in Murie (1951). Severinghaus and Cheatum (1956)

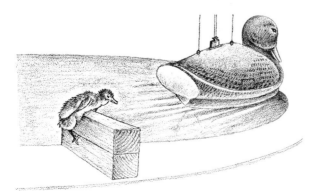

FIGURE 11-4. Ducklings can be imprinted to mechanical objects. (From "'Imprinting' in Animals" by E. H. Hess. Copyright © 1958 by Scientific American, Inc. All rights reserved.)

report that a Conservation Department employee observed a white-tailed doe chasing a red fox, striking at it with its front hooves as it ran.

The strength of family ties in wild ruminants is difficult to assess because of the few instances in which several generations have been tagged for field identification. It appears to be generally true that males are seldom a part of a family group; the bond between mother and young is much stronger. Further, the association between mother and young deteriorates during the breeding season. Nursing usually stops altogether, although the young may remain in close proximity to the mother. Moose cows seem to tolerate the presence of their previous young for a longer time than deer do, the young remaining until driven out by the pregnant moose cow just prior to calving.

The longer period of social attachment between mother and young for moose compared with deer is probably a reflection of the earlier breeding age of deer. Female white-tailed fawns can breed at the age of six months if they have been on a high nutritional plane, so they are capable of giving birth to their first young at the age of about one year. Moose do not reach sexual maturity until they are over a year old, and many do not breed until they are over two years old. This is also true for elk. Thus the completion of the weaning process, both nutritionally and socially, seems to be related to the sexual development of the young, but it is influenced to some extent by the reproductive condition of the mother.

11-3 MOVEMENT PATTERNS

The home range of any wild animal depends both on its behavioral characteristics, especially toward members of its own species, and on the physical and chemical characteristics of the habitat. Behavioral characteristics are difficult to quantify. Their effects can be measured, and some of them may be related to the density of the population. Seasonal changes occur in the reproductive condition of the animals and in the physical and chemical characteristics of the habitat.

The sizes of the home ranges of different wild ruminants vary widely. Some ruminants are migratory, while others remain in the same general area throughout the year or even for their entire lives. White-tailed deer seem to have one of the smallest home ranges of all the wild ruminants. The summer range of each individual is about one square mile, with little tendency for the animals to group together at that time. The doe and her fawn(s) form a family unit that is very close-knit both socially and nutritionally. They remain in close nutritional association until weaning and in social association until parturition approaches. Since the young of the previous year may have their own fawns if the range is in good condition, the tendency for separation of the doe and fawn is reinforced by parental instincts in both age groups.

Male white-tailed deer are quiescent during the summer. The antlers are developing, and the males seem content to forage, rest, and escape the flies and other insects that can be very bothersome. Relatively few observations of bucks in velvet are reported; they seem to be quite inconspicuous.

Seasonal movements of white-tailed deer vary considerably. Deer tagged in Minnesota have been relocated forty miles from the original site of tagging

(Blankenship 1957). Hawkins, Klimstra, and Autry (1971) observed that yearling bucks dispersed most widely from original trapping sites, with November being the principal time of dispersal.

White-tailed deer sometimes move several miles to a winter area of concentration. The triggering mechanism for the gregarious behavior is unknown. It is possible that reduced visibility causes deer to gather into groups (Moen 1966). Observers in prairie and agricultural habitats report that winter herds did form in certain areas, but up to 30% of the population remained dispersed throughout the winter (Sparrowe and Springer 1970). Some observers relate the yarding instinct to cold weather, high winds, or snow. The number of deer in a winter concentration area seems to be related to snow conditions inasmuch as deer may disperse during the winter if snow conditions do not inhibit travel. Large deer herds can be seen in late winter in many areas.

The daily activity patterns of white-tailed deer in Illinois were studied with the aid of radio telemetry (Montgomery 1963). The deer were active for 1 to 2 hours before sunset and continued to be active for an hour or so after sunset. Progressively more deer bedded until the peak of bedding occurred just before dawn in the summer. In the winter, with long nights, the peak of bedding occurred about 4 hours after sunset, following the activity period, with another bedding period before dawn. Undisturbed deer frequently bedded near their food supply. This was also observed by the author in Western Minnesota [(Moen 1966) and unpublished data] where the deer bedded in open fields near standing corn. Subzero temperatures did not cause the animals to bed in more sheltered areas.

Mule deer migrate between summer and winter ranges. The fall migration from the summer range at high elevations to the winter range at lower elevations seems to be coupled with a combination of weather factors. Storms and an accumulation of snow seem to be closely associated with the fall migration in some areas. The spring migration to higher elevations follows the snowline.

Elk are migratory and have a tendency to herd. Annual behavior patterns vary from the calving period when the cows are more or less scattered, through summer groups that become progressively larger and have a greater variety of individuals as the summer progresses, fall groups that are partly determined by the ability of the bull to maintain a harem, and winter herds that can include several hundred animals or more, depending on the winter range. The home range occupied by these different groups changes with time. In general, the area covered by elk is larger than that used by white-tailed or mule deer. This is reasonable because both the individual animal and the herd size is generally larger for elk.

The size of the home range of moose has been discussed by several writers and summarized in Peterson (1955). He suggests that an individual animal will likely remain within a radius of 2 to 10 miles for an entire lifetime, but that ecological conditions have an effect on the actual size of the home range for a given individual. It is fairly clear that moose are not migratory, and there seems to be little indication of territoriality. The fact that the animal has a large home range reduces the need for territorial instincts since the number of contacts with other members of the same species is small.

The widest-ranging and most gregarious of the wild ruminants is the barren-ground caribou. Caribou use forested winter ranges and barren-ground summer ranges. Kelsall (1960) describes the movement of one large herd over a period of 17 months. The daily and seasonal movements of the herd cover many miles, with individuals in a constant state of flux as they assemble, disperse, and reassemble in different locations with different individuals. Individual animals may shift from one herd to another as they travel, with some of the separations caused by reproductive conditions. Nonherding females and the males may travel farther during the calving season, with animals of all ages rejoining in herds for the autumn migration.

Snow seems to have a major influence on the movements of caribou. The animals move from areas of high snow density to low snow density, from high snow hardness to low snow hardness, and from greater to lesser snow depths. Wind is of considerable importance when insects bother the animals; Kelsall noted that the caribou took advantage of the wind to escape harassment. After the insect season, a pronounced dispersal was observed, with the herd spread over an area of 50,000 square miles in August. This dispersal included calves as well. Family ties were broken and new herds were formed as the animals herded up for movements to the winter range.

11-4 FEEDING BEHAVIOR

Wild ruminants forage for their food, combining this behavior with an alertness that is necessary for survival in the wild. Since they can feed at will, they do not reach the appetite levels of domestic ruminants from which feed is withheld until a certain time. However, daily feeding patterns for wild ruminants are fairly regular. Generally speaking, the animals feed actively in the early morning and in the evening. Feeding periods for white-tailed deer have been observed at noon or shortly thereafter, and there are also one or two feeding periods during the night.

Wild ruminants do not seem to be dependent on the availability of open water at all times. Snow is used as a substitute in the winter. During spring and summer, ingested forage is usually succulent and water requirements are partially satisfied by succulent vegetation. In addition, the vegetation is often covered by dew. The drinking periods seem to coincide generally with the daily feeding and activity cycle.

After eating and drinking, a ruminant usually lies down and ruminates. The food is regurgitated and chewed over again while the animal remains alert. Within groups of white-tailed deer, individuals usually lie in different orientations so that the group as a whole can see in all directions.

11-5 REST

The question of whether wild ruminants sleep has been debated often, but the answer seems fairly clear if the problem of semantics is avoided. If sleep is considered to be a resting mode with a marked reduction in alertness, then white-tailed deer "sleep." This conclusion is based on my own field observations in which deer were seen placing their heads alongside their bodies and remaining unalert. Measurements of physiological parameters by telemetry have shown that the deer do indeed become unalert and sleep, especially in cold weather.

Murie (1951) describes an instance in which he walked up on a male Alaskan caribou that was resting quietly, its head drooping lower and lower until an antler rested on the ground. On another occasion a sleeping female caribou was actually captured and tied with a rope. He did not observe sleep in elk, but suggests that it probably occurs.

11-6 PLAY

Young animals play at times; this seems to be a normal part of the life of wild ruminants. Play activity in black-tailed deer was observed to decrease as the density of the population increased (Dasmann and Taber 1956). This may be an indication of social stress within the population. White-tailed fawns at the Bio-Thermal laboratory often start running in late afternoon, tearing around recklessly until they are panting heavily. Collisions do occur, but their agility is remarkable. Murie (1951) describes play activity in elk. A shallow pond is enjoyed by all ages, and erratic running has been observed in elk cows as well as calves.

11-7 SOCIAL ORDER

Social order in wild populations has been recognized by behaviorists for some time. The article by Watts and Stokes (1971) on wild turkeys in Texas is a lucid account of a very rigid social structure. The social structure of the Rio Grande turkey at the Welder Refuge is more rigid than that of the Eastern wild turkey inhabiting the Atlantic coastal states. Behavioral differences may be attributed to habitat differences; the authors suggest that habits of game birds of North America indicate that woodland species such as the ruffed grouse and spruce grouse are widely dispersed except during the mating season, while grassland brush inhabitants such as the prairie chicken, sharp-tailed grouse, and sage grouse live in large flocks. "The Welder turkeys . . . follow the grassland pattern of social organization, whereas the Eastern wild turkey, living in woodlands, favors small social units." Welty (1962) discusses various aspects of social behavior in birds, including songs, territoriality, and reproductive behavior. Social order in ruminants is well defined, particularly at certain times of the year.

DOMINANCE PATTERNS.   The social structure of a deer herd is generally dominated by the adult doe. Family groups at the Crab Orchard National Wildlife Refuge are generally matriarchal, with the most common group consisting of an adult doe, her yearling daughter, and two fawns belonging to the older doe (Hawkins and Klimstra 1970). Doe dominance at feeding sites has been observed many times by several investigators, and often a larger number of does are trapped during winter trapping operations because the dominant doe moves into the trap first to feed. A study in Maine, designed to expose the effects of different types of cover on the physical condition of deer, showed that the final condition of an individual deer seemed related to its position of dominance in the pen.

There are seasonal differences in the dominance pattern. Adult bucks exert a physical dominance during the breeding season. Male deer tend to form groups after the breeding season is over. This is apparently related to the reduction in the reproductive condition of the deer. It does not appear to be related to the time of the antler drop. Males in western Minnesota held their antlers well into March, but groups of 3 to 7 were common from January through the rest of the winter (Moen, unpublished data).

Elk tend to be more gregarious, and the social structure of a herd does not seem to be quite as separated as that of deer. The bulls assume a position of dominance during the breeding season, at least within their own sex group. Moose are quite independent and have little tendency to congregate. The size of the animal, with its concomitant high requirement for food, precludes the possibility of herding unless the animal migrates to new feeding areas as elk do.

Deer fawns are often thought of as physiologically inadequate to survive severe winters, but this may be due to their subdominant behavioral position. Verme (1965) states that ". . . exposure to icy gales can be fatal to nearly famished, completely unprotected fawns. A similar exposure of well-fed animals had little effect on their physical condition." He also states that deer often behave as though inadequate protection from bad weather is a greater peril than a lack of proper food. These observed responses of deer suggest that a very basic analysis of their energy exchange, such as that described in Chapters 6, 13, and 14, may provide a physiological basis for analyzing deer behavior in the winter.

11-8 RADIO TELEMETRY AND BEHAVIORAL ANALYSES

The description of behavioral characteristics has been greatly enhanced in the last decade by the use of radio telemetry. One of the earliest reports on the use of radio transmitters to locate free-ranging animals (woodchucks, *Marmota monox,* in this case) was published in 1958 by Le Munyan et al. Marshall (1962) used a radio-positioning technique for determining certain aspects of the activities of porcupines (*Erethizon dorsatum*). Both daytime and nighttime activities were observed using radio telemetry. Marshall and Kupa (1963) discuss the development of radio-telemetry techniques for ruffed grouse (*Bonasa umbellus*) studies, pointing out the potential of this approach for the analysis of an animal's reactions to weather conditions, food supplies, cover conditions, and other ecological considerations. Additional accounts of early work in bio-telemetry may be found in Slater (1963); species studied and described there include grizzly bears, woodchucks, porcupines, ruffed grouse, fish, monkeys, birds, insects, reptiles, and others. Additional references may be found in *BioScience* (1965; vol. 15).

An automatic radio-tracking system was developed at the Cedar Creek Natural History Area, University of Minnesota (Cochran et al. 1965), permitting continual surveillance of the locations of animals. Signal characteristics can be interpreted to reveal something about their activity and behavior. This was used to advantage by Marshall (1965) when the movements of ruffed grouse caused the antenna to "whip," which resulted in corresponding changes in pitch in the audio signal. Observations of birds carrying transmitters permitted the field biologists on the project to equate signal variations with resting, walking, running, flying, feeding, and drumming activities. Loop antennae in collars show less signal variation than whip antennae, but many activities can still be determined from signal characteristics.

Radio telemetry systems can be designed to perform specific functions, including positioning by triangulation (Cochran et al. 1965), activity determination by signal interpretation, bioelectric potentials such as electrocardiograms and electroencephalograms (Mackay 1968), and physical factors associated with the animals such as sounds, light intensity, pressure, and many others.

Commercial instrumentation designed for the transmission of physiological data from humans can often be adapted to wild species. Equipment is currently being used on white-tailed deer to transmit heart rates and breathing rates. Close attention is paid to the activity and behavior of the deer during tests so that the signal can be interpreted properly (Moen and Jacobsen, unpublished data).

The suggestion that radio telemetry equipment be used in an ecological study often leads to the conclusion that the experimental work is now easier and less demanding inasmuch as the investigator can sit back and receive the data by remote means. Actually, quite the opposite is true. Radio telemetry permits much more efficient work in the field by providing a clue to an animal's location or physiological condition. Observations become much more meaningful, and the time spent in the field increases rather than decreases. So many new insights can be gained from field work with the *aid* of radio telemetry that the investigator tends to enjoy it much more, too!

A good example of the usefulness of radio telemetry is shown by the work of Mech et al. (1971) on the timber wolf. This wide-ranging animal is difficult to study because of its habits and life-style. Very fruitful studies of wolves have been conducted at Isle Royal National Park where the island naturally limits their range. On the mainland, radio telemetry becomes even more useful because of the greater possible variation in movement.

Their study in Minnesota combined visual observation with radio-locations. Visual sightings did not always result in day-to-day identification because most of the wolves were the same color. Five wolves carried transmitters emitting different frequencies, permitting individual recognition. Directional receiving antennae were attached to the wing struts of an airplane, permitting the investigator to locate the wolf by maneuvering the plane within visual sighting distance. The five wolves were located within time spans of 47 to 84 days before technical problems resulted in loss of signal.

Between 9:00 A.M. and 6:00 P.M., the wolves spent 62% of the time resting, 28% traveling, and 10% feeding. Resting was observed most often about noon. The wolves ranged several miles on most days, with the longest recorded straight-line distance between two locations being 12.8 miles. This is an underestimation of the actual distance traveled, of course. The winter range of the five wolves bearing transmitters varied considerably. Some stayed in a relatively small area most of the winter while others moved to different areas. About 8% of the wolves were observed alone, with 92% of them sighted in a group.

The use of radio telemetry with aircraft tracking and observation is efficient if an animal such as a wolf, which may travel several miles a day is being studied. The technique is obviously limited to daylight hours, and it is also expensive. The cost per unit of information gained may be low, however.

Another study using radio telemetry techniques is described in Schneider, Mech, and Tester (1971), who used an automatic radio tracking system (Cochran et al. 1965) for tracking racoons (*Procyon lotor*). This resulted in a significant amount of information that could not be obtained by direct observation since the racoon is chiefly a nocturnal animal. The activity of the racoon was divided into three major phases: (1) from the end of winter dormancy until parturition; (2) from parturition until the cubs could travel with their mother; and (3) from the beginning of cub travel until winter denning. A fourth phase, covering the denning period, should also be recognized.

The telemetry technique was particularly successful in the racoon study because the young could be radio tagged, resulting in information on the social characteristics of the family group. An interesting aspect was the presence of temporary cub-cub and adult-cub liaisons that were repeatedly formed and dissolved. As the time for winter denning approached, however, the strength of the family bond began to increase. Family members bedded together more often, and in the last few days before denning they moved as a family unit. The entire family denned for the winter in the same tree or in a group of trees that were close together.

Recent advances in instrumentation make it possible to study free-ranging animals in much more detail than before. The student of analytical ecology should not feel that the use of instruments is necessary for successful ecological analyses, however. The significant part of any scientific investigation is the analysis of data and the use of the data in discovering ecological interactions. Work out these relationships through the use of simple models that depict the roles of several principal characters in the ecological theatre, whether they are insects, song birds, predators, game birds, or any other organism. As these relationships are worked out within the fundamental framework of matter and energy interactions, more use can be made of the ecological details that can be obtained with the aid of advanced instrumentation.

LITERATURE CITED IN CHAPTER 11

Blankenship, L. H. 1957. Deer management includes tagging. *Conserv. Volunteer* **20**(119): 53–55.

Cochran, W. W., D. W. Warner, J. R. Tester, and V. B. Keuechle. 1965. Automatic radio-tracking system for monitoring animal movements. *BioScience* **15**: 98–100.

Dasmann, R. F., and R. D. Taber. 1956. Determining structure in Columbian black-tailed deer populations. *J. Wildlife Management* **20**(1): 78–83.

Hawkins, R. E., and W. D. Klimstra. 1970. A preliminary study of the social organization of white-tailed deer. *J. Wildlife Management* **34**(2): 407–419.

Hawkins, R. E., W. D. Klimstra, and D. C. Autry. 1971. Dispersal of deer from Crab Orchard National Wildlife Refuge. *J. Wildlife Management* **35**(2): 216–220.

Hochbaum, H. A. 1944. *The canvasback on a prairie marsh*. Washington, D. C.: American Wildlife Institute, 201 pp.

Kelsall, J. P. 1957. Continued barren-ground caribou studies. Wildlife Management Bulletin. Series 1. No. 12. *Can. Dept. Northern Affairs Natl. Resources, Natl. Parks Branch, Can. Wildlife Serv.* 148 pp.

Kelsall, J. P. 1960. Cooperative studies of barren-ground caribou, 1957–1958. Wildlife Management Bulletin. Series 1. No. 15. *Can. Dept. Northern Affairs Natl. Resources, Natl. Parks Branch, Can. Wildlife Serv.* 145 pp.

Lorenz, K. Z. 1937. The companion in the bird's world. *Auk* **54**: 245–273.

LeMunyan, C. D., W. White, E. Nyberg, and J. J. Christian. 1959. Design of a miniature radio transmitter for use in animal studies. *J. Wildlife Management* **23**(1): 107–110.

Mackay, R. S. 1968. *Bio-medical telemetry*. New York: Wiley, 388 pp.

Marshall, W. H. 1962. Radio-tracking of porcupines and ruffed grouse. In *Bio-telemetry*, ed. L. E. Slater. New York: Macmillan, pp. 173–178.

Marshall, W. H., and J. Kupa. 1963. Development of radio-telemetry techniques for ruffed grouse studies. *Trans. North Am. Wildlife Nat. Resources Conf.* **28**: 443–456.

Marshall, W. H. 1965. Ruffed grouse behavior. *BioScience* **15**(2): 92–94.

Martinka, C. J. 1969. Population ecology of summer resident elk in Jackson Hole, Wyoming. *J. Wildlife Management* **33**(3): 465–481.

Mech, L. D., L. D. Frenzel, Jr., R. R. Ream, and J. W. Winship. 1971. Movements, behavior, and ecology of timber wolves in Northeastern Minnesota. In *Ecological studies of timber wolf in Northeastern Minnesota,* ed. L. D. Mech and L. D. Frenzel, Jr. USDA Forest Service Research Paper NC-52, pp. 1–35.

Moen, A. N. 1966. Factors affecting the energy exchange of white-tailed deer, western Minnesota. Ph.D. dissertation, University of Minnesota, 121 pp.

Montgomery, G. G. 1963. Nocturnal movements and activity rhythms of white-tailed deer. *J. Wildlife Management* **27**(3): 422–427.

Müller-Schwarze, D. 1971. Pheromones in black-tailed deer (*Odocoileus hemionus columbionus*). *Animal Behaviour* **19**: 141–152.

Murie, O. J. 1951. *The Elk of North America*. Harrisburg, Pennsylvania: The Stackpole Co., 376 pp.

Peterson, R. L. 1955. *North American moose*. Toronto: University of Toronto Press, 280 pp.

Schneider, D. G., L. D. Mech, and J. R. Tester. 1971. Movements of female racoons and their young as determined by radio-tracking. *Animal Behaviour Monographs* **4**(1): 43 pp.

Severinghaus, C. W., and E. L. Cheatum. 1956. The white-tailed deer. In *The Deer of North America,* ed. W. P. Taylor. Harrisburg, Pennsylvania: The Stackpole Co., pp. 57–186.

Slater, L. E., ed. 1963. *Bio-telemetry*. New York: Macmillan, 372 pp.

Sowls, L. K. 1955. *Prairie Ducks*. Harrisburg, Pennsylvania: The Stackpole Co., and Washington, D.C.: Wildlife Management Institute, 193 pp.

Sparrowe, R. D., and P. F. Springer. 1970. Seasonal activity patterns of white-tailed deer in eastern South Dakota. *J. Wildlife Management* **34**(2): 420–431.

Verme, L. J. 1965. Swamp conifer deer yards in Northern Michigan. *J. Forestry* **63**(7): 523–529.

Watts, C. R., and A. W. Stokes. 1971. The social order of turkeys. *Sci. Am.* **224**(6): 112–118 (Offprint No. 1224).

Welty, J. C. 1962. *The life of birds.* Philadelphia: Saunders, 546 pp.

IDEAS FOR CONSIDERATION

What energy and matter limitations are imposed on an animal that are reflected in its home-range or territorial characteristics? What is the relative importance of energy and matter compared with intraspecific interactions? Does food limit the size of the home range? Does the size of the home range vary in relation to the abundance of food? If food is plentiful, what minimum size can be tolerated in terms of intraspecific relationships? What seasonal variation in size is there in relation to variation in the abundance of food? How do the home ranges of predators depend on the abundance of prey?

SELECTED REFERENCES

Ables, E. D. 1969. Home range studies of red foxes (*Vulpes vulpes*). *J. Mammal.* **50**(1): 108–120.

Aleksiuk, M. 1968. Scent-mound communication, territoriality, and population regulation in beaver (*Castor canadensis* Kuhl). *J. Mammal.* **49**(4): 759–762.

Anderson, R. K. 1971. Orientation in prairie chickens. *Auk* **88**(2): 286–290.

Brander, R. B. 1967. Movements of female ruffed grouse during the mating season. *Wilson Bull.* **79**(1): 28–36.

Bromley, P. T. 1969. Territoriality in pronghorn bucks on the National Bison Range, Moiese, Montana. *J. Mammal.* **50**(1): 81–89.

Carthy, J. D., and F. J. Ebling, eds. 1964. *The natural history of aggression.* New York: Academic Press, 159 pp.

Cowan, I. McT., and V. Geist. 1961. Aggressive behavior in deer of the genus *Odocoileus.* *J. Mammal.* **42**(4): 522–526.

Darling, F. F. 1937. *A herd of red deer: a study in animal behaviour.* London: Oxford University Press, 215 pp.

Davis, D. E. 1966. *Integral animal behavior.* New York: Macmillan, 118 pp.

Etkin, W. 1964. *Social behavior and organization among vertebrates.* Chicago: University of Chicago Press. 307 pp.

Fletcher, I. C., and D. R. Lindsay. 1968. Sensory involvement in the mating behaviour of domestic sheep. *Animal Behaviour* **16**(4): 410–414.

Fox, M. W., ed. 1968. *Abnormal behavior in animals.* Philadelphia: Saunders, 563 pp.

Fraser, A. F. 1968. *Reproductive behavior in ungulates.* New York: Academic Press, 202 pp.

Geist, V. 1968. On delayed social and physical maturation in mountain sheep. *Can. J. Zool.* **46**(5): 899–904.

Geist, V. 1971. *Mountain sheep: a study in behavior and evolution.* Chicago: University of Chicago Press, 383 pp.

Godfrey, G. A., and W. H. Marshall. 1969. Brood break-up and dispersal of ruffed grouse. *J. Wildlife Management* **33**(3): 609–920.

Gullion, G. W. 1967. Selection and use of drumming sites by male ruffed grouse. *Auk* **84**(1): 87–112.

Hafez, E. S. E., ed. 1969. *The behaviour of domestic animals.* Baltimore: Williams & Wilkins, 647 pp.

Haugen, A. O., and D. W. Speake. 1957. Parturition and early reaction of white-tailed deer fawns. *J. Mammal.* **38**(3): 420–421.

Herring, D. S., and A. O. Haugen. 1969. Bull bison behavior traits. *Proc. Iowa Acad. Sci.* **76**: 245–262.

Hinde, R. A. 1966. *Animal behavior: a synthesis of ethology and comparative psychology.* New York: McGraw-Hill, 534 pp.

Hornocker, M. G. 1969. Winter territoriality in mountain lions. *J. Wildlife Management* **33**(3): 457–464.

Johnsgard, P. A. 1965. *Handbook of waterfowl behavior.* Ithaca, New York: Cornell University Press, 378 pp.

Jones, R. E., and A. S. Leopold. 1967. Nesting interference in a dense population of wood ducks. *J. Wildlife Management* **31**(2): 221–228.

Keeler, C., and E. Fromm. 1965. Genes, drugs and behavior in foxes. *J. Heredity* **56**(6): 288–291.

Klopfer, P. H. 1962. *Behavioral aspects of ecology.* Englewood Cliffs, New Jersey: Prentice-Hall, 171 pp.

Klopfer, P. H., and J. P. Hailman. 1967. *An introduction to animal behavior.* Englewood Cliffs, New Jersey: Prentice-Hall, 297 pp.

Lent, P. C. 1965. Rutting behaviour in a barren-ground caribou population. *Animal Behaviour* **13**(2/3): 259–264.

Lewin, V., and J. G. Stelfox. 1967. Functional anatomy of the tail and associated behavior in woodland caribou. *Can. Field Naturalist* **81**(1): 63–66.

Linsdale, J. M., and P. Q. Tomich. 1953. *A herd of mule deer.* Berkeley and Los Angeles: University of California Press, 567 pp.

Linsday, D. R., and I. C. Fletcher. 1968. Sensory involvement in the recognition of lambs by their dams. *Animal Behaviour* **16**(4): 415–417.

Lowe, C. H., D. S. Hinds, P. G. Lardner, and K. E. Justice. 1967. Natural free-running period in vertebrate animal populations. *Science* **156**(3774): 531–534.

Marler, P., and W. J. Hamilton, III. 1966. *Mechanisms of animal behavior.* New York: Wiley, 771 pp.

Pack, J. C., H. S. Mosby, and P. B. Siegel. 1967. Influence of social hierarchy on gray squirrel behavior. *J. Wildlife Management* **31**(4): 720–728.

Ralls, K. 1971. Mammalian scent marking. *Science* **171**(3970): 443–449.

Robinette, W. L. 1966. Mule deer home range and dispersal in Utah. *J. Wildlife Management* **30**(2): 335–348.

Rongstad, O. J., and J. R. Tester. 1971. Behavior and maternal relations of young snowshoe hares. *J. Wildlife Management* **35**(2): 338–346.

Samuel, W. M., and W. C. Glazener. 1970. Movement of white-tailed deer fawns in south Texas. *J. Wildlife Management* **34**(4): 959–961.

Schneider, D. G., J. D. Mech, and J. R. Tester. 1971. Movements of female raccoons and their young as determined by radio-tracking. *Animal Behaviour Monographs* **4**(1): 1–43.

Stokes, A. W. 1967. Behavior of the bobwhite, *Colinus virginianus*. *Auk* **84**(1): 1–33.

Tinbergen, N. 1965. *Social behaviour in animals, with special reference to vertebrates.* London: Methuen, 150 pp.

Thomas, J. W., J. G. Teer, and E. A. Walker. 1964. Mobility and home range of white-tailed deer on the Edwards Plateau in Texas. *J. Wildlife Management* **28**(3): 463–472.

Vince, M. A. 1964. Social facilitation of hatching in the bobwhite quail. *Animal Behaviour* **12**(4): 531–534.

Woolf, A., T. O'Shea, and D. L. Gilbert. 1970. Movements and behavior of bighorn sheep on summer ranges in Yellowstone National Park. *J. Wildlife Management* **34**(2): 446–450.

Woolpy, J. H. 1968. The social organization of wolves. *Nat. Hist.* **77**(5): 46–55.

Wynne-Edwards, V. C. 1962. *Animal dispersion in relation to social behaviour.* Edinburgh: Oliver & Boyd, 653 pp.

INTERSPECIES INTERACTION

Interactions between populations of different species can take many forms, including the following: (1) *neutralism*—a population is not affected by association with another; (2) *competition*—populations adversely affect one another in the struggle for food, living space, or other common needs; (3) *mutualism*—populations benefit one another's growth and survival and cannot survive under natural conditions without one another; (4) *protocooperation*—populations benefit by associating with one another, but relations are not obligatory; (5) *commensalism*—one population benefits but the other is not affected; (6) *amensalism*—one population is inhibited and the other not affected; (7) *parasitism* and (8) *predation*—a population adversely affects another by direct attack but is dependent on it (Odum 1959). These are summarized in relation to their effect on population growth and survival in Table 12-1.

These kinds of interactions are generally considered to be between animals, although some of them can exist between plants and animals. Amensalism, for example, occurs if free-ranging animals consume poisonous plants on the range, the consumer being inhibited by the action of the ingested toxin and the plant being unaffected except for losing that which has been consumed. The plant as a whole survives and may even be stimulated by the cropping.

This chapter deals primarily with predation, competition, and parasitism. These have been of interest to the wildlife biologist, although it may be true that other, more subtle, factors have equally as great an impact on the population dynamics of any organism.

TABLE 12-1   ANALYSIS OF TWO-SPECIES POPULATION INTERACTIONS

Type of Interaction	Effect on Population Growth and Survival of Two Populations, A and B				General Result of Interaction
	When Not Interacting		When Interacting		
	A	B	A	B	
Neutralism (A and B independent)	0	0	0	0	Neither population affects the other
Competition (A and B competitors)	0	0	−	−	Population most affected eliminated from niche
Mutualism (A and B partners or symbionts)	−	−	+	+	Interaction obligatory for both
Protocooperation (A and B cooperators)	0	0	+	+	Interaction favorable to both, but not obligatory
Commensalism (A commensal; B host)	−	0	+	0	Obligatory for A; B not affected
Amensalism (A amensal; B inhibitor or antibiotic)	0	0	−	0	A inhibited; B not affected
Parasitism (A parasite; B host) Predation (A predator; B prey)	−	0	+	−	Obligatory for A; B inhibited

SOURCE: Odum 1959.
Note: + indicates population growth increased; − indicates population growth decreased; 0 indicates population growth not affected.

12-1   PREDATOR-PREY RELATIONSHIPS

Predator-prey interactions have been studied in controlled situations (e.g., fish tanks as miniature aquatic systems) and historical ones, such as the attempts to correlate the populations of predators with populations of prey species. These populations have often been estimated by the take of trappers and hunters, and lags between the peak abundance of the predator and prey have been observed. The lynx–hare relationship is perhaps the best known example; another is the fox–pheasant.

Analytically, the ecologist is interested in the mechanisms of predation, in the factors regulating the amount of predation, and in the effect of predation on population structures and levels through time. Before considering some of the mechanisms, let us review some general characteristics of predation that have been observed among free-ranging animals.

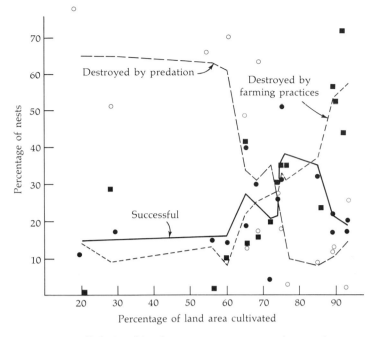

FIGURE 12-1. Relationships between percentages of nests destroyed by predation and by farming practices, total nesting success, and the percentage of land area under cultivation. The lines were drawn from three-point moving averages. (From Wagner, Besadny, and Kabat 1965.)

Upland game birds and waterfowl are preyed upon by a variety of predators. The individual birds are killed by fox, coyote, house cats, raccoon, mink, lynx, bobcat, hawks, owls, and other predators in local areas. Nests are destroyed by skunks, badgers, crows, magpies, and others. Many studies of nest destruction have been made, with destruction rates varying from nearly zero to three-quarters or more of the nests. In Wisconsin, pheasant nest losses varied from 3% to 78%, with the losses inversely correlated with losses due to farming activity and the percentage of land under cultivation (Figure 12-1). Note that the losses due to predation were high under low-intensity cultivation when few nests were destroyed by agricultural practices. As cultivation increases, the rate of nest loss due to predation decreased, but that due to farming practices increased. A reversal in the rate of loss due to predation occurred when over 80% of the land area was cultivated. The percentage of successful nests was highest when 60% to 90% of the land area was cultivated, a characteristic unique to the pheasant and certainly not characteristic of other game-bird species such as the prairie chicken.

Sowls (1955) summarizes data from a variety of studies and presents data for the Delta Marsh, too. Ducklings are subject to predation by fish and turtles. Carnivorous fish are often not present in the shallow brood marshes, but ducks living on larger lakes and rivers are susceptible.

Larger animals such as deer and elk have fewer predators than do the upland game birds. Fawns and calves can be taken by coyotes, and wolves prey on deer and moose in areas such as northern Minnesota. Wolves prey almost exclusively on moose on Isle Royal. Very high mortality of white-tailed deer fawns owing to coyote predation has been observed in Texas, with losses of about 72% of the fawns within two months (Cook et al. 1971). Mountain lions, grizzlies, and black bear prey on elk calves.

One important point to consider in evaluating predation is that intrusion by the biologist in the process of marking animals, attaching radios, checking nests, counting eggs, and so forth may result in an increase in the amount of predation. The effect of human intrusion is dependent on cover conditions and on the abilities of the predators to use human artifacts to locate prey. Crows have been known to locate nests by watching biologists working at nest sites. Merely flushing a bird from a nest leaves the eggs exposed to crow predation. Tags attached to the backs of ruffed grouse seemed to be responsible for significantly accelerated losses (Gullion and Marshall 1968). The attachment of radios, even though they may weigh but a few grams, may affect the susceptibility of the carrier bird to predation. They usually are quite inconspicuous, however, so if their weight is sufficiently low they may have less effect than colored back tags.

It is usually very difficult to estimate accurately the rate of predation on adult animals because the chances of finding the remains and determining the cause of death are very slim. One of the advantages of using a radio telemetry system to aid field work lies in the ability to locate the transmitter regularly, in order to check on the condition of the carrier animal. Sometimes the signal changes location more rapidly than the carrier animal can travel; a snowshoe hare traveling one-half mile in a minute or so clearly indicates that something is carrying it at a speed unknown to hares!

The usefulness of radio telemetry in studying predation was demonstrated by Marshall and Kupa (1963) with grouse. Mech (1967) describes the use of telemetry in predation studies of snowshoe hares (*Lepus americanus*) and cottontail rabbits (*Sylvilagus floridanus*). In one case, a snowshoe hare was killed by a red fox (*Vulpes fulva*) who was also carrying a radio transmitter. The photographed signals of both transmitters were the same because the radios were being shaken in a similar manner. The episode ended with the fox burying part of the hare and the transmitter in 10 inches of snow. Several telemetry techniques are designed specifically for the detection of mortality in free-ranging animals [see Stoddart (1970)]; temperature sensors reveal a loss of body temperature.

12-2   FACTORS AFFECTING PREDATION RATES

The factors affecting the amount of predation are exceedingly complex. Leopold (1933) has listed five variables that affect annual mortality due to predators:

1. The density of the game (prey) population.
2. The density of the predator population.
3. Food preferences of the predator.

4. The physical condition of the game and the escape cover available.

5. The abundance of "buffers" or alternative foods for the predator.

None of these factors are stable over time. The density of the game population varies, depending on the conditions affecting productivity and on the reproductive potential of the species. Quail and pheasants have greater annual variation in population density than do large mammals such as deer, elk, and moose. Densities vary with the season also; many more birds are present in a given habitat just after hatching than during the incubation period. Further, the mobility of the birds is considerably less during incubation than after hatching. Incubating females spend most of the time on the nest where they are usually concealed by cover. The chance of a predator locating such a bird is probably less than when the bird is actively rearing the brood. Thus there is an actual density (this is difficult if not impossible to determine) and there is an apparent density, or a population level that appears to exist based on direct or indirect observations of the individuals. Keep in mind that the density apparent to a human driving along a road or spending a few hours in the field is different from that apparent to a predator living in the habitat. Further, the visual environment of the predator is different from that of a human since it may be looking through vegetative cover from a lower height or it may be looking down into the vegetation as it flies. Also, predators can detect the presence of prey by scents and sounds that humans are not aware of. These kinds of considerations were discussed in Chapter 2.

The food preferences of a predator are a complex summation of the previous experiences of the predator, the chemical interactions through taste and the digestive processes, the abundance of any food that could be eaten, and other factors that work together with any or all of these. It is difficult to determine the relative importance of these factors; the anlytical ecologist should first concentrate on studying the effects of different combinations of foods in the diet rather than the causes of these combinations.

The physical condition of the prey species and the escape cover available vary greatly on a seasonal basis. Further, they frequently vary together, with a decline in the physical condition of the animal coinciding with a decline in the amount of cover available owing to snow accumulations, plant decadence, and other factors that change the mechanical, thermal, and optical characteristics of cover. A reduction in the density of the cover may force an animal to travel farther to reach sufficient cover. The distance to this cover becomes more critical as the animal's physical capacity for travel is reduced, thereby compounding the interaction between cover and predation. As the vegetation resumes growth in the spring, the animal's condition improves, too, increasing its ability to escape.

The importance of "buffers" or alternative foods available to a predator is closely tied to Leopold's third variable. If there is a variety of prey species or alternate foods available (such as berries for fox and coyote), the predator's diet is likely to be varied. As hunting conditions worsen, the predator is forced to eat fewer kinds of foods and more of each kind.

The term "buffer" species has generally been applied to species other than the prey under consideration. From a functional biological point of view, a buffer can be a member of the same species as the prey under consideration. If the prey species is polygamous, such as the pheasant, the shooting of cocks only, for example, is not considered detrimental to the total pheasant population. If, however, there is a shortage of other buffer species available, then the cocks may become buffers for the hen population. This may not be of much significance during periods in which populations are large, but it could be important when the pheasant population is at or below a minimum threshold for a rapid rate of increase. Leopold suggests that this kind of relationship is worth considering by pointing out that different prey-predator abundance ratios warrant different considerations.

Rates of predation alone do not permit an analysis of the effects of predation in an analytical ecological context. In the absence of sound biological information on rates and their effect in relation to the total ecology of a species, it is beneficial to evaluate the effect of different predation rates in relation to different reproductive rates by simulation, resulting in an understanding of the balance between productivity and mortality of animals in different age, weight, and reproductive classes. There are realistic limits within which to confine the analysis; predation, for example, can vary from zero to 100% of the prey population. Biologically reasonable rates are found somewhere between those two extremes, and an analysis of rates at intervals between zero and 100% in relation to reproductive rates at intervals between zero and the maximum reproductive potential of the animal will provide insight into the relative importance of predation to different population densities. If a time element is added to this analysis, both seasonally and over several generations, further insights are gained into long-range trends. These considerations are discussed in Chapter 19. They are a part of a population model that starts with the smallest possible working model and progresses toward the most "real" model possible. This is different from a common procedure in population ecology in which masses of numbers are accumulated in the field and an equation(s) is derived without an understanding of the important factors causing those numbers to vary.

12-3   ENERGETIC CONSIDERATIONS

Life was defined in Chapter 1 as a process consisting of the orderly rearrangement of matter with the expenditure of energy. Everything an animal does "costs" something in terms of energy, and predation is no exception. There are some generalizations about predation that provide background for understanding the variations in rates of predation by different predator-prey combinations. A large predator must consume more very small prey than larger prey. In other words, a fox needs to consume more mice than pheasants to satisfy its nutritional requirements. This assumes that the prey are eaten, and killing is not just a form of displacement behavior. The cost of catching the prey becomes an important consideration when evaluating the predation rate necessary for meeting the nutritional requirements, since these requirements vary in part in relation to the cost of catching. Thus if it is harder to catch mice than pheasants, relatively more mice must be caught to meet the extra energy demand. If mice are easy to catch, the predator may be better off energetically by eating several small mice rather than spending more energy trying to catch one large pheasant.

The spatial distribution of prey is an important factor in determining the cost of predation. Game birds are widely distributed during the nesting season, for example, and will most likely be more difficult to locate and capture than when a group of them is confined to a small patch of cover in the winter. After hatching, there is a high density of birds within the brood radius, and it is likely to be less costly for a predator to locate and catch one member of the brood than if only single birds were available.

Weather conditions affect the behavior patterns of predator and prey differently. Storms affect the distribution of both predator and prey, the condition of each, and the rate of predation. Snow as a mechanical barrier affects the usefulness of cover. Deep snow covers the living area of many potential prey such as mice. Snow is a barrier to travel, although frequently it is a greater barrier to the larger predator than the smaller prey. Some prey are larger than the predator, however; moose are larger than wolves, for example. The effectiveness of snow as a barrier varies in relation to both of these animals. Loose fluffy snow may be a mechanical barrier to wolves but not to moose. Deep snow with a crust strong enough to support a wolf may not support a moose, and the prey are much more easily caught. Very cold weather may cause white-tailed deer to reduce their activity, but it may not affect wolves. Pheasants may not feed in open fields during high winds, but fox may be less affected.

Mech et al. (1971) describe several accounts of the hunting behavior of wolves. The most interesting characteristic observed was a fairly short chase after white-tailed deer. The wolves appeared to have as much or more difficulty in snow than the deer at times, and the wolves would stop to rest after only several hundred yards of chase. After the radio-tagged wolves were successful in killing a deer, they remained in the vicinity of the kill for one to seven days, depending on how recently they had eaten.

Predators are frequently credited with catching the weaker animals. This has been demonstrated in predator-prey relations between wolves and moose, caribou, bison, and other prey [see Mech (1970)]. The recent study by Mech and Frenzel (1971) on the age, sex, and condition of deer killed by wolves in northern Minnesota shows that wolves tend to prey on old, debilitated, or abnormal deer. This generalization should not necessarily be applied to all predator-prey relationships since the manner in which the predator hunts may affect the opportunity of the prey to utilize its physical capacity for escape.

The difficulty in catching a prey depends on its ability to escape owing to its physical condition and abilities as well as the condition of the cover. Thus it may be harder for a predator to catch a rabbit in thick brush than a pheasant in a field if both stay on the ground. If the pheasant flies when it detects the predator, it may become immediately unavailable if the predator is confined to the ground.

Some birds can fly faster than others, however; it may be more efficient for the predator to pursue a grouse, which has a shorter flight distance than a pheasant. The behavior of predators may change with variations in hunting conditions. According to Nellis and Keith (1968), lynx (*Lynx canadensis*) had daily cruising distances of 3.0 miles and 5.5 miles in two successive winters in Alberta, and their success in catching snowshoe hares dropped from 24% to 9%, respectively. The authors attributed this change to different snow conditions. No significant differences were observed in the number of kills, although the kill per mile traveled was lower in the second of these two winters. This indicates that they tried harder when hunting conditions were poorer because of snow conditions. They caught less food during the less successful (relatively) year, which may indicate a higher utilization of each kill under poor hunting conditions.

Differential survival of individual prey (ruffed grouse in this case) that can be related to behavioral or physiological characteristics has been noted by Gullion and Marshall (1968). Male grouse survived four months longer when they used transient drumming logs not used by a predecessor than those that replaced earlier drummers on perrenial logs. This indicates that the raptorial predators focused on a physical object in the habitat that elicited a perennial behavioral response, waiting in ambush for the grouse to use it. The actual attack occurs when the grouse are en route to or from the log. The few males observed to live to an old age in the Cloquet Forest (Minnesota) all occupied transient logs. The relationship between body size and predation habits has been analyzed by Rosenzweig (1966; 1968). He has pointed out that larger predatorial carnivores (secondary consumers) tend to eat larger prey. The exceptions to this include the secondary consumers that are also primary consumers, such as the black bear. They are larger than mountain lions or wolves, but they prey on smaller animals. The difference can be explained by their consumption of a considerable amount of vegetation also. Thus the body size of a predatory carnivore appears to be related to both the size of the prey and the level of utilization of nonprey foods.

The whole predator-prey complex of interactions is very challenging to the analytical ecologist. The approach to predator-prey analyses should be made within the confines of natural laws governing the energetics of the interactions, including both nutritive and thermal considerations. Such an approach will provide insight into the mechanisms operating, after which behavioral constraints and other characteristics that have developed over time can be considered.

TROPHIC LEVELS. The idea of trophic levels in interspecies interactions is a valid one. Herbivores eat plants, carnivores eat herbivores, and secondary carnivores eat the carnivores that feed on herbivores. Different levels of efficiency can be attached to the different trophic levels [see Odum (1959)]. There is a "loss" of energy in going from a lower to a higher trophic level. This results in "pyramids" that depict reductions in numbers, biomass, or energy from the lower to the higher level. These representations are fine for depicting general principles, but the drama associated with spatial and temporal factors is lost. None of the animals living in an ecosystem can depend on the average biomass or energy transfer from one trophic level to another to survive; each must meet its needs on a day-to-day basis. Some of these needs may be met by the mobilization of body reserves, or they may be met by daily feeding and other regulatory behavior.

As Rosenzweig (1968) points out, some species are both primary and secondary consumers, so they are at two different trophic levels at the same time. This characteristic permits an animal to "regulate" its energy efficiency by going to the lower trophic level, that is, toward the primary productivity level when food is scarce and to levels above primary productivity when hunting conditions are favorable. Changes in weather conditions, seasonal changes in animal behavior and activity, and seasonal changes in plant phenology can cause these shifts to occur.

The timing of the shifts may be of considerable significance to both predator and prey. The arrival of spring may make a difference to the predator as it shifts from one prey species to another, and changes in phenological characteristics may result in an abundance of alternative foods that could replace the prey. The latter seems to be the situation in Texas where the timing and abundance of the berry crop has a significant effect on the amount of coyote predation on fawns (Knowlton, personal communication).

Differences in population and biomass parameters for moose on Isle Royale have been presented by Jordan, Batkin, and Wolfe (1971) (Figure 12-2). Note that the population can decrease but the standing crop can increase during part (April–August) of that time. The transfer, or total live weight of animals dying each month and hence available to predators, remains fairly constant. While these dynamics through time are occurring, moose of different ages play distinctly different roles in population and biomass dynamics (Figure 12-3). The population in July includes more animals that are one year old or younger than any other age class, but the highest standing crop can be attributed to the six-year class. The transfer by death is highest in the one-year class, lowest in the three-year class, and rises to a secondary peak in the twelve-year class. Figures 12-2 and 12-3 combined illustrate how multidimensional the population characteristics are in space and time. The averaging of all these factors into single values for each year and all moose would result in the loss of considerable information.

When factors affecting the productivity of both predator and prey fluctuate greatly, a greater departure from the average is noted and productivity increases or decreases greatly, depending on whether the combination of circumstances has a positive or negative effect. It is these kinds of fluctuations that result in population fluctuations over an ecologically short time span of just a few years, but a long enough time span to cause significant changes in legislation regulating hunting and the activities of game biologists and managers. Thus it is particularly important for the biologist and the manager to understand the cause of short-term fluctuations in order to predict and explain the biology of the current situation. It is hoped that this would result in the institution of appropriate policy that would eliminate undue political and economic pressure, which often results in wresting decision-making authority from resource biologists and managers and placing it in the hands of political groups.

Predator control has always been a controversial topic among biologists and sportsmen. Disagreements have often been caused by a lack of understanding of the impact of predators on game populations in relation to social and economic factors. Predator removal *can* result in higher prey populations according to Chesness, Nelson, and Longley (1968). They found that the cost of predator control per pheasant chick hatched on a controlled versus an uncontrolled area was $4.50, a rather high figure. The effectiveness of predator control was demonstrated by Balser, Dill, and Nelson (1968) also; 60% more ducklings from about one to three weeks of age were counted in the area with intensive predator control.

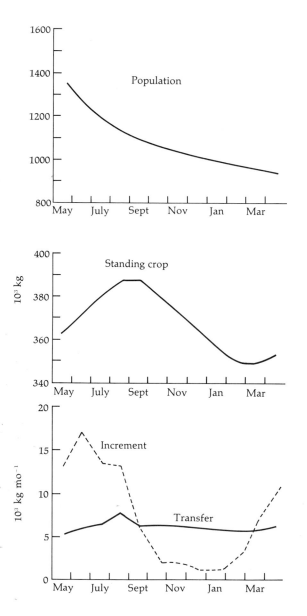

FIGURE 12-2. Monthly levels of population and of biomass parameters computed for the Isle Royale moose herd. Standing crop is the live weight of the population; increment is the total live weight gained each month; and transfer is the total live weight of animals dying each month. (From Jordan, Batkin, and Wolfe 1971.)

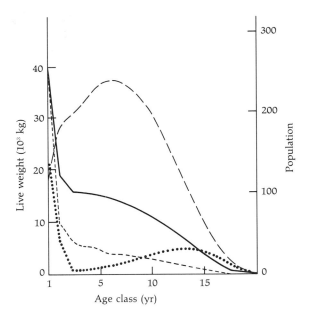

FIGURE 12-3. Numbers and biomass parameters computed for each age class of the Isle Royale moose herd. Population (solid line) is the number present in July; standing crop (long-dashed line) is total live weight in July of the population; annual increment (short-dashed line) is the year's total of live-weight gains; and annual transfer (dotted line) is the year's total of live weight on moose dying. (From Jordan, Batkin, and Wolfe 1971.)

The entire predator-prey complex needs to be studied with considerable attention given to the roles of all potential predators over time. The biology of predator-prey interactions should be studied first, after which social and economic considerations can be appropriately added.

12-4 MAN AS A PREDATOR

Man as a biological species is at different trophic levels. Different cultures and societies are oriented more toward being either primary consumers or secondary consumers, and some individuals are entirely vegetarian or primary consumers. The hunting instinct is strong in man, and the hunting tradition is present even after intensive agriculture reduces the need for hunting as a source of food. The ecological role of man at the present time may necessarily include hunting, if some populations of animals are to be maintained in balance with the supply of resources.

The incompatibility of man and wolf or man and mountain lion, for example, has resulted in a lack of natural predators necessary to maintain a balance between the consumer and the supply of resources. The prohibition of hunting in national parks has demonstrated the need for natural predators, since populations of elk, for example, have increased beyond the ability of the resources to support them. Programs to reduce the number of elk have been attempted for many years in some areas because the herd is too large for the food supply. Trapping is often resorted to when a serious food shortage develops on a range, but the removal and release of weakened animals is usually not a desirable way to handle the problem. Trapping and transporting wild ruminants is costly and often results in mortality due to shock and the effect of other stresses on the animals.

A hunted animal is more wary than an unhunted one; Behrend and Lubech (1968) have observed this for white-tailed deer in the Adirondacks. Increased wariness is a desirable thing ecologically, but resort owners and their clientele often prefer the tamer, more abundant animals. Thus a social problem is superimposed on the basic biology of the situation. The important point to remember is that man should be primarily responsible to the biology of the organism and of the ecosystem, and social, economic, and political considerations should remain secondary.

12-5 PARASITES AND PATHOGENS

A large number of parasites and pathogens have been found living on and in free-ranging animals. Many short reports describing lesions, tumors, aberrent behavior, and other abnormalities have been published. Diagnostic work is usually done coincidentally with the collection of other biological information such as that gathered at checking stations during the hunting season. Treatment of free-ranging animals is not feasible because they are not under the control of man. The control of diseases in wild populations is usually by natural attrition when the number of carriers reaches such a low point that transmission is interrupted. Disease control by natural means can be accelerated by reducing the carrier population through hunting, trapping, or other means.

Several publications have appeared that describe various characteristics of the epidemiology of different parasites and diseases. A small book entitled *Diseases of Free-living Wild Animals* by McDiarmid (1962) is organized according to the taxonomy of the pathogens, with short descriptions of diseases due to bacteria, fungi, viruses, protozoa, rickettsiae, and true neoplasms of viruses associated with tumor formation. Different hosts known or suspected to be carriers are discussed for each kind of pathogen.

A book on diseases transmitted from animals to man (Hull 1963) includes discussions of the role of both domestic and wild animals as carriers. Thirty-two authors contributed to this edition of the book, each discussing their subject within a format of their choosing.

Another book (McDiarmid 1969) is a collection of papers on a variety of diseases that were considered at a symposium sponsored by the Zoological Society of London. In his foreword, Dr. McDiarmid points out how general world interest in diseases of wild animals has been amply demonstrated in the 1960s but that some countries still do not devote much attention to diseases in relation to wildlife ecology.

Davis and Anderson (1971) have edited a book on parasitic diseases of wild mammals. It is interesting to note that this book is also a collection of papers by several authors. It appears that no single ecologist with a deep interest in wildlife diseases has written a comprehensive treatise on the role of diseases in wildlife ecology. The analytical treatment of diseases as a factor affecting productivity, complete with models relating the effect of pathogens and parasites to productivity, is an exciting prospect in wildlife ecology.

Individual ecologists may often ignore diseases as a vital part of ecological relationships. This may be because ecologists are generally not prepared to detect the effect of diseases on productivity. The ecological effects of diseases on the structure of a population owing to relative effects on different members of that population, the changes in productivity because of changes in population characteristics, and the effect of diseases and parasites on the physiology and behavior of an individual as well as its relation to other species deserves careful analytical attention.

A recent but pioneering attempt to use the systems approach in the analysis of a host-parasite interaction is described by Ractliffe et al. (1969). Their model has three principal components, including (1) a parasite control mechanism, (2) a parasite population, and (3) a mechanism for haematocrit regulation. Host-parasite interactions were simulated, using data from field observations, which illustrated how such a simulation model can provide the basis both for selecting the specific variables that should be studied and for estimating the quantitative nature of the variables.

Parasites (and pathogens) may affect an animal directly by diverting energy through the parasite system or by causing a metabolic constraint that upsets the metabolic efficiency of the host. The energy requirements of *Haemonchus contortus*, a roundworm found in the abomasum of sheep, deer, and other ruminants, were calculated and compared with the energy requirements of deer at different levels of productivity (Moen, unpublished data). These calculations showed that the number of parasitic worms necessary to cause an energy drain equal to that of normal productive functions was too great for the space available in the abomasum. Thus the effect of the energy drain was not the dominant cause of death but contributed to it by increasing the total energy requirements at a time when energy supplies were limited.

A more significant effect of *Haemonchus* and other parasites is their effect on the metabolic functions of various systems of the body. *Haemonchus,* for example, is apparently related to erythrocyte levels (Whitlock and Georgi 1968), which in turn are related to oxygen transport efficiency and the susceptibility of the host to a variety of other decimating factors. Productivity functions, the maintenance of homeothermy in cold environments, and sustained activity are three obvious metabolic functions that require oxygen transport.

Another example of the physiological effect of parasites may be seen in the relationship between deer, moose, and the roundworm *Pneumostrongulus tenuis.* This parasite is carried by white-tailed deer with little or no known effects. However, infected moose suffer a neurological disease that results in aberrant behavior, including reduced fear of man and a decrease in muscular coordination, which is manifested externally but caused internally. The result is death for the moose; large moose populations may not be found in areas with large deer populations because deer are carriers of a parasite that is potentially lethal for moose.

12-6   COMPETITION

The most powerful forces in the life of an animal may be exerted by members of its own species, but other species interact with the animal in several ways. They compete directly or indirectly for the available material and energy within the space that they share. The extent of this competition depends on how much their operational environments overlap. Long-term adaptations often tend to minimize direct competition. For example, subtle differences in food preferences may develop.

The ranges of wild ruminants often overlap. Mule deer, white-tailed deer, elk, and moose can be found in the same general area. There may not be much competition for food, however. In the summer, for example, deer eat a mixture of browse and herbaceous plants, while elk are primarily grazers. Moose and deer feed on both aquatic and terrestrial plants, but the larger size of the moose permits it to feed in areas that deer cannot reach.

The restricted winter range is often the site of the most direct competition among species. Elk may forage on woody browse at that time, competing directly with deer. Moose and deer forage heavily on woody browse during the northern winters, competing directly for those species that both prefer.

Snow depths can affect the amount of competition for food. Moose are not bothered much by depths of 20 inches, a critical depth for the deer. Snow characteristics that are related to cover types seem to affect the distribution of moose and deer in the winter; Telfer (1970) observed little competition between the two species because of the different effects of snow cover on their distribution in New Brunswick. Only 1% of the area was shared by the two species.

Direct competition for food may exist between wild animals and domestic stock. In general, competition is less intense when both the wild and the domestic animal populations are low, permitting each to select its most preferred forage. As populations increase, competition between the groups increases.

Indirect competition is a more subtle type of interspecies interaction. Small mammals such as mice, rabbits, prairie dogs, ground squirrels, gophers, and others affect the growth of forage without competing directly with livestock or game animals for some of the aerial parts. These animals consume the roots and early growth of grasses and forbs, and their absence can result in higher forage yields.

Another form of indirect competition takes place between livestock and game birds and waterfowl. Overgrazed pastures are of little value as nesting cover. Stock ponds with muddy banks devoid of vegetation are not suitable for waterfowl. This kind of competition can be reduced by carefully regulating the number of livestock using an area, with fences for confining access to water to specially prepared areas with proper slope and substrate.

12-7   CONCLUSION

The relationships between an individual free-ranging animal and members of its own species and other species are very labile, varying in both time and space. The keen observer recognizes many of these relationships, mentally integrating them into a system that may be biologically realistic but difficult to explain. After all, the human mind is a kind of computing system, but without a single common program format. An analysis of behavioral interactions is difficult because of a lack of definitive units to quantify these relationships. Thus this chapter and Chapters 10 and 11 have been somewhat general, with brief indications of how such analyses might be approached.

Having progressed from physical to physiological to behavioral interactions thus far, let us now turn to an analysis of interactions involving all of these. The analyses described in the remainder of this text are representative of what can be done in analytical ecology. Many other analyses can be made, of course, and students with backgrounds in basic sciences are urged to develop further analyses of additional interactions present in the ecosystem.

LITERATURE CITED IN CHAPTER 12

Balser, D. S., H. H. Dill, and H. K. Nelson. 1968. Effect of predator reduction on waterfowl nesting success. *J. Wildlife Management* **32**(4): 669–682.

Behrend, D. F., and R. A. Lubech. 1968. Summer flight behavior of white-tailed deer in two Adirondack forests. *J. Wildlife Management* **32**(3): 615–618.

Chesness, R. A., M. M. Nelson, and W. H. Longley. 1968. The effect of predator removal on pheasant reproductive success. *J. Wildlife Management* **32**(4): 683–697.

Cook, R. S., M. White, D. O. Trainer, and W. C. Glazener. 1971. Mortality of young white-tailed deer fawns in south Texas. *J. Wildlife Management* **35**(1): 47–56.

Davis, J. W., and R. C. Anderson, eds. 1971. *Parasitic diseases of wild mammals.* Ames: The Iowa State University Press, 374 pp.

Gullion, G. W., and W. H. Marshall. 1968. Survival of ruffed grouse in a boreal forest. *The Living Bird* **7**: 117–167.

Hull, T. G., ed. 1963. *Diseases transmitted from animals to man.* 5th ed. Springfield, Illinois: Charles C Thomas, 967 pp.

Jordan, P. A., D. B. Batkin, and M. L. Wolfe. 1971. Biomass dynamics in a moose population. *Ecology* **52**(1): 147–152.

Leopold, A. 1933. *Game Management.* New York: Scribner, 481 pp.

Marshall, W. H., and J. J. Kupa. Development of radio-telemetry techniques for ruffed grouse studies. *Trans. North Am. Wildlife Nat. Resources Conf.* **28**: 443–456.

McDiarmid, A. 1962. *Diseases of free-living wild animals.* Rome: Food and Agriculture Organization of the United Nations, 119 pp.

McDiarmid, A., ed. 1969. *Diseases in free-living wild animals.* New York: Academic Press, 332 pp.

Mech, L. D. 1967. Telemetry as a technique in the study of predation. *J. Wildlife Management* **31**(3): 492–496.

Mech, L. D. 1970. *The wolf: the ecology and behavior of an endangered species.* New York: Natural History Press and Doubleday, 384 pp.

Mech, L. D., and L. D. Frenzel, Jr. 1971. An analysis of the age, sex, and condition of deer killed by wolves in Northeastern Minnesota. In *Ecological studies of the timber wolf in Northeastern Minnesota,* ed. L. D. Mech and L. D. Frenzel, Jr. USDA Forest Service Research Paper NC-52, pp. 25–51.

Mech, L. D., L. D. Frenzel, Jr., R. R. Ream, and J. W. Winship. 1971. Movements, behavior, and ecology of timber wolves in Northeastern Minnesota. In *Ecological studies of the timber wolf in Northeastern Minnesota*, ed. L. D. Mech and L. D. Frenzel, Jr. USDA Forest Service Research Paper NC-52, pp. 1–35.

Nellis, C. H., and L. B. Keith. 1968. Hunting activities and successes of lynxes in Alberta. *J. Wildlife Management* **32**(4): 718–722.

Odum, E. P. 1959. *Fundamentals of ecology.* Philadelphia: Saunders, 546 pp.

Ractliffe, L. H., H. M. Taylor, J. H. Whitlock, and W. R. Lynn. 1969. Systems analysis of a host-parasite interaction. *Parasitology* **59**: 649–661.

Rosenzwieg, M. L. 1966. Community structure in sympatric carnivores. *J. Mammal.* **47**(4): 602–612.

Rosenzwieg, M. L. 1968. The strategy of body size in mammalian carnivores. *Am. Midland Naturalist* **80**(2): 299–315.

Sowls, L. K. 1955. *Prairie ducks.* Harrisburg, Pennsylvania: The Stackpole Co., and Washington, D.C.: Wildlife Management Institute, 193 pp.

Stoddart, L. C. 1970. A telemetric method for detecting jackrabbit mortality. *J. Wildlife Management* **34**(3): 501–507.

Telfer, E. S. 1970. Winter habitat selection by moose and white-tailed deer. *J. Wildlife Management* **34**(3): 553–559.

Wagner, F. H., C. D. Besadny, and C. Kabat. 1965. *Population ecology and management of Wisconsin pheasants.* Technical Bulletin No. 34. Madison: Wisconsin Conservation Department, 168 pp.

Whitlock, J. H., and J. R. Georgi. 1968. Erythrocyte loss and restitution in ovine *Haemonchosis*. III. Relation between erythrocyte volume and erythrocyte loss and related phenotype displays. *Cornell Vet.* **58**(1): 90–111.

SELECTED REFERENCES

Altman, M. 1958. The flight distance in free-ranging big game. *J. Wildlife Management* **22**(2): 207–209.

Boddicker, M. L., E. J. Hugghins, and A. H. Richardson. 1971. Parasites and pesticide residues of mountain goats in South Dakota. *J. Wildlife Management* **35**(1): 94–103.

Bue, I. G., L. Blankenship, and W. H. Marshall. 1952. The relationship of grazing practices to waterfowl breeding populations and production on stock ponds in western South Dakota. *Trans. North Am. Wildlife Nat. Resources Conf.* **17**: 396–414.

Croll, N. A. 1966. *Ecology of parasites.* London: Heinemann, 136 pp.

Davis, J. W., A. L. Karstad, and D. O. Trainer. 1970. *Infectious diseases of wild mammals.* Ames: Iowa State University Press, 420 pp.

Errington, P. L. 1934. Vulnerability of bobwhite populations to predation. *Ecology* **15**: 110–127.

Errington, P. L. 1967. *Of predation and life.* Ames: Iowa State University Press, 277 pp.

Fox, M. W. 1969. Ontogeny of prey-killing behavior in Canidae. *Behaviour* **35**(3/4): 259–272.

Montgomery, G. G. 1968. Rate of tick attachment to a white-tailed fawn. *Am. Midland Nat.* **79**(2): 528–530.

Pearson, O. P. 1964. Carnivore-mouse predation: an example of its intensity and bioenergetics. *J. Mammal.* **45**(2): 177–188.

Pimlot, D. H. 1967. Wolf predation and ungulate populations. *Am. Zool.* **7**(2): 267–278.

Rosenzweig, M. L., and R. H. MacArthur. 1963. Graphical representation and stability conditions of predator-prey interactions. *Am. Naturalist* **97**(895): 209–223.

Schnell, J. H. 1968. The limiting effects of natural predation on experimental cotton rat populations. *J. Wildlife Management* **32**(4): 698–711.

Schranck, B. W. 1972. Waterfowl nest cover and some predation relationships. *J. Wildlife Management* **36**(1): 182–186.

Skovlin, J. M. P. J. Edgerton, and R. W. Harris. 1968. The influence of cattle management on deer and elk. *Trans. North Am. Wildlife Nat. Resources Conf.* **33:** 169–181.

Urban, D. 1970. Raccoon populations, movement patterns, and predation on a managed waterfowl marsh. *J. Wildlife Management* **34**(2): 372–382.

Walther, F. R. 1969. Flight behaviour and avoidance of predators in Thompson's gazelle. *Behaviour* **34**(3): 184–221.

Courtesy of Paul M. Kelsey
New York State Department of Environmental Conservation

PART

ENERGY FLUX AND
THE ECOLOGICAL ORGANIZATION
OF MATTER

Discussions of weather and food must ultimately be synthesized into a dynamic energy exchange for an organism if the fundamental life processes that permit survival and production are to be understood. This is true for an individual, and when a collection of individuals is regarded as a community, the community assumes a degree of organization that is regulated by basic energy flow plus the added behavioral interactions that are superimposed on the energy framework. Much can be said about basic energy exchange because of the recent interest in this approach to ecology. Given this background information for the organism, there is a rich potential for analytical studies of plant and animal communities. This part of the book includes fairly detailed descriptions of the energy relationships of an individual, with insights into the factors affecting the organization of communities.

THERMAL ENERGY EXCHANGE
BETWEEN ORGANISM
AND ENVIRONMENT

13-1 THERMAL ENERGY EXCHANGE

The effects of weather are manifested in an organism-environment relationship through the medium of thermal or heat exchange. Meteorological parameters alone may be quite meaningless when used in interpreting animal responses because weather instruments do not respond to the thermal regime of the atmosphere in the same way that a living organism does. Further, weather instruments are often placed in standard weather shelters at standard heights so that as many variables as possible are eliminated. A living organism, however, is exposed to changing weather conditions, and it also has its own physiological and behavioral variables. Knowledge of the basic principles of heat exchange enables the biologist to understand the functional relationships between weather and the living organism.

An organism is *coupled* to energy-exchange processes by certain specific properties of its own (Gates 1963). The amount of heat exchange by radiation, convection, conduction, or evaporation depends on the thermal characteristics of the atmosphere and substrate, such as soil, rock, or snow, and the thermal characteristics of an organism. If a surface is highly reflective to radiant energy, then there can be little thermal effect from radiation. A cylinder with a very small diameter is a very efficient convector and slight air movement can result in a large amount of heat loss. Conversely, a large cylinder is a poor convector. An object covered with a layer of good insulation loses little heat by conduction, and an object with no water or other fluid that can be vaporized can have no heat loss by evaporation.

The thermal characteristics of an organism are related to its physiological and behavioral characteristics also, and these may change drastically in a short period of time. A deer, for example, may be bedded quietly until frightened, when it literally leaves its bed on the run! This results in very abrupt changes in the thermal regime.

It has been traditional to categorize animals as either homeothermic or poikilothermic, but the distinction between the two is not entirely clear. Reptiles can regulate their body temperatures somewhat by behavioral thermoregulation, including changes in activity, location, and posture.

Mammals do not have the same body temperature throughout their entire bodies. Deep body temperature is quite constant, with the temperatures of appendages more closely coupled to the external thermal environment. Hibernating mammals regulate their body temperatures by increasing heat production when their body temperatures approach the freezing point. They are in a sense homeothermic, but at a lower set point.

13-2 THE CONCEPT OF HOMEOTHERMY

Warm-blooded or homeothermic animals are usually described as animals that maintain a constant body temperature. This very simple idea is often presented to students in elementary grades, and it is commonly said that the body temperature of humans is 98.6°F. All humans do not have the same body temperature though, and all parts of the human body are not at the same temperature.

Careful consideration of the basic concept of homeothermy leads to a conclusion that is much more basic than the simple statement that the body temperature remains "constant." Heat energy is produced when food is "burned," and heat is lost to the environment when the environment is colder than the animal. Homeothermic animals regulate the balance between heat production and heat loss. The *effect* of this balance is a relatively constant body-core temperature.

The actual heat exchange between a homeothermic animal and its environment is very complex. The concept is simple, however, and can be illustrated by the example of a heated home in the winter (Figure 13-1). Fuel is burned in the furnace, and the heat energy is distributed throughout the house by water or steam pipes or by air ducts. The amount of heat energy distributed throughout the house is controlled by a thermostat that can be set at a desired temperature.

The fuel in the furnace (oil, gas, coal, or wood) is a source of energy like the food eaten by an animal. The water pipes, steam pipes, or air ducts are like the blood vessels. The thermostat can be compared to the hypothalamus in an animal—a part of the brain that regulates heat production. If the weather gets colder, the fire in the furnace will burn longer in order to maintain a balance between the amount of heat energy released from the furnace and the amount of heat lost from the house. The net result is the maintenance of a constant house temperature. The Astrodome in Houston, Texas, operates in a similar manner, with air conditioning to keep it cool in the summer and furnaces to keep it warm in the winter. There are instruments or "nerve endings" on the outside surface

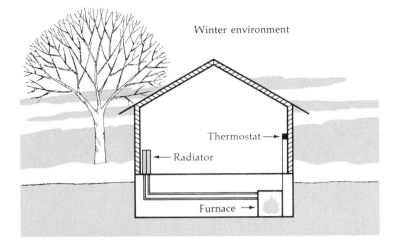

FIGURE 13-1. A heated home is like a homeothermic animal; a balance between the heat energy produced in the heat source (furnace) and the heat energy lost to the heat sink (the winter environment) is regulated by a thermostat so that the house temperature remains quite constant.

of the Astrodome that perceive the amount of energy striking it from the sun, just as a person feels heat energy on the skin surface in bright sunlight!

REGIONS OF THERMAL EXCHANGE.   The exchange of heat between animal and environment occurs in three distinct regions, including the internal thermal region, the boundary region, and the external thermal region. The external thermal region includes both the atmosphere and the substrate (Figure 13-2). The internal thermal region includes all of the body tissue except the hair. Heat flow through body tissue is primarily by conduction and convection, with additional heat exchange by evaporation in the respiratory tract.

The rate of conduction in this region depends on the thermal conductivity of the different kinds of body tissue, including muscle, fat, bones, and other tissue. The circulation of blood results in the distribution of heat within the body by conduction and convection. The sites of active metabolism may have an excess of heat energy, which can be removed by the flow of blood and dissipated in areas in which the body tissue is cooler. Thus the circulatory system is a *thermal transport system* as well as an oxygen and carbon dioxide transport system.

The boundary region is a thermally active region in which the influence of the animal's surface is exerted, both physically and thermally, on the external thermal region. Temperature differences between the internal thermal region of animals and the external thermal region (the atmosphere) often exist, resulting in temperature gradients that characterize the boundary layer. Further, the boundary region surrounding birds and mammals includes the feather-air and hair-air interfaces, respectively.

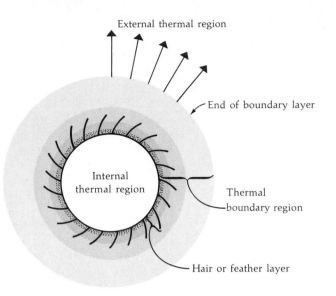

FIGURE 13-2. The three regions of heat exchange between
animal and environment: internal thermal region, boundary
region, and external thermal region.

The hair-air interface is a fibrous and porous medium with a density gradient
from the base to the tips of the hairs, while a feather-air interface has a density
gradient and a considerable amount of lateral overlap (Figure 13-3). These physical
characteristics result in a very complex heat transfer that is made even more
complex by the pliability of the hair or feather coat. Movement of the hairs or
feathers increases the rate of heat transfer because it disturbs the hair-air or
feather-air geometry.

The relative importance of each mode of heat transfer varies across the interface
layer. Conduction is the most important mode at the base of the hair because
the hairs are tightly packed together and there is little air movement. Heat is
conducted along the shafts of the hairs (or feathers) and through the air that is
trapped between them. The space between the hairs in the outer portions of the
hair-air interface is greater and the hair shafts are more exposed to the environment. Thus there is radiation exchange between the surfaces of the hair shafts
and the surroundings. The decreased density of the hair permits a larger amount
of air movement so convection processes accelerate. Each hair functions as a little
convection cylinder exposed to both free and forced convection, and the proportion of free and forced convection is dependent on the air motion at the site of
convection and the amount of penetration by the wind. The overlapping feathers
function in a similar manner, with some modifications due to differences in
physical morphology.

Evaporative heat exchange through the interface results if energy is involved
in the conversion of secretions of the skin or of rain or snow from a liquid to
a gaseous phase. The rate at which this change of phase occurs is dependent on

the movement of water molecules in the interface. A very dense hair layer will trap water molecules at the skin surface and will also prevent the penetration of rain or melted snow. The oily characteristic of an animal's coat further enhances its protective characteristics. The destruction of normal feather structure by oil on the surface of ducks destroys its insulative and waterproof characteristics, resulting in a high rate of heat flow through the feather-air interface.

ANALYSES OF HOMEOTHERMIC RELATIONSHIPS. A basic concept that emerges from the consideration of all of the thermal relationships between animal and environment is that the thermal exchange of an animal is centered on the heat flow through the boundary region. This region, including hair or feather insulation, is a barrier to both heat absorption and heat dissipation. If the functional thermal characteristics of the boundary region are known, then the calculated heat flow through that layer will be in balance with the heat production of the animal and with the heat loss from its surface, with the additional consideration of the heat lost by respiratory evaporation.

There are two approaches to the study of homeothermy. One is to measure the heat production of an animal in a chamber, which can be assumed to be equal to the heat loss as long as the body temperature remains stable. The other is

FIGURE 13-3. A schematic representation of the structural characteristics of hair and feathers that determine the types of heat transfer through the hair or feather layer.

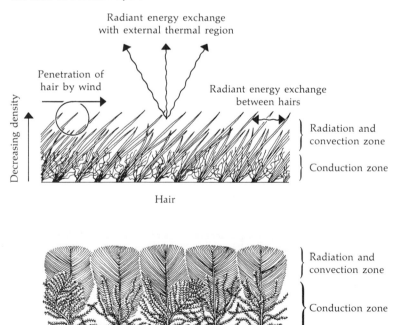

to measure the heat loss, which can be assumed to be equal to the heat production if body temperature remains stable.

The use of heat-production measurements as analogs to heat loss is valid, but the means by which heat is lost are usually not considered in this type of analysis. An animal held in a refrigerated metabolism chamber exhibits an increase in heat production as ambient conditions in the chamber become colder. This provides very useful information on the metabolic potential of the animal, but the chamber temperature does not represent outdoor conditions in which so many other thermal factors are present. The use of a mask on an animal held outdoors provides a more realistic exposure of the animal to natural weather conditions, but the limitations imposed by the mask prevent the animal from exhibiting normal behavioral responses.

The calculation of heat loss is very complex owing to the complicated geometry of the animal and the very labile nature of the thermal energy regime that is commonly referred to as weather. The heat loss must be equivalent to the heat production except for transient fluctuations over short periods of time.

The synthesis of heat-production and heat-loss measurements into a single unified concept of heat exchange is the most logical approach to the analysis of the energy requirements of a free-ranging animal. The animal and its thermal environment are a thermal system, a homeostatic one; and analyses of the relationships between the two should include the recognition of every system component.

13-3 MEASUREMENT OF THERMAL PARAMETERS

CONDUCTION COEFFICIENTS. A thermal conduction coefficient (k) is an expression of the rate of heat flow through an insulating medium per unit depth for each degree of difference in temperature. A perfect insulator would have a k value of zero. The conduction coefficient is a necessary parameter for the calculation of heat flow by conduction. The thermal conductivity coefficients of wild ruminants reported in the literature are shown in Table 13-1.

Conduction coefficients cannot be expressed as single values for each species, however, since the density of the hair or feather layer, the compression and wetness of the hair or feathers, inclination of the shafts, air temperature, radiation, and wind all affect the conductivity. Hammel (1953) has measured the conductivity of summer and winter coats of several mammals, and the greater insulation value of the winter coat seems to be primarily a function of depth.

The rate of heat flow through fibrous material is high when the fiber density (ratio of fiber material to air) is low. It decreases as the density of the fiber increases up to a certain point, and it increases again as the density increases further. Convection and radiation exchange are also important, especially at the lower densities, increasing the total heat flow. Berry and Shanklin (1961) found that cattle with denser hair coats had a higher heat transfer; the hair shafts are better conductors than the air between the hairs. There is less air space in denser hair and in compressed hair, and the amount of air in the hair layer is a very important characteristic affecting conductivity (Hammel 1953).

TABLE 13-1 THE INSULATION VALUES OF FURS

Pelt	Thickness of Fur (cm)	Conductivity per Cm of Hair Layer (kcal m^{-2} hr^{-1} $°C^{-1}$)	Total Conduction
Virginia deer no. 1* (early fall, fresh pelt)	1.7	3.32, 3.30	1.95, 1.94
Caribou no. 1* (fresh winter pelt)		3.29, (3.30)	1.94, 1.94
(air 7°)	2.5	2.88	1.15
Caribou parka* (air 24°)	1.5	3.26	2.17
Caribou† (winter)	3.28	3.81	1.18
Caribou† (summer)	1.19	3.13	2.66
Caribou† (thin summer)	0.74	3.06	4.17
Deer† (winter)	2.39	3.06	1.28

* Data from Hammel 1953.
† Data from Moote 1955.

The compression of fibrous insulation such as hair or feathers is an important factor in determining the rate of heat loss by conduction from the animal. A bedded deer, for example, compresses the hair on its legs and trunk. The snow under the deer melts as heat is conducted from the body through the compressed hair. Direct measurements of heat flux under deer indicate that this heat loss is important when considering the energy requirements of the deer (Moen, unpublished data).

The amount of moisture in the coat has a marked effect on the rate of heat flow through the conduction zone in the hair layer. This was indicated by the large increase in the metabolism of infant caribou when their coats were wet (see Table 7-7).

The inclination of the hair or feather shafts affects the rate of conduction through these media. The lowest rate of conduction occurs when the fibers are parallel to the plane of the insulation and perpendicular to the mean direction of heat flow, and the highest when the fibers are perpendicular to the plane of the insulation and parallel to the direction of heat flow (Figure 13-4).

Hair in normal lie falls between the parallel and perpendicular extremes. It has its maximum insulation per unit depth when in normal inclination. Piloerection results in a decrease in the insulation value of the fur per unit depth, but Hammel (1953) has data that indicate that the increase in depth compensates for the change in hair inclination. The total insulation of the erect fur was slightly greater than the total insulation of fur in normal position. These relationships have also been observed by Berry and Shanklin (1961) for cattle. The inclination

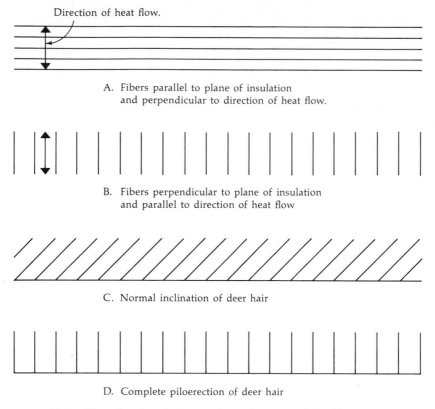

Direction of heat flow.

A. Fibers parallel to plane of insulation
   and perpendicular to direction of heat flow.

B. Fibers perpendicular to plane of insulation
   and parallel to direction of heat flow

C. Normal inclination of deer hair

D. Complete piloerection of deer hair

FIGURE 13-4.   Heat flow in relation to the inclination of the fibers (arrows in-dicate direction of heat flow): A is the best insulator, B the poorest. Normal inclination of deer hair is between A and B.

of the hair had more influence on hair-coat conductivity than other measured parameters. Increasing depth-to-length ratios increased the conduction of heat, but the increase in hair depth caused a decline in the conduction loss through the entire length of the hair coat.

The temperature of the air around the fur affects the conduction rate through the fur layer itself. The experiments of Hammel (1953) show that the insulation of dog, hare, and fox pelts was greater in colder temperatures. This is to be expected since the conductivity of air has a major influence on conduction through the hair-air interface, and air has a lower thermal conductivity at lower tempera-tures. This was expressed in equation (6-8).

The rate of heat flow through the hair-air interface is affected by both the velocity of the wind and the angle at which the wind strikes the hair or feather surface. The effective angle is not a single value determined from the mean wind direction, but is an extremely variable parameter that is dependent on the geome-try of the animal in relation to the mean wind direction and on the turbulent wind-flow characteristics at the animal's surface.

The insulation value of fibrous material decreases as the wind velocity increases and the angle at which the wind strikes the surface increases from zero or parallel flow. Experiments on merino sheep fleece show this effect (Allen et al. 1964), and higher conduction or lower insulation values for rabbit fur exposed to higher wind velocities were reported by Tregear (1965). Tracy (1972) discusses conductivity through an animal's hair coat, pointing out that wind speed is an important consideration in determining the overall conductance of an animal. A similar conclusion has been reached for white-tailed deer also (Stevens 1972) and for sharp-tailed grouse (Evans 1971).

TEMPERATURE-PROFILE MEASUREMENTS.   Measurements of the thermal characteristics of the hair-air layer from the trunks of white-tailed deer have been completed in the Thermal Environment Simulation Tunnel (TEST) (Figure 13-5) at the BioThermal Laboratory. Simulations (Figure 13-6) are used in the TEST to reduce the experimental conditions to a physical system devoid of biological variability. This permits an analysis of the mechanisms of heat flow according to the basic principles of thermal engineering. These measurements are then compared with results of similar measurements of live deer with their biological variability superimposed on the effects of uncontrolled weather factors, including wind, radiation, and temperature.

The validity of results from a simulation is often questioned. My own research at the University of Minnesota indicated agreement between measurements of simulations and those of live deer, and this has been confirmed in current experiments. The agreement appears to result from the dominant effect of the hair layer

FIGURE 13-5. The TEST is used to measure the thermal characteristics of animal simulators exposed to controlled wind and radiation conditions.

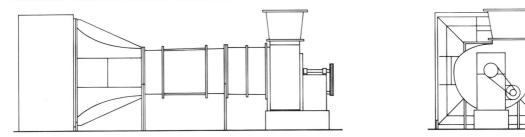

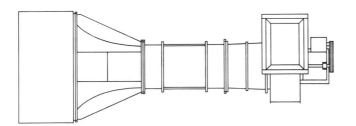

0 _____ 5 Feet

as a thermal barrier. This barrier is primarily a physical one and is therefore subject to the physical laws of heat transfer with little regulation by the animal of the thermal characteristics of the hair layer itself. Further, simulations are used to understand the mechanisms of heat flow under controlled conditions, and not necessarily to provide actual data that are numerically equal to values for live animals.

TEMPERATURE PROFILES IN FREE CONVECTION.   The first consideration of the thermal characteristics of the deer simulator is the nature of the temperature profile under stable radiation conditions and no wind (Figure 13-7). Temperature measurements with thermocouples at different depths indicate that the boundary layer of temperatures extends from the skin surface to about 4 cm. The temperature field within this 4-cm zone is a function of the thermal characteristics of the simulator, and this effect extends beyond the hair itself. At lower air temperatures the gradient is steeper, at higher temperatures the gradient is less steep, and at 38°C the gradient line would be vertical. The very systematic shape to these curves indicates that the relationship between air temperature and the temperature profile is a predictable one.

TEMPERATURE PROFILES IN FORCED CONVECTION.   Wind has an effect on the distribution of temperatures in the boundary region. A wind velocity of 6 mi hr^{-1} results

FIGURE 13-6.   A schematic drawing of a deer simulator used in the TEST.

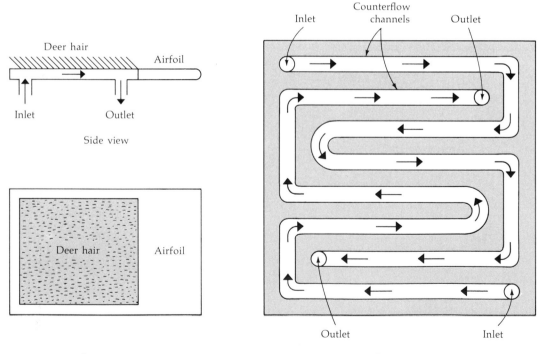

Side view

Front view

Rear view

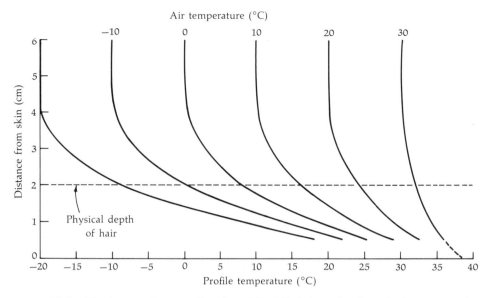

FIGURE 13-7. The temperature profile of a white-tailed deer simulator in free convection in the TEST. (Data from Stevens 1972.)

in a compression of the temperature field in deer hair to a depth of about 3 cm (Figure 13-8). The effect of wind on the compression of the temperature field at an air temperature of $+20°C$ is less than at $-20°C$. As velocities increase, the profile is compressed more (Figure 13-9).

A distinctly nonlinear effect of wind on the temperature profile and surface temperature of deer hair has been observed in our experiments on white-tailed deer and is also reported by Tregear (1965) for rabbits, horses, and pigs. The lower wind velocities have a proportionately greater effect than high velocities, and the temperature depression in the boundary layer is also related to the density of the hair. The nonlinear effects of velocity on the temperature field in deer hair are evident in Figure 13-9. At an air temperature of $0°C$ an increase in wind velocity from 0 to 6 mi hr^{-1} results in a $3.8°C$ decrease in surface temperature (from point A to point C), but an increase from 6 to 14 mi hr^{-1} causes an additional decrease in surface temperature of only $1.5°C$ (from point C to point E).

Simulator measurements of sharp-tailed grouse show a pattern very similar to that for deer (Figure 13-10). The temperature profile is steepest in the feather layer, with the end of the boundary region at about 7 cm.

Figures 13-7 through 13-10 show the characteristics of the temperature profile of the simulators when the wall infrared temperatures in the experimental tunnels are in equilibrium with air temperature. This is similar to the thermal conditions in the field when an animal is under heavy coniferous cover or the sky is overcast at night.

TEMPERATURE PROFILES UNDER RADIANT ENERGY LOADS. Absorbed radiant energy is important in the maintenance of homeothermy. This energy becomes a part

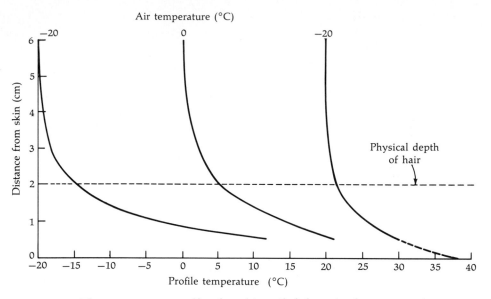

FIGURE 13-8. The temperature profile of a white-tailed deer simulator exposed to a wind velocity of 6 mi hr^{-1} in the TEST. (Data from Stevens 1972.)

FIGURE 13-9. Temperature gradients in the thermal boundary region of a deer simulator in an air temperature of 0°C. A, B, C, D, and E indicate radiant surface temperature. (Data from Stevens and Moen 1970.)

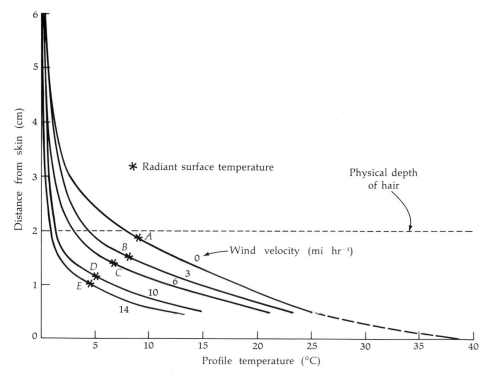

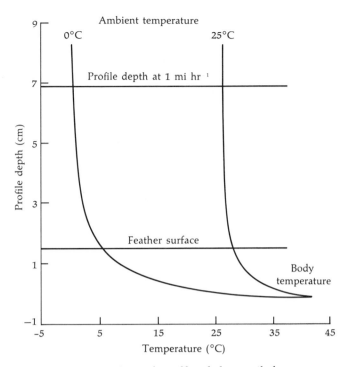

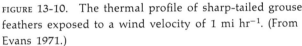

FIGURE 13-10.   The thermal profile of sharp-tailed grouse
feathers exposed to a wind velocity of 1 mi hr^{-1}. (From
Evans 1971.)

of the thermal regime of the animal, and it can be dissipated by any one of the
four modes of heat transfer. The amount of energy absorbed by the hair surface
of a mammal depends on the spectral characteristics of the hair and the angle
at which the solar energy strikes the surface. The absorption coefficients for cattle
have been measured by Riemerschmid and Elder (1945) and were shown in Figure
6-3. White coats absorb less solar energy and reflect more; black coats absorb
the most solar energy. The greatest amount of energy is absorbed when the solar
radiation strikes perpendicular to the surface, with no absorption when the rays
are parallel to the surface. Since animals are not plane surfaces but have a rather
complicated geometry, the absorption characteristics of a whole animal include
all angles from 0 to 90 degrees. Inclination of the hair, the smoothness or curliness
of the coat, and seasonal changes in the characteristics of the coat change the
absorptivity by less than 2%. The distribution of solar radiation on the surface of an animal is called the solar-radiation profile, which was discussed in
Chapter 6.

The infrared emissivities of most biological materials are close to 1.0. Several
hair surfaces have been tested, with measured emissivities ranging from 0.92 to
1.0 for several species (Table 6-2). An emissivity of 1.0 is usually a satisfactory
approximation; the error in using this approximation increases as the temperature
difference between two surfaces exchanging radiant heat increases. The error is

usually small in the range of temperatures experienced by animals in natural habitats.

Experiments in the TEST with an infrared lamp as a radiant energy source of 0.5 cal cm^{-2} min^{-1} show that at an air temperature of 20°C, the temperature profile in deer hair has an inversion at wind velocities of 0 and 1 mi hr^{-1} (Figure 13-11). As wind velocity increases, the effect of radiation on the temperature profile is reduced. Radiant energy absorbed in the outer portions of the hair layer is removed by convection so that the net effect on the animal's thermal regime may be considerably less than the measurements of radiation flux alone would indicate. This convection process is very efficient at the hair surface since each hair acts as a tiny convection cylinder and all of the hairs together expose a very large surface area to the wind.

The same general pattern in the temperature profile is observed when the air temperature is −20°C, except that the inversions do not occur at the low wind velocities because the relative effect of the radiation is less at low air temperatures. At 14 mi hr^{-1} the effect of radiation is negligible; the profiles are essentially the same as those in Figures 13-7 through 13-10.

13-4 RADIANT SURFACE TEMPERATURE RELATED TO AIR TEMPERATURE

There is a predictable relationship between the radiant temperature (T_r) of an animal and the air temperature if radiation and convection conditions are held constant. The difference between the radiant temperature and air temperature is greatest when the air is cold and the wind velocity high. This was illustrated for both deer and grouse in Figures 6-8, 6-9, and 6-10.

FIGURE 13-11. The temperature profile of a white-tailed deer simulator at two air temperatures and five wind velocities when exposed to thermal radiation in the TEST. (Data from Stevens 1972.)

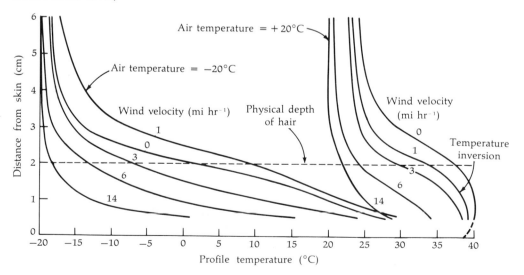

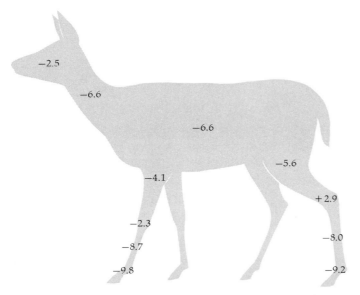

FIGURE 13-12. Radiant temperatures (in °C) on eleven body parts
of white-tailed deer exposed to an air temperature of −20°C.
(From Stevens 1972.)

Radiant temperature measurements taken at fifty-one points on deer have been
completed throughout two winters at the BioThermal Laboratory. A "thermal
profile" composed of regression equations expressing the radiant surface temperature in relation to air temperature has been calculated (see Appendix 4). An
illustration of radiant temperature distribution on eleven body parts at an air
temperature of −20°C is shown in Figure 13-12. Note that at an air temperature
of −20°C the radiant surface temperature of the deer varied from a high of 2.9°C
on the hock to a low of −9.8°C on the lower part of the front leg. These temperature differences can be explained by the anatomy and physiology of the circulatory
system, differences in hair depth on different parts of the body, wind flow, and
other variables.

Maximum and minimum trunk temperatures of deer during each measurement
period are expressed with regression equations and compared with the regression
equation for a simulator in Figure 13-13. A slightly greater variability in the results
for a live deer (compare the r values) is expected since more biological factors
are involved, and also the wind conditions in the outdoor pen are more variable
than those in the TEST.

Radiant temperature measurements on sharp-tailed grouse show a similar
general pattern, with much less variability over their surfaces because of their
simplified geometry. The data points and correlation coefficient for the sharp-
tailed grouse tests at 1 mi hr^{-1} are shown in Figure 13-14.

A final generalization on the effect of wind and radiation on the thermal
boundary region is that the effect of radiant energy is greatest at high air temperatures and low wind velocities and least at low temperatures and high wind

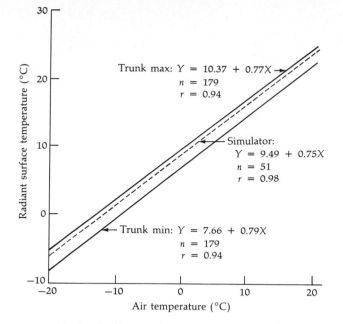

FIGURE 13-13. Radiant surface temperature in relation to air temperature for a flat-fur simulator and the body trunk of live deer in free convection. Maximum and minimum values are presented for the deer. (From Stevens 1972.)

FIGURE 13-14. Effect of air temperature on feather surface temperature of sharp-tailed grouse at a wind velocity of 1 mi hr^{-1}. (From Evans 1971.)

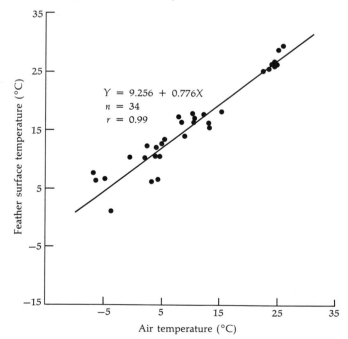

velocities. This suggests that the reduction of wind velocities is one of the most important benefits of cover in the winter. The thermal benefits due to different levels of radiation from different types of cover can be evaluated only if free or forced convection is considered also.

13-5 THE CONCEPT OF THERMAL DEPTH

One concept that has emerged from the experiments in the TEST is that of thermal depth (d_t). This concept is apparent in that obvious changes take place in the characteristics of the thermal boundary region without corresponding changes in the physical depth of the hair. Changes in the shape of the temperature profile are related to air temperature, absorbed radiation, and wind velocities. The radiant temperature also changes.

A temperature profile shows the temperature in the hair-air interface at any given depth. The radiant temperature is an integration of the vertical and horizontal distribution of the temperature field exposed to the environment. This integration results in an average radiant temperature, which can be compared with the temperatures on the profile. The distance between the point on the profile at which the radiant temperature equals the profile temperature and the base of the hair can be determined, and this distance can be considered the thermal depth.

The effective thermal depth of the hair on the trunk of a deer is shown in Figure 13-15. As wind velocities increase, there is a reduction in d_t, equaling nearly 50% of the physical depth when the wind velocity is 14 mi hr^{-1}. A similar pattern is observed for sharp-tailed grouse, although the thermal depth is relatively less

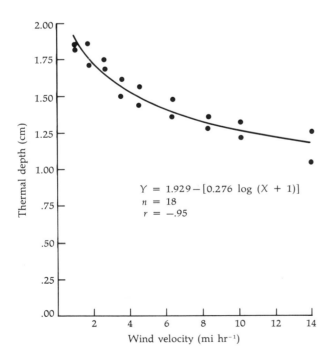

$$Y = 1.929 - [0.276 \log (X + 1)]$$
$$n = 18$$
$$r = -.95$$

FIGURE 13-15. Effective thermal depth of deer hair on the simulator when exposed to different wind velocities. (From Stevens 1972.)

than the physical depth throughout the entire range of wind velocities (Figure 13-16). Current experiments at the BioThermal Laboratory indicate that the relationship between d_t and wind velocities depends on several factors, including hair structure and wind turbulence; the data in Figures 13-15 and 13-16 are applicable only to the flow characteristics during those tests.

13-6 GEOMETRY AND SURFACE AREA

Surface area is one of the most obvious parameters necessary for the calculation of the heat loss from an animal. The measurement of this parameter is exceedingly difficult, however, because of the complex geometry of an animal's body and its hair or feather surface. Physiologists have used many different methods, including skinning the animal and measuring the flat hide, using rollers across the animal's surface, and direct measurement with a tape.

Direct measurements with a fiberglass tape have been completed on 309 white-tailed deer ranging in age from a few days to several years (Moen, unpublished data). Twenty-two linear and circumferential measurements of each animal have been made (Figure 13-17), and the areas have been calculated using equations for cylinders, frustums of cones, and rectangles (see Appendix 5). The regression equations expressing the total surface area in relation to body weight for male and female deer are shown in Figure 13-18, with data points plotted

FIGURE 13-16.   Effect of wind speed on thermal depth of breast and back feather tracts of sharp-tailed grouse. (From Evans 1971.)

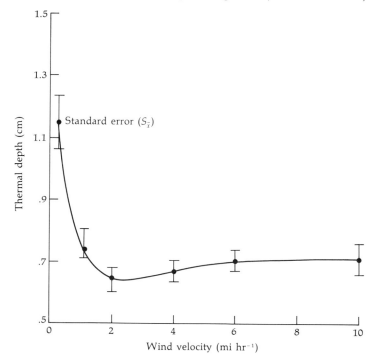

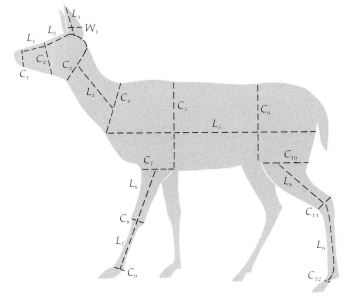

FIGURE 13-17.   Twenty-two measurements (C = circumference, L = length, W = width) have been used to calculate the surface area of white-tailed deer. The subscripts are identified in Appendix 5.

FIGURE 13-18.   The relationship between surface area and body weight of white-tailed deer. The data points are for females only.

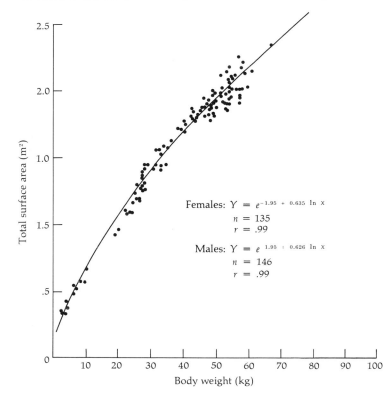

for females only. Variations between individual deer are fairly small; the correlation coefficients are high. Measurements taken by different people have also been found to be quite consistent. Equations for the head, neck, ears, trunk, upper front leg, lower front leg, upper hind leg, and lower hind leg for female white-tailed deer are listed in Appendix 5.

Measurements of the geometric components of sharp-tailed grouse have been made by Evans (1971). The equation is

$$\text{Surface area (cm}^2) = 7.46 \ W^{0.652}$$

Similar measurements have been made for pheasants up to 16 weeks of age at the BioThermal Laboratory (Figure 13-19). Regression equations for the surface area of the beak, head, neck, body, upper leg, metatarsus, and toes of pheasants are found in Appendix 5.

The several geometric portions of an animal's physical surface participate differently in heat exchange. The thin legs exchange heat at different rates than the trunk, for example. This suggests that it is necessary to consider a thermal profile that is based on the thermal characteristics of different body parts in

FIGURE 13-19.   Total surface area of ring-necked and Japanese green pheasants in relation to weight. Replicates are indicated by circles.

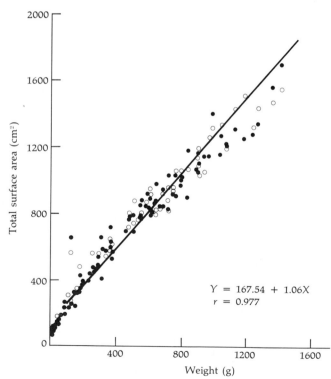

$$Y = 167.54 + 1.06X$$
$$r = 0.977$$

relation to their surface area, orientation, and other thermal characteristics and the distribution of thermal energy in the environment. This was discussed in Chapter 6 (see Figure 6-8) with reference to solar radiation.

13-7 THE CALCULATION OF HEAT LOSS

The complexity of heat exchange between animal and environment is beyond the capabilities of real-time analyses. The alternative is the use of a model in which some of the variables in the thermal equation are held constant. Posture is a variable, and two model postures—standing and bedding—can be compared. Wind velocities are continuously variable, and discrete velocities can be used in an analysis of wind effects.

The hair-air interface is the thermal barrier through which heat is conducted from the heat source to the heat sink. The metabolic heat lost from the surface of an animal passes through the hair-air interface at a rate that is dependent on the conduction characteristics of the hair layer. When the conduction characteristics are known and the radiation characteristics have been identified for particular weather conditions, the heat lost by convection can be determined as a residual. This basic concept is expressed in equation (13-1).

$$Q_k = Q_r + Q_c + Q_v \qquad (13\text{-}1)$$

where

Q_k = heat lost by conduction through the hair-air interface
Q_r = heat lost by radiation from the thermal surface of the animal
Q_c = heat lost by convection from the thermal surface of the animal
Q_v = heat lost by evaporation

Specific values used in equation (13-1) are dependent on the particular model being analyzed. Appropriate geometric data for the model selected are used for the calculation of convection. A deer in a standing posture is composed mostly of cylinders and frustums of cones. Convection coefficients for these geometries have been calculated using the measured dimensions of deer. The vertical distribution of air flow past a deer in a standing posture varies, with the lower portions of the animal's body exposed to lower mean air velocities than the upper parts. The wind profile can be calculated as illustrated in Chapter 6. Evaporative heat loss is not considered in this model because its quantity appears to be quite small from the surface of a deer in cold weather.

Heat loss by conduction through the hair-air interface of a 60-kg deer standing in an open field with uniform radiation is shown in Figure 13-20. The heat loss by nonevaporative means is much greater at $-20°C$ than at $+20°C$. The increasing effect of wind at $-20°C$, indicated by the steeper slope of the line, illustrates the greater importance of wind when the air temperature is low.

Note the convex curvature of the lines. This is due to the reduced effect of wind as velocities increase. This was illustrated for cylinders in Chapter 6 (Figure 6-17). The complex geometry of both the hair and the entire animal precludes the measurement of local convection losses directly, but they can be determined as a residual when the other factors in equation (13-1) are known. When the convection losses are known, the "convective diameter" of the deer can be calculated.

The convective diameter (Figure 13-21) is an expression of the effective thermal diameter of the deer in relation to the efficiency of the convective process. Note that there is a marked increase in the convective diameter as the wind velocity increases from 0 to 1.5 mi hr^{-1}. This velocity range includes a transition from free to forced convection; it also indicates a rapid decrease in the efficiency of convection inasmuch as the tiny hairs become quickly linked to convection processes but the larger body cylinders are not yet efficient convectors. This was illustrated in Chapter 6 (Figure 6-17); the larger diameters are less efficient convectors, especially at the low wind velocities. The smaller thermal diameter observed when the air temperature is −20°C illustrates the effect of increased convective losses at low air temperatures. The deer behave thermally like a smaller cylinder in the colder air.

The curves shown for each air temperature in Figure 13-21 reflect the separate curves that might be projected for the hair and the body separately, as indicated by the dashed lines (+20°C T_a). Although these considerations may seem a bit detailed, they deserve some thought because of the indication that hair and body structure may play a more important role in heat loss than conductivity comparisons alone would indicate. Tracy (1972) comes to the same conclusion with respect to body size and the effect of wind on the conduction of heat through the hair layer of a mammal.

Further documentation of the importance of the role of the hairs in relation to the body is provided by the experiments on the interaction between radiation and convection discussed earlier in this chapter. The simulated cold night sky

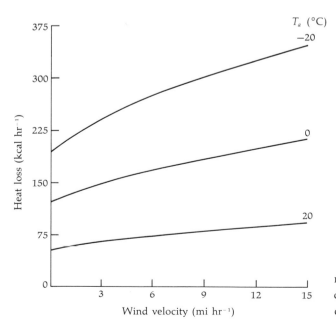

FIGURE 13-20. The calculated heat loss by conduction through the hair of a 60-kg deer. (From Stevens 1972.)

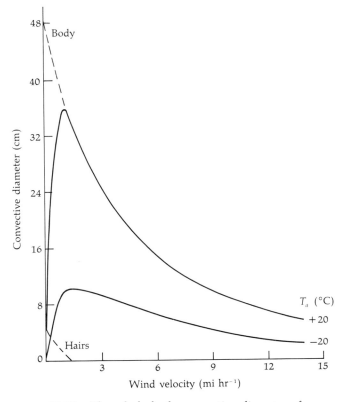

FIGURE 13-21. The whole body convective diameter of a
60-kg deer, expressing the relationship between gross body
diameter and hair diameter. (From Stevens 1972.)

was effective in reducing the radiant temperature of the hair surface only when
the wind velocity was less than 1 mi hr^{-1} (Moen, unpublished data). At velocities
greater than 1 mi hr^{-1}, the heat-sink effect was not observed, indicating that the
radiant temperature of the hairy surface was linked to air temperature rather
than radiant sky temperature. Since the clear sky at night is a heat sink, the effect
of wind flow across the hairy surface is "convection in reverse" inasmuch as heat
energy is added to the hair layer from the air in motion over the hairy surface.
This process is called advection.

The heat lost by conduction through the hair-air interface equals the sum of
the radiation and convection losses from the thermal surface of the animal (Figure
13-22). Convective losses increase at higher wind velocities. Radiation losses
decrease at higher wind velocities, because the cooling effect of wind reduces the
radiant temperature of the animal's thermal surface.

The relative importance of convection and radiation expressed in Figure 13-22
indicates that the effect of wind is much more important than the effect of radiant
energy in the cold, winter habitat. This suggests that the primary value of overhead
cover is its effect on wind velocities rather than its higher radiation flux. Early

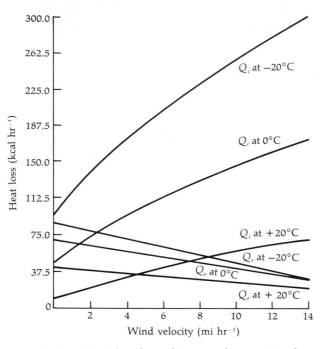

FIGURE 13-22.   Heat loss by radiation and convection from the thermal surface of a 60-kg white-tailed deer. Q_c = convection loss; Q_r = radiation loss. (From Stevens 1972.)

FIGURE 13-23.   Percentages of the total heat loss attributed to convection from a 60-kg deer in a standing posture at night. (From Stevens 1972.)

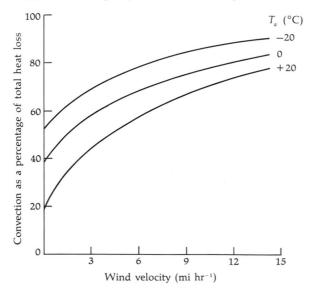

stages in succession—such as the shrub stage—may be very effective in reducing wind velocities and thus may be as good a thermal cover as a dense canopy.

The percentage of nonevaporative heat loss at three air temperatures that can be attributed to convection is shown in Figure 13-23. As wind velocities increase, convection losses increase in a nonlinear fashion, indicating a reduction in the efficiency of convection at the higher velocities. Convection is more important at low air temperatures, with free convection making up over 50% of the heat loss at −20°C, and rising to about 90% at that air temperature with a wind velocity of 14 mi hr^{-1}. These data are for a 60-kg deer; a smaller deer would have higher convective losses.

Calculating the percentage of the convective loss from different geometric components of the deer is interesting because theory suggests that the smaller cylinders, such as the lower parts of the legs, should be significant contributors to the convective heat loss. The actual percentages at different wind velocities are highest for the trunk (Table 13-2), but this is due to the larger surface area of the trunk and to the position of the trunk at a higher point in the vertical wind profile. The legs, with their smaller diameters, show a convective loss that is relatively greater than their surface-area percentages, thus conforming to theory on the effects of cylinder size.

One interesting characteristic of Table 13-2 is that the convection losses from the lower legs change less than the losses from the trunk at different wind velocities. This is due to the small cylinder size—the legs become linked quite quickly to low wind velocities—and to their position in the lower part of the wind velocity profile.

TABLE 13-2 PERCENTAGE OF HEAT LOSS BY CONVECTION AND RADIATION FROM DIFFERENT BODY PARTS OF A 60-KG DEER IN A STANDING POSTURE IN AN OPEN FIELD AT NIGHT WITH AN AIR TEMPERATURE OF −20°C

Body Part	Percentage of Surface Area	Percentage of Heat Loss at −20°C and Different Wind Velocities (mi hr^{-1})									
		Convection					Radiation				
		0	2	4	8	14	0	2	4	8	14
Head	6.6	4.6	4.8	4.8	4.7	4.5	8.2	8.9	9.8	11.9	17.1
Ears	2.4	9.4	7.9	7.2	6.4	5.8	3.1	3.4	3.7	4.6	6.5
Neck	7.6	3.2	4.2	4.7	5.1	5.3	7.5	7.2	7.0	6.4	5.2
Trunk	56.5	35.2	39.9	41.8	43.6	44.6	54.9	53.9	52.7	49.8	43.3
Upper front legs	5.0	7.4	6.6	6.2	5.8	5.5	5.7	6.1	6.5	7.5	10.0
Lower front legs	4.5	18.5	15.9	14.9	14.2	13.9	3.6	3.4	3.2	2.5	0.8
Upper hind legs	9.7	4.9	5.5	5.8	6.1	6.3	9.9	10.2	10.5	11.2	12.8
Lower hind legs	7.7	16.5	14.7	14.2	13.8	13.8	6.6	6.5	6.3	5.7	3.9

SOURCE: Data from Stevens 1972.

The percentages of the radiant heat loss from each geometric part show that the major portion of radiation loss is from the trunk. This is reasonable because of the trunk's sluggish response to convective forces and its large surface area. The trunk has a higher radiant surface temperature than the legs, and radiation loss is proportional to this radiant temperature.

The previous illustrations have included considerations of the radiant temperature profile of the animal, the thermal depth of the hair, the vertical wind profile, and the geometry of the animal in a standing posture. Convective losses have been treated as a residual, and the results are in agreement with basic theory. Knowledge of the convective loss permits the calculation of convection coefficients for the *whole* animal as illustrated in Figure 13-24, with weights from 20 to 100 kg and an air temperature of −20°C. The smallest deer loses less heat by convection than the largest one.

Expression of the convection coefficients on a unit-area basis indicates the effect of body size on the efficiency of heat loss (Figure 13-25). More heat is lost by convection from a square meter of surface of a small deer than from a square meter of surface of a large deer. The point of inflection is at a weight of 35 to 40 kg, indicating that animals over that weight become relatively less efficient as convectors and, conversely, more efficient in the conservation of heat energy. This weight range is similar to the average weight of fawns on many ranges in late autumn. These calculations lend substantiative evidence to the idea that fawns entering the winter period at below-average weights have certain physical laws operating against their chances of survival.

FIGURE 13-24.   Convection coefficients for the total surface area of white-tailed deer at an air temperature of −20°C. (From Stevens 1972.)

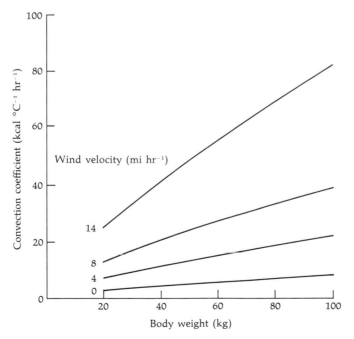

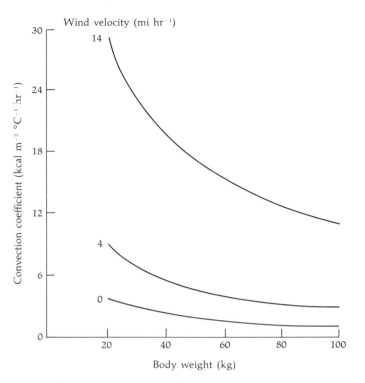

FIGURE 13-25. Convection coefficients calculated on a whole-body basis but expressed as kilocalories per square meter of surface area per hour. The air temperature is −20°C. (From Stevens 1972.)

LITERATURE CITED IN CHAPTER 13

Allen, T. E., J. W. Bennett, S. M. Donegan, and J. C. D. Hutchinson. 1964. Moisture in the coats of sweating cattle. *Proc. Australian Soc. Animal Prod.* **5:** 167–172.

Berry, I. L., and M. D. Shanklin. 1961. Environmental physiology and shelter engineering with special references to domestic animals; physical factors affecting thermal insulation of livestock hair coats. *Mo. Agr. Exp. Res. Bull.* **802:** 1–30.

Birkebak, R. C., R. C. Birkebak, and D. W. Warner. 1963. Total emittance of animal integuments. Paper #63-WA-20, presented at the Winter Annual Meeting, Philadelphia, Nov. 17–22, of the American Society of Mechanical Engineers.

Evans, K. E. 1971. Energetics of sharp-tailed grouse (*Pedioecetes phasianellus*) during winter in western South Dakota. Ph.D. dissertation, Cornell University, 169 pp.

Gates, D. M. 1963. The energy environment in which we live. *Am. Scientist* **51**(3): 327–348.

Hammel, H. T. 1953. A study of the role of fur in the physiology of heat regulation in mammals. Ph.D. dissertation, Cornell University, 105 pp.

Hammel, H. T. 1956. Infrared emissivities of some arctic fauna. *J. Mammal.* **37**(3): 375–378.

Moote, I. 1955. Insulation of hair. *Textile Res. J.* **25:** 832–837.

Riemerschmid, G., and J. S. Elder. 1945. The absorptivity for solar radiation of colored hairy coats of cattle. *Onderstepoort J. Vet. Sci. Animal Ind.* **20**(2): 223–234.

Stevens, D. S. 1972. Thermal energy exchange and the maintenance of homeothermy in white-tailed deer. Ph.D. dissertation, Cornell University, 231 pp.

Stevens, D. S., and A. N. Moen. 1970. Functional aspects of wind as an ecological and thermal force. *Trans. North Am. Wildlife Nat. Resources Conf.* **35:** 106–114.

Tracy, C. R. 1972. Newton's law: its application for expressing heat losses from homeotherms. *BioScience* **22**(11): 656–659.

Tregear, R. T. 1965. Hair density, wind speed, and heat loss in mammals. *J. Appl. Physiol.*
20(4): 796–801.

SELECTED REFERENCES

Alderfer, R. G., and D. M. Gates. 1971. Energy exchange in plant canopies. *Ecology* **52**(5):
855–861.

Billings, W. D., and R. J. Morris. 1951. Reflection of visible and infrared radiation from
leaves of different ecological groups. *Am. J. Bot.* **38**: 327–331.

Birkebak, R., and R. Birkebak. 1964. Solar radiation characteristics of tree leaves. *Ecology*
45: 646–649.

Fuchs, M., and C. B. Tanner. 1966. Infrared thermometry of vegetation. *Agron. J.* **58**:
597–601.

Gates, D. M. 1962. *Energy exchange in the biosphere.* New York: Harper & Row, 151 pp.

Gates, D. M. 1965. Energy, plants, and ecology. *Ecology* **46**: 1–13.

Gates, D. M. 1971. The flow of energy in the biosphere. *Sci. Am.* **224**(3): 88–100 (Offprint
No. 664).

Gates, D. M., and C. M. Benedict. 1963. Convection phenomena from plants in still air.
Am. J. Bot. **50**: 563–573.

Gates, D. M., E. C. Tibbels, and F. Kreith. 1965. Radiation and convection for ponderosa
pine. *Am. J. Bot.* **52**: 66–71.

Hadley, E. B., and L. C. Bliss. 1964. Energy relationships of alpine plants on Mt. Washington, New Hampshire. *Ecol. Monographs* **34**(4): 331–358.

Hart, J. S. 1956. Seasonal changes in insulation of the fur. *Can. J. Zool.* **34**: 53–57.

Herreid, C. F., II, and B. Kessel. 1967. Thermal conductance in birds and mammals. *Comp.
Biochem. Physiol.* **21**: 405–414.

Hutchinson, J. C. D., and G. D. Brown. 1969. Penetrance of cattle coats by radiation. *J.
Appl. Physiol.* **26**(4): 454–464.

Idso, S. B., D. G. Baker, and D. M. Gates. 1966. The energy environment of plants. *Advan.
Agron.* **18**: 171–218.

Joyce, J. P., and K. L. Blaxter. 1964. Effect of air movement, air temperature, and infrared
radiation on the energy requirements of sheep. *Brit. J. Nutr.* **18**(1): 5–27.

Knoerr, K. R., and L. W. Gay. 1965. Tree leaf energy balance. *Ecology* **46**: 17–24.

Miller, P. C. 1971. Sampling to estimate mean leaf temperatures and transpiration rates
in vegetation canopies. *Ecology* **52**(5): 885–889.

Morrison, P. R., and W. J. Tretz. 1957. Cooling and thermal conductivity in three small
Alaskan mammals. *J. Mammal.* **38**(1): 78–86.

Raschke, K. 1960. Heat transfer between the plant and the environment. *Ann. Rev. Plant
Physiol.* **11**: 111–126.

Tibbols, E. C., E. K. Carr, D. M. Gates, and F. Kreith. 1964. Radiation and convection
in conifers. *Am. J. Bot.* **51**: 529–538.

Turrell, F. M., and S. W. Austin. 1965. Comparative nocturnal thermal budgets of large
and small trees. *Ecology* **46**: 25–34.

Vogel, S. 1968. "Sun leaves" and "shade leaves": differences in convective heat dissipation. *Ecology* **49**(6): 1203–1204.

PHYSIOLOGICAL, BEHAVIORAL, AND GENETIC RESPONSES TO THE THERMAL ENVIRONMENT

14-1 THE ROLE OF THE ANIMAL IN THERMAL ENERGY EXCHANGE

A homeothermic animal maintains a balance between its heat production and the physical processes of heat exchange with its environment by regulating its physiology and behavior. The complexity of heat transfer is beyond comprehension, but the characteristics of heat transfer for different animals in different habitats can be determined through analytical procedures. For example, evaporation losses are more important at high temperatures when vapor-pressure deficits are high, but little research has been done on evaporative loss from wild animals.

A series of thermal analyses, mostly for white-tailed deer, are presented in this chapter. It is difficult to illustrate dynamic heat exchange in two-dimensional graphs, but insight into the relative importance of different modes of heat exchange contributes to our understanding of the basic energy relationships between an animal and its environment.

The heat loss described in Chapter 13 must be balanced by an animal's heat production if homeothermy is to be maintained. An animal can either increase its heat production or reduce its heat loss to maintain homeothermy. These are regulated in several ways, both behaviorally and physiologically. The predicted energy balance, using a model, is discussed in this chapter, along with responses and adaptations described by others studying homeothermy and energy exchange. Before going on to specific responses, however, let us consider the concept of a "critical thermal environment" in relation to thermal energy exchange.

14-2   THE THERMAL REGIME AND THE CRITICAL THERMAL ENVIRONMENT

A homeothermic animal is in a continual state of dynamic equilibrium between heat production and heat loss. The continual adjustment of physiological and behavioral responses to the changing energy flux in the environment results in short-term temperature changes in the animal, but this is a normal part of life for a homeotherm. A *critical thermal environment* (Moen 1968) exists when the animal must make a response in order to maintain homeothermy. When the heat production of an animal is greater than the heat loss, even though the animal may attempt to maintain a balance, the animal is in a *critical hyperthermic environment.* If heat loss is greater than heat production, the animal is in a *critical hypothermic environment.*

Physiologists have defined a thermoneutral zone as the range of temperatures that do not cause a metabolic response to maintain homeothermy during basal metabolic measurements. This has been a useful definition for the establishment of a laboratory standard for comparative work, but it is quite inadequate for ecologists who are concerned with an animal in its natural habitat. This animal lives at an "ecological metabolic rate," and it may spend a considerable amount of time outside the physiologically defined thermoneutral range, yet it survives and reproduces.

Situations can develop in which the many factors that make up the thermal regime, including both heat-production and heat-loss factors, may result in a critical thermal environment that is partly independent of existing weather conditions. A deer being chased by dogs on a cool autumn day, for example, may be in a critical hyperthermal environment because running has caused a great increase in heat production. The weather conditions are not critical; running is the critical factor. A deer may be in a critical hypothermal environment during a spring rain when its winter coat is being shed and its thin new summer coat becomes soaked with water. The effect of this rain might be critical during this molt period, but would have been quite unimportant when the animal was still in winter coat. Thus the critical factor is the stage of the molt in combination with the rain.

Any factor that is a part of the thermal regime could become critical if other thermal conditions were near critical. In a high wind, for example, there could be a critical orientation of an animal that would result in a balance between heat production and heat loss. A critical wind velocity, critical radiation level, critical activity level, critical posture, or any other critical variable is meaningful only when all other thermal factors are identified.

Because there are so many different combinations of thermal factors for free-ranging animals, the traditional idea of upper and lower critical temperatures to identify the limits of the thermoneutral zone is entirely inadequate in field ecology. It has served a useful purpose in the laboratory where chamber conditions are quite well represented by a temperature measurement, but data from these simplified laboratory experiments cannot be applied to the dynamic thermal regimes of free-ranging animals. The disparate descriptions given in the literature of the effects of weather on different species of wild ruminants suggests that it is necessary to analyze basic thermal relationships between an animal and its environment so that the physiological and behavioral responses to the thermal environment can be separated from the responses to social interaction between animals or to other stimuli that are not related in any way to weather factors. Thus the ecological definition of the thermoneutral range is a concept that requires an understanding of the principles of heat exchange instead of a precise set of factors with definite limits.

14-3 HEAT LOSS IN RELATION TO HEAT PRODUCTION

The control mechanisms available to an animal for maintaining a balance between heat loss and heat production can be represented conceptually with input and output energy. Food contains energy, and solar and infrared radiation also add to the heat load. Output energy includes the energy that goes into growth and reproduction, as well as the waste energy that has never been a useful part of the system (Figure 14-1). Note that there are several variables in this system, including vascular control over physical heat exchange, respiratory and perspiratory control over evaporation losses, behavioral control over physical heat exchange, and rate of heat production. The rate of production depends on the impact of such environmental forces as disease, social factors, and others on the physiological efficiency of the animal.

Predicted heat losses per hour for deer exposed to different wind velocities at an air temperature of $-20°C$ are shown in Figure 14-2. The smaller deer lose less heat on an absolute basis, but the greater efficiency of heat conservation by

FIGURE 14-1. Regulatory and control mechanisms for the maintenance of a balance between heat production and heat loss.

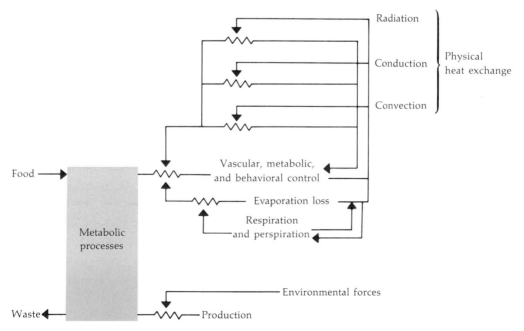

larger deer is indicated by the fractional exponent. Note that the exponents, ranging from 0.600 to 0.617, are less than the exponent used to express the relationship between heat production and weight (0.75). This indicates that a deer's efficiency in conserving heat is greater than its efficiency in producing it.

This illustration of heat loss is based on a standing model in an open field at night, which has been used because it represents the coldest habitat for a deer in the winter. Thermoregulatory mechanisms, such as vascular control in the extremeties, metabolic responses, behavioral thermoregulation, feeding, activity, and others, can be used by deer and other species to regulate heat production in relation to heat loss. The complexity of these dynamic, constantly changing physiological and behavioral responses is beyond mathematical representation, but the use of models for specified sets of conditions provides an insight into the relative values of different responses.

14-4 PHYSIOLOGICAL RESPONSES TO THE THERMAL ENVIRONMENT

An animal's physiological responses to the thermal environment can be either (1) heat-producing or thermogenic or (2) heat-conserving. Three categories of an animal's metabolic responses to variations in the thermal environment over a

FIGURE 14-2. Total heat loss (*HL*) per hour from white-tailed deer expressed as a multiple of basal metabolic rate (2.9 $W_{kg}^{0.75}$). Air temperature is $-20°C$.

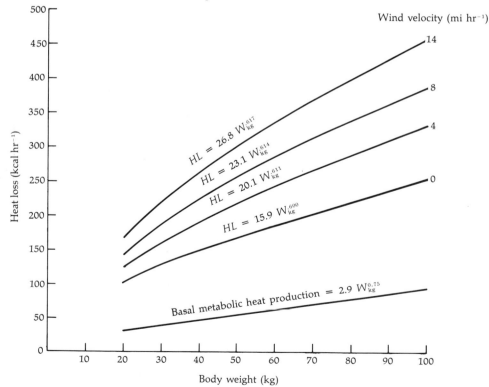

period of several days to several generations have been suggested by Folk (1966, citing Eagan):

1. "Genetic" adaptation: Used for alterations which favor survival of a species or of a strain in a particular environment, which alterations have become part of the genetic heritage of the particular species or strain. This is the same as acclimatization of the race.

2. Acclimatization: The functional compensation over a period of days to weeks in response to a complex of environmental factors, as in seasonal or climatic changes.

3. Acclimation: The functional compensation over a period of days to weeks in response to a single environmental factor only, as in controlled experiments.

These categories cover the range of possibilities over large increments of time, but what about the responses to short-term fluctuations that affect an animal's choice of bedding site, posture, or orientation with respect to the vegetative cover, the position of the sun, or wind characteristics? An animal is constantly employing heat-producing responses and heat-conserving responses in order to regulate body temperatures within certain limits.

Heat production is an active response involving the metabolism of absorbed nutrients or of body tissue. Heat conservation is accomplished by regulating an aspect of the animal-to-environment gradient, resulting in alterations in the dissipation of heat energy. In reality, these two processes occur together so it is the *net* effect—whether the heat production is greater or less than the heat loss—that is important in homeothermy.

THERMOGENIC RESPONSES. An animal may increase its heat production by increasing gross body activity or by increasing the rate of metabolism in specified organs or tissues. Raising the level of gross body activity is not 100% efficient since greater body movement results in greater heat dissipation due to the effect of movement in the hair-air interface. Heat exchange by convection is also increased as the animal moves through the air, increasing the effective wind velocity at its surface. More body activity results in an increase in the energy requirement also.

An elevated metabolic rate cannot be sustained without development of an oxygen debt and muscle fatigue, so sustained body activity is not possible during extended periods of cold weather. The daily metabolic rate of a free-ranging animal may result in an energy expenditure that exceeds the net energy in the food ingested because the animal has reduced food intake during the rutting season, because of a depleted food supply on the range, or for other reasons. In any case, energy stored in the body must be utilized, and a weight loss will ensue. Thus, an increase in body activity to maintain homeothermy is not a feasible short-term solution because of metabolic limits nor is it a feasible long-term solution because of limits imposed on the amount of weight loss that can be tolerated physiologically.

Less obvious but very important thermogenic responses are found in metabolic potentials that do not involve overt muscular activity. Animals conditioned to cold weather [acclimatization (Folk 1966)] are more capable of withstanding cold than animals that have had no previous exposure to cold. Rabbits and rats that were conditioned to cold temperatures did not become hypothermic or suffer frostbite, but unconditioned animals did when both groups were exposed to −50°C for 8 hours. Four yearling sheep that were exposed to temperatures as low as −20°F in a chamber had higher heart rates and lower rectal temperatures than four that were maintained in 55°–65°F temperatures (Hess 1963).

A general relationship has been observed between heart rate, oxygen consumption, and heat production; the increased heart rate is indicative of a higher heat production. Heart rate in relation to oxygen consumption in sheep was studied by Webster (1967) who found a close relationship between heart rate and energy expenditure when the energy expenditure was increased by cold stress or increased levels of food intake. This general relationship between heart rate and metabolism shows considerable variation among individual animals, however, so it is necessary to calibrate each animal to determine the curve relating its heart rate to oxygen consumption.

The relationship between heart rate and oxygen consumption does not seem to hold true under hot conditions. Suggs (1965) did not find an increase in oxygen consumption in human subjects exposed to additional radiation, but this may be due to the necessity to dissipate heat, so increases in cardiac output were effective in elevating the amount of vascularization.

One metabolic response that has received considerable attention by physiologists is nonshivering thermogenesis (NST), which involves an increase in heat production that results from the metabolism of brown fat. Metabolic rates up to six times the basal rate have been reported for rats. Cold-adaptive nonshivering thermogenesis is not common to all mammals, however. Some species have a high capability for NST when young but much less when adult, and in larger species such as cattle it seems to be lacking altogether (Brück, Wünnenber, and Zeisberger 1969). They formulated a general rule, documented also by Jansky et al. (1969), that the bigger the animal, the less brown fat it possesses. This conclusion was reached after careful consideration of their own work and that of others reported in the literature. It appears that wild ruminants do not rely on NST for the maintenance of homeothermy.

The metabolic rate of white-tailed deer confined to a cold respiration chamber rises as the chamber temperature is lowered (Figure 14-3). The increase for four deer in winter coat is less than the increase for two deer in summer coat in May and June. The rate of heat production in May and June is greater than in the winter. The proportional increase in heat production as the chamber temperature is lowered from +20°C to −20°C is 1.7 times in the winter and 2.6 times in the summer coat. Thus the rise in heat production for deer in summer coat appears to be a function of the decreased insulation value of the summer coat and of changes in the metabolic characteristics of the animal.

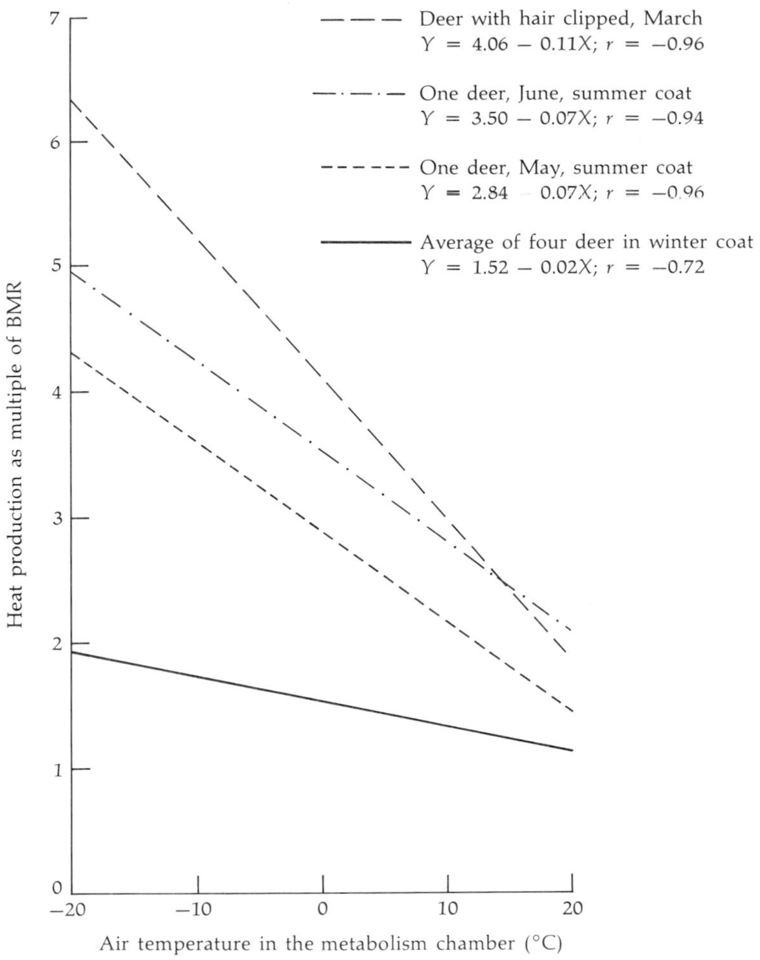
Deer with hair clipped, March
Y = 4.06 − 0.11X; r = −0.96
One deer, June, summer coat
Y = 3.50 − 0.07X; r = −0.94
One deer, May, summer coat
Y = 2.84 − 0.07X; r = −0.96
Average of four deer in winter coat
Y = 1.52 − 0.02X; r = −0.72
Heat production as multiple of BMR
7
6
5
4
3
2
1
0
−20
−10
0
10
20
Air temperature in the metabolism chamber (°C)

FIGURE 14-3. The relationships between heat production and air temperature in the metabolism chamber at the University of New Hampshire. (From Silver et al. 1971.)

The high metabolic potential of the deer (6 × BMR) while confined to a chamber is shown in the results for the clipped deer. Since a metabolic rate of over six times BMR cannot be maintained because of physiological constraints, the measured rates of heat production in cold chambers cannot be applied directly to temperature data in the field. The experiments are useful, however, for determining the metabolic potential of the animal without a significant increase in gross body activity or for determining the effect of diet on heat production.

Another metabolic process that is important in thermogenesis in ruminants is the effect of the heat of rumen fermentation. The heat energy released during the exothermic fermentation process may contribute to the maintenance of homeothermy. The absorption and metabolism of nutrients that follows is an additional source of heat, attributed to nutrient metabolism. These heat increments are eliminated during basal metabolism measurements by imposing the standard condition of post-absorptive digestion.

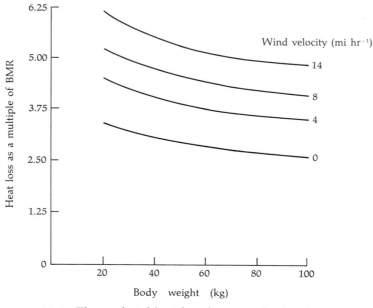

FIGURE 14-4. The predicted heat loss from a 60-kg deer in a standing posture at an air temperature of −20°C. (From Stevens 1972.)

Experiments on sheep in chambers show that the critical temperature based on the chamber environment is higher when the sheep are at a low feeding level, lower at a medium feeding level, and lowest at the highest feeding level (Graham et al. 1958; Graham 1964). This pattern can be expected, but it must be emphasized that an animal has several alternative pathways to maintain heat production or regulate heat loss. In the absence of sufficient food, for example, fat catabolism occurs under cold conditions so that the heat production is maintained at a level necessary for the maintenance of homeothermy. New-born lambs, not yet developed as homeotherms, that were confined in a room at 23°C survived longer than those confined at 9°C, indicating that a lower heat production was possible because the warmer room resulted in a reduced rate of heat loss from the animal (Alexander 1962a). The rate of depletion of the fat reserve is an important consideration under these conditions; fat and carbohydrate reserves appeared to be exhausted when the animal was near death from starvation.

The predicted heat loss from a deer at an air temperature of −20°C is shown in Figure 14-4. The wind velocity of zero is most nearly like the conditions in the respiration chamber at the University of New Hampshire. The predicted results show the heat loss to be about three times the basal rate, although the measured results shown in Figure 14-3 are about two times the basal rate at −20°C. This discrepancy can be explained by the fact that the model used to predict heat loss did not include vascular responses. Thus any reduction in the blood flow to the legs or ears would reduce the heat loss below the predicted level. A discussion of simulations of these vascular controls follows.

THERMOREGULATORY RESPONSES.   The homeothermic animal maintains a balance between heat production and heat loss by distributing the heat produced during normal life processes, increasing the heat production when the animal is in a critical hypothermal environment, and increasing the heat loss when in a critical hyperthermal environment. Heat production by metabolic processes takes place in specific areas of the body. Muscle metabolism and rumen fermentation, for example, are two localized exothermic processes. The heat energy released is distributed throughout the body by the circulatory system, which functions in thermal transport. This thermal transport is physically a passive process until the animal exhibits control over blood flow in order to maintain a thermal balance.

Another passive mechanism (Figure 14-5A) that has been suggested as one adaptation by which caribou maintain sharp temperature gradients in their extremities is a heat exchanger (Irving and Krog 1955). This arrangement permits the exchange of heat without an exchange of arterial and venous blood. The heat flow is simply a conduction process from the warm arterial walls through the body tissue to the cooler venous blood that is returning from the extremities. Hart (1964) has data for rabbit ears that indicate the possible importance of this mechanism for the conservation of heat. Five calories per milliliter of blood flow were lost to the environment, with 22 calories returned to the body via the heat

FIGURE 14-5.   Mechanisms for the conservation of heat energy: (A) passive; (B) active.

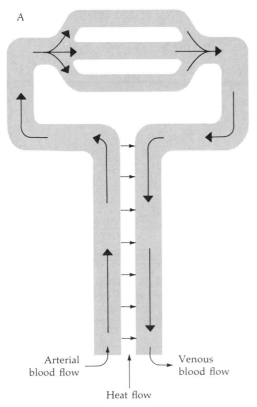

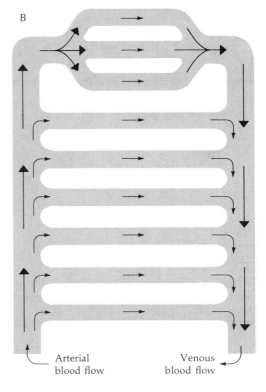

A

Arterial
blood flow

Venous
blood flow

Heat flow

B

Arterial
blood flow

Venous
blood flow

exchanger. This results in a colder tissue temperature in the extremities, but it seems to be an adaptation that does not inhibit the normal functions of the tissue.

A vascular shunt is an active mechanism for the distribution of heat energy as the blood is shunted from an artery to a vein, diminishing the blood supply in the terminal capillary beds (Figure 14-5B). This results in a reduction in heat loss from the extremities since the blood flow is reduced, but it also causes a reduction in the oxygen supply.

Regulation of heat loss by the control of blood flow is not precise enough to result in a constant temperature at the extremities. Ear temperatures of caribou, for example, showed phasic changes from lows of nearly freezing (0°C) to 15°C, back to 0°C, and so forth (Irving, Peyton, and Monson 1956). This indicates that the different parts of a homeotherm's anatomy are not in thermal balance.

Henshaw, Underwood, and Casey (1972) have studied the peripheral thermo-regulation in Arctic foxes (*Alopex lagopus*) and gray wolves (*Canis lupus*), two species that spend much of their lives in subzero ambient temperatures. These Arctic canines did not show phasic rewarming of their foot pads when immersed in a −35°C bath or standing on extremely cold snow. Without phasic circulation to prevent freezing, continuous temperature regulation is necessary, and these animals maintained a foot temperature just above the freezing point.

Vasoconstriction can serve to reduce heat loss by reducing blood flow to an extremity, thus reducing the loss of blood-borne heat. This physiological response has a lower limit of effectiveness, however, since hyperconstriction would result in tissue freezing. The predicted effects of vasoconstriction in the legs of deer are illustrated in Figure 14-6. As the air temperature becomes colder and the wind velocity increases, the importance of vasoconstriction increases. These data are for a standing deer; bedded deer with vasoconstriction could reduce the heat loss even further.

The importance of vascular control of heat flow is related to the insulation characteristics of an animal's coat, since heat energy dissipated from metabolic tissue must pass through this insulative layer. The amount of vascular control in the extremeties is relatively more important because of the shorter, sparser hair there than on the rest of the body. The insulation quality of deer hair on the body trunk is illustrated by the steep temperature gradient between the skin temperature and the base of the hair; I found that at an air temperature of −20°C, the skin temperature of deer was 37°C, a gradient of 57° over a distance of 2 cm. The importance of blood flow and hair insulation is further modified by the distribution of subcutaneous fat. A fat layer reduces the heat loss from an animal's surface since fat tissue is a good insulator.

Piloerection of hair or fluffing of feathers has been considered a useful adaptation for conserving heat. The piliary system is ". . . a variable thermal resistance which accomodates a basic tissue heat production conditioned by animal size to a seasonal factor," according to Herrington (1951). The effectiveness of piloerection in reducing heat loss from deer has been analyzed by Stevens (1972). Changes in the physical depth (d_p) of 1.5 and 1.8 times the depth of the hair in normal lie cause a reduction in insulation of 15% in each case. The effectiveness of

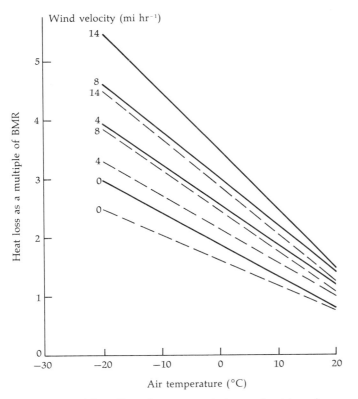

Wind velocity (mi hr⁻¹)

FIGURE 14-6.   The effect of vasoconstriction on heat loss from a 60-kg deer: dashed line indicates vasoconstriction; solid line, no vasoconstriction.

piloerection as a heat conservation mechanism is greater at low air temperatures and high wind velocities (Figure 14-7).

Piloerection has traditionally been considered a heat-conservation mechanism because of the increase in the depth of the insulation layer. This increase is partially offset by the more open characteristic of the hair coat. The loss of heat through a more open hair layer is greater because of increased penetration by wind and subsequent convective heat loss. Thus the decrease in thermal conductivity due to the greater depth of erected hair is counteracted by the increase in heat flow through the more open erect hair.

An important function of piloerection in the maintenance of homeothermy that has not been investigated is its value as a heat-producing adaptation. This theory is illustrated in Figure 14-8. Under warm conditions, normal muscle tonus results in a small amount of heat production from chemical processes involved in muscular contraction. A small temperature gradient exists between the external surface of the skin and the subcutaneous muscle layer. As the environment gets colder, heat loss increases and the temperature gradient from skin to muscle increases in magnitude as the temperature decreases. Thermal receptors in the skin may then sense the colder conditions, the central nervous system relates the signal

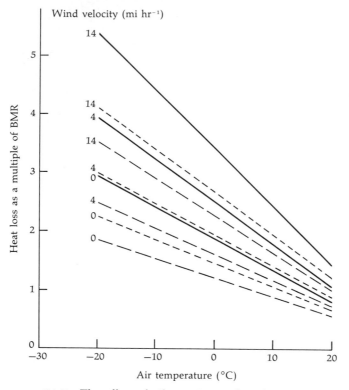

FIGURE 14-7. The effect of piloerection on heat loss from a 60-kg deer: solid line indicates normal piloerection; short-dashed line, 1.5; long-dashed line, 1.8.

to the muscle tissue beneath the skin, and the muscle tissue may contract with a concomitant release of heat energy. This heat energy reduces the temperature gradient and raises the temperature in that area of the skin. As the muscles contract and release heat, the hair is erected! Piloerection is then a secondary effect, and the exothermic muscular contraction is a heat-producing process of primary importance in altering the temperature gradients and subsequent flow of heat.

Evaporation of body fluids is a thermoregulatory response that often occurs under hot conditions. The balance between heat loss and heat production in hot environments can be regulated by a reduction in nutrient intake and an increase in evaporative losses. Under these conditions, a large amount of water is consumed during hot weather. When the effective environmental temperature is equal to or greater than body temperature, evaporative heat loss is the only kind of heat loss possible. Radiation, conduction, and convection contribute to an increase rather than a reduction in body temperature.

There are two sources of evaporation: the moist surface of the respiratory system and skin moistened by perspiration. The relative importance of each source depends on current thermal conditions and the species. Some animals such as cattle do not perspire very much, although horses perspire profusely.

In sheep with and without sweat glands, evaporative losses from respiration are more important than those from sweating, although sweating is advantageous to the shorn animal (Brook and Short 1960). Alexander and Williams (1962) conclude that newborn lambs do sweat, but respiratory water loss is more important in a hot environment.

The evaporative heat loss from respiratory surfaces can be increased by panting. This has often been considered an inefficient process because exothermic reactions take place as muscles are used for panting. This increases heat production, which must be offset by an increase in heat loss by evaporation. There are few data available that provide an indication of the relative magnitude of the two opposing forces. Hales and Findlay (1968) suggest that it may be more efficient than is generally supposed.

The loss of heat by evaporation is a desirable physiological response only when there is a sufficient supply of water. Two large African antelopes, the eland and the oryx, can survive with little or no free water. The oryx is capable of withstanding high internal temperatures by the presence of a countercurrent system involving respiratory cooling in the nasal passages that results in a brain temperature that is lower than the rest of the body. The eland has metabolic control that

FIGURE 14-8.   A schematic representation of the heat-production function of piloerection.

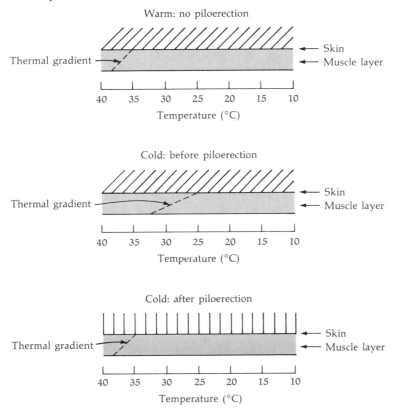

parallels the thermal characteristics of the environment, seeking shade during the heat of the day to minimize a rise in body temperature (Taylor 1969).

There are times when the wetting of an animal's surface results in an undesirable increase in heat loss by evaporation, which may be counteracted by an increase in heat production. For example, wet lambs have a higher heat production than dry ones (Alexander 1962b). This relationship was also shown in Table 6-6 for infant caribou. Calves with wet fur had a heat production as high as ten times the basal rate. Such heat losses due to evaporation have been suspected to be a cause of mortality in the newborn of several species. Analytical studies of newly hatched game birds exposed to prolonged wet weather may reveal a considerable amount of information on mortality in the wild at this critical age.

Several other physiological responses are observed when animals are exposed to thermal regimes that are beyond the zone of thermoneutrality for any length of time. Body growth and production is retarded in hot environments, and reproduction is generally less successful. Critical hyperthermal environments for a few hours or days can result in mortality since the body tissue of most species is not adapted to the maintenance of metabolic processes at temperatures much over 2°C above normal.

The physiological effects of cold environments are frequently less critical than those of hot environments. The effect of cold environments can be compensated for by higher planes of nutrition that permit a level of heat production up to the maximum metabolic potential of the animal. It is extremely important to recognize that food is the ultimate source of energy for metabolic processes; radiant and atmospheric energy present during warm weather can only serve to reduce thermal gradients.

A response to a cold environment exhibited by white-tailed deer and possibly other animals considered to be homeothermic is a reduction in body temperature as an animal becomes lethargic in cold weather. This response has been observed in our current experiments using physiological telemetry equipment. It is a useful response for a free-ranging animal because it reduces its energy requirements on an already restricted winter range, which may be very important for survival and reproduction.

14-5 BEHAVIORAL RESPONSES

Animals respond behaviorally to the thermal regime, altering the balance between heat loss and heat production by changes in orientation, posture, activity, or the selection of cover. This behavioral thermoregulation has an effect on the physiological responses that result from thermoregulation, since behavioral responses cannot be made without the contraction of muscles. Thermoregulatory behavior—of both an individual and a group—includes thermogenic responses and heat-conservation responses, and it is the latter that are discussed in this section.

INDIVIDUAL RESPONSES. Each individual animal can assume a posture that will result in the conservation or dissipation of heat. For example, the surface area of an animal that is in contact with the snow or soil is an important consideration in determining the amount of conductive heat loss. A standing animal has little area in contact with the substrate, so conduction losses are small. A deer bedded in the snow may have about 30% of its surface area in contact with the snow, and heat loss by conduction is an important part of the total heat loss. Figure 14-9 shows the amount of heat lost from a 60-kg deer in standing and bedded postures. The difference between standing and bedding is clear, with small differences due to the position of the head while bedded. The effectiveness of bedding posture as a heat conservation mechanism in the total thermal regime is relatively greater at higher wind velocities and colder temperatures. The predicted heat loss of a bedded deer is about two times basal at an air temperature of −20°C, so heat-production mechanisms, such as activity, may be employed to compensate. An alternative is for the deer to allow heat loss to exceed heat production, resulting in a drop in body temperature as the animal progresses toward a lethargic condition.

Many descriptions of thermoregulatory behavior are found in the literature. Severinghaus and Cheatum (1956) report that deer will remain in a bed for one to three days after a storm, usually under low-hanging conifers or windfalls. This observation may be related to the lethargic responses we have seen in our current experiments. In subzero weather in Maine, beds were found under conifer branches that were bent down and covered with snow, or under hardwoods that had retained their leaves (Hosley 1956). On very cold nights, deer in the Adirondacks were observed moving slowly on the trails, and Severinghaus and Cheatum

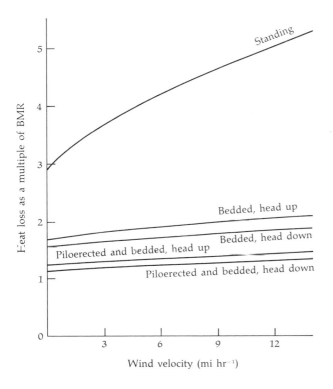

FIGURE 14-9. The effect of posture on heat loss from a 60-kg deer at an air temperature of −20°C.

(1956) suggested that it was too cold for the deer to remain bedded. They also cite several cases in which deer moved to seek shelter from cold winds.

Sharp-tailed grouse can exhibit thermoregulatory control by altering their posture. The heat loss from the head is much higher *per unit area* than that from the body (Figure 14-10), indicating that thermal benefits would accrue if the head and neck were withdrawn or curved back under a wing. Predicted thermal benefits from this postural change in different air temperatures and wind velocities are shown in Figure 14-11. Note that the least amount of head exposure is most beneficial at 0 mi hr^{-1}; 1.5 × BMR could be maintained with 5% head exposure down to about −8°C.

All calculations of the thermal balance between heat production and heat loss in different thermal regimes indicate that wind is the most important factor at low air temperatures. Calculations by Robbins (1971) indicate the importance of wind in altering the heat loss in different cover types (Figure 14-12). Wind velocities of 10–20 mi hr^{-1} in an open field raise the predicted heat loss above realistic limits, whereas the effect of a conifer canopy in reducing wind velocities puts a 30-kg deer in a much more stable thermal balance. This effect and the effect that a conifer canopy has on snow depths and structure are benefits that must be considered in evaluating different cover types. It must be pointed out,

FIGURE 14-10. Effect of wind speed, air temperature, and body part on predicted heat loss through the feather layer.

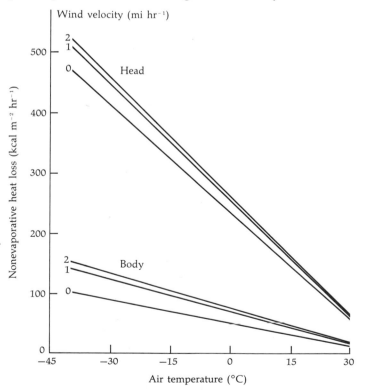

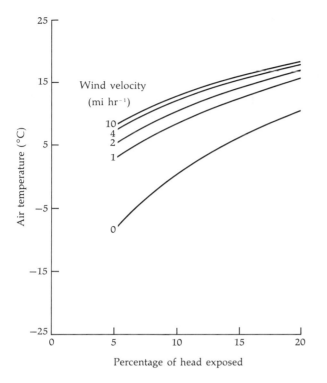

FIGURE 14-11. Effect of wind velocity and head exposure at ambient temperature at which predicted nonevaporative heat loss by a 1000-gram grouse would be 1.5 × BMR.

however, that a conifer canopy reduces food production so that a very essential source of energy is reduced. Further, topography and herbaceous and shrubby vegetation in more open habitats may reduce the wind effects nearly as much as a conifer canopy while producing a considerably greater quantity of food.

The relative benefit of herbaceous vegetation from 0 to 100 cm in height for deer of different weights is shown in Figure 14-13. The smallest deer benefits the most from the reduction in wind because it is in the lower part of the wind profile where the vegetation has its greatest effect on wind velocities. Small animals such as rabbits and hares, fox, game birds, and others can benefit from the use of low vegetation to reduce the wind velocity past their bodies; effective winter cover is not necessarily a dense coniferous overstory!

Henshaw (1968) noted that caribou bedded in areas of irregular topography during continued high winds. Their bodies were generally broadside to the wind. No apparent discomfort due to a low temperature with little or no wind could be detected. Observations of white-tailed deer in the cold, continental climate of western Minnesota also indicated that cold temperatures without wind had little or no apparent effect on their behavior. (Moen 1966).

An animal can exhibit many energy-conserving responses at one time, and no single order of their occurrence has been observed. Suppose that a 60-kg deer were standing in an open field at night, the coldest possible postural thermal model. At a wind velocity of 14 mi hr^{-1}, a negative thermal balance could be predicted, if the deer had a metabolic rate of two times basal, at 15°C (Figure

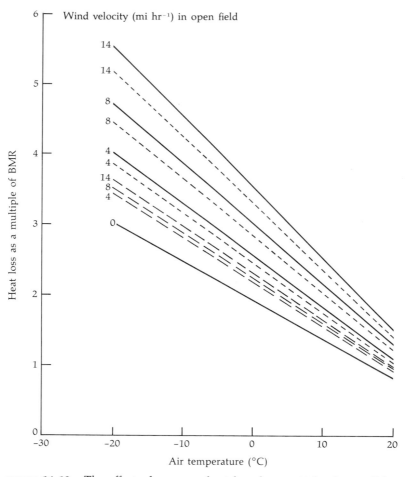

FIGURE 14-12.   The effect of cover on heat loss from a 60-kg deer: solid
line indicates open field; short-dashed line, hardwood; long-dashed line,
conifer. (Data from Stevens 1972.)

14-14). A sequence of energy-conserving responses might be as follows: vasoconstriction, piloerection, both vasoconstriction and piloerection, bedding, the
head could be placed down alongside the body, and finally at −40°C several
mechanisms could be employed and the deer could still be losing heat at a rate
less than 2 × BMR. This indicates that deer have a considerable potential for
withstanding cold weather if they have a source of energy for heat production,
and further capabilities if they can become lethargic during periods of cold
weather.

One aspect of the heat-production–heat-loss relationship currently under
investigation, using physiological telemetry techniques, is the reduction in heart
rate, respiration rate, and body temperature when deer are exposed to cold

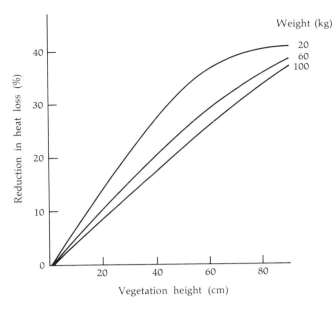

FIGURE 14-13. Reduction in total heat loss of a standing deer by vegetation, expressed as the percentage of decrease in heat loss from that of a deer in vegetation that is 0.75 cm in height with a 14 mi hr^{-1} wind at 2 meters.

FIGURE 14-14. Possible physiological and behavioral responses for maintaining heat loss below 2 × BMR for a 60-kg deer in a 14 mi hr^{-1} wind.

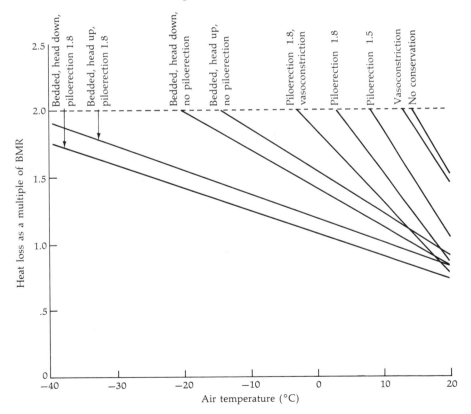

weather. These reductions indicate that heat production must also be going down. Thus the dynamic response of the animal seems to be one of heat conservation rather than the more energetically costly heat-production response.[1]

GROUP RESPONSES. Groups of wild ruminants seem to respond to general weather conditions in somewhat predictable ways. Caribou have been observed to be more gregarious during cold weather (Henshaw 1968). White-tailed deer move to or are confined to yards, especially during a winter with deep snow. In western Minnesota they form larger groups during periods of reduced visibility without apparent relation to cold weather (Moen 1966). Elk migration from the summer to winter range seems to be triggered by weather changes. Heavy snowstorms cause mule deer to migrate to a winter range in the fall whereas the migration back to the summer range is related to plant growth (Russell 1932). Moose bed in soft snow, which may reduce the energy requirements of the animals because snow is a good insulator (Des Meules 1964).

Studies have been done to evaluate the different directions of slopes that are used, and the usual conclusion is that southern exposures are used more than northern ones. There is a greater energy flux on south slopes than on north slopes because of the distribution of solar radiation. Northerly winds are also more common in the winter, resulting in generally harsher conditions on north slopes. The preference for south slopes may not be related only to current weather conditions. Snow depths are frequently less on south exposures, especially in late winter and early spring when melting begins. A decrease in snow cover results in an increase of available food, and this may have a considerable effect on the distribution of animals.

The grouping of animals has potential benefit in the reduction of heat loss. Animals that are huddled together exchange heat with each other, thus conserving it within the group. Caribou move in bands when feeding and the water vapor released by respiration sometimes condenses and forms a cloud that can reduce heat loss by radiation. However, this "cloud cover" may have little real benefit since the heat production of the animals is higher when they are active and the cloud is likely to be unnecessary for the maintenance of homeothermy. The physiological benefits from grouping may be social, with heat conservation benefits only incidental.

The distribution of energy in field habitats can be interpreted in a meaningful way only when the physiological benefits are related to environmental energy. Further, it is difficult to separate the physiological causes of behavioral responses from social causes. The physiological benefits from behavioral responses can be quantified more easily than the social benefits can because the basic unit of energy, the calorie, can be used. No such unit exists for the quantification of social traits, and interpretation is much more subjective.

---

[1] These responses will be discussed in detail in a Ph.D. dissertation in preparation by Nadine L. Jacobsen, graduate assistant at the BioThermal Laboratory.

14-6 GENETIC RESPONSES

All animals exhibit individual variation owing to genetic differences. The obvious characteristics, such as coat color, body size, and so forth, are often the only characteristics considered by field biologists in studying particular populations. Since energy metabolism is vital to the survival of an individual animal, genetic factors that affect its energetic efficiency may have a significant bearing on its survival, especially when conditions of stress begin to appear.

Evans and Whitlock (1964) discuss the genetic adaptation of sheep to cold environments. A subtle but perhaps vitally important genetic characteristic that has been isolated in sheep is the inheritance pattern of the hemoglobin type. They point out that sheep breeds indigenous to cold environments tend to have high gene frequencies for hemoglobin A. This in turn is associated with a high efficiency of oxygen transport.

Hemoglobin is a blood protein with four iron-containing heme groups attached to it. Oxygen unites reversibly by combining with one of four iron atoms in the hemoglobin (HbO_2) molecule. Thus a fully saturated hemoglobin molecule contains four oxygen atoms, and is called oxyhemoglobin. Since oxygen is required for energy metabolism the efficiency of an animal's oxygen transport system may have a direct influence on its survival when exposed to cold conditions. Further, sheep with a low erythrocyte volume and related oxygen transport efficiency have a smaller chance of surviving a natural challenge by the parasite *Haemonchus contortus*, a helminth parasite of ruminants usually found in the abomasum, than does an animal with a greater metabolic efficiency. This parasite has been found in deer. The combined effects of cold weather and a parasite load may require a highly efficient metabolic system if a deer is to survive.

Owing to the effects of natural selection, different deer populations may have genetic adaptations—such as hemoglobin types—that result in higher survival rates. One complicating factor, however, in the process of natural selection is the effect of man on the distribution of wild ruminants. The Adirondacks of New York and the coniferous forests in northern Minnesota, for example, were not inhabited by deer populations prior to the settlement of North America by the white man. Woodland caribou and moose were present, but the opening of the forests by timber-cutting operations caused a decline in the number of caribou and moose and an increase in the number of white-tailed deer. Thus the white-tails are now found in areas that have weather, soil, and forage characteristics that may be quite different from those in presettlement deer ranges. The normal patterns of genetic evolution have been further complicated by the transplanting of deer from one geographical region to another. The period of time since settlement and the transplanting of deer is but a fleeting moment on the evolutionary scale, and the genetic characteristics of present populations may not result in maximum metabolic efficiency.

There are daily and seasonal metabolic cycles that are inherent in different species. Generally, the metabolic rate is higher during the day for diurnally active animals, and higher at night for those that are nocturnal. Many animals have seasonal metabolic rhythms—hibernation is an extreme example. Deer appear to exhibit a seasonal rhythm with lower metabolism in the winter (Helenette Silver, personal communication; Moen and Jacobsen, unpublished data).

Basal metabolic rates are similar for tropical and arctic animals, but the thermal characteristics of the hair coat are vastly different. It is generally recognized that the insulation of the coat is the major adaptation of northern species. There are marked differences in the depth and structure of summer and winter coats of white-tailed deer. Their summer coats of finer hair are not nearly as thick as their winter coats. The depth of winter hair at several points on whitetails is shown in Figure 14-15. The hairs in the longer winter coat are also hollow and crinkled, with an underfur that consists of very fine hairs that are about as numerous as the longer hairs (Figure 14-16).

Coat color is a genetic characteristic of interest in analyzing the radiation exchange of an animal, especially because of seasonal changes in color, thickness, and depth. From a thermal point of view, it would be advantageous for an animal to have a dark coat in the winter so that a maximum amount of solar radiation could be absorbed. The infrared-radiation exchange is not related to the visible color of the coat, however. Many arctic animals are white in winter and dark in the summer, so other factors such as protective coloration must have had a greater influence on their genetic characteristics than did any thermal benefits from color. The slight differences in coat color of different ruminants in northern climates appear to have little relationship to the rigorous thermal regime.

Two ecological rules that relate body geometry to climate have become firmly entrenched in the ecological literature. Bergman's rule states that northern members of a species have a larger body size than southern members. This is inter-

FIGURE 14-15.   The physical depth (cm) of winter hair at several points on white-tailed deer. (From Stevens 1972.)

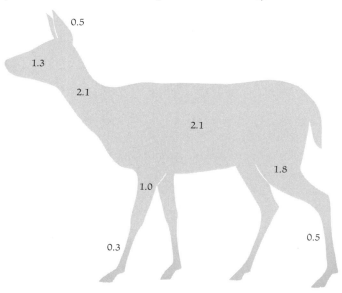

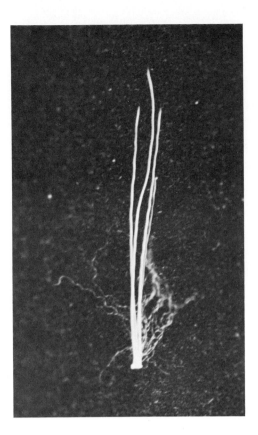

FIGURE 14-16.   The long, guard hairs and the fine underfur of the winter coat of a white-tailed deer.

preted as a genetic adaptation to cold since a larger body has a higher volume-to-surface-area ratio than a smaller one. Allen's rule states that northern species have smaller appendages than southern ones. This is interpreted as an adaptation for the conservation of heat in the north with less surface area on legs and ears, and for the dissipation of heat in southern climates as the appendages act as cooling fins.

Several ecologists have recently questioned the validity of interpreting Bergman's rule on the basis of a climatic gradient. The rule is logical in theory, but the number of factors participating in thermal exchange and the compensatory effects of interaction between physiological and behavioral factors indicate that differences in body size are quite insignificant compared with other thermoregulatory mechanisms. Further, the younger members of each species are usually smaller than the mature animals in either north or south, so the rule is violated by each individual during its life span.

The functional basis for rules such as Bergman's can be successfully analyzed using a thermal-engineering approach. Work at the BioThermal Laboratory has indicated that differences in body size do appear to be significant but that deer seem to have other behavioral and physiological capabilities that compensate in part. Wathen, Mitchell, and Porter (1971) have studied the energy exchange of jackrabbit (*Lepus californicus*) ears, using both field and laboratory techniques.

Convection from the ears was large enough to account for the dissipation of all of the animal's metabolic heat production at an air temperature of 30°C, indicating that genetic adaptations such as large appendages may play a significant role in the maintenance of homeothermy.

14-7 SUMMARY

Careful analyses of energy relationships between an animal and its environment result in the conclusion that animals in northern climates are quite able to survive periods of cold weather if they can exhibit metabolic responses for the release of heat energy from ingested food or body tissue and physiological and behavioral responses for the conservation of heat energy. An adequate supply of energy and other nutrients is necessary for basal metabolic processes, for growth, and for production. The assimilation of nutritive considerations into a large ecological consideration of the concept of carrying capacity reveals some very interesting relationships, especially with respect to the importance of the characteristics of different age and weight classes within a population. This is discussed in Part 6.

LITERATURE CITED IN CHAPTER 14

Alexander, G. 1962a. Energy metabolism in the starved newborn lamb. *Australian J. Agr. Res.* **13**(1): 144–164.

Alexander, G. 1962b. Temperature regulation in the newborn lamb. IV. The effect of wind and evaporation of water from the coat on metabolic rate and body temperature. *Australian J. Agr. Res.* **13**(1): 82–99.

Alexander, G., and D. Williams. 1962. Temperature regulation in the newborn lamb VI. Heat exchanges in lambs in a hot environment. *Australian J. Agr. Res.* **13**(1): 122–143.

Brook, A. H., and B. F. Short. 1960. Regulation of body temperature of sheep in a hot environment. *Australian J. Agr. Res.* **11**(3): 402–407.

Brück, K., W. Wünnenber, and E. Zeisberger. 1969. Comparison of cold-adaptive metabolic modifications in different species, with special reference to the miniature pig. *Federation Proc.* **28**(3): 1035–1041.

Des Meules, P. 1964. The influence of snow on the behavior of moose. *Quebec Serv. Faune Rappt.* **3**: 51–73.

Evans, J. V., and J. H. Whitlock. 1964. Genetic relationship between maximum hematocrit values and hemoglobin type in sheep. *Science* **145**(3638): 1318.

Folk, G. E., Jr. 1966. *Introduction to environmental physiology.* Philadelphia: Lea & Febiger, 308 pp.

Graham, N. M. 1964. Influences of ambient temperature on the heat production of pregnant ewes. *Australian J. Agr. Res.* **15**(6): 982–988.

Graham, N. McC., F. W. Wainman, K. L. Blaxter, and D. G. Armstrong. 1959. Environmental temperature, energy metabolism, and heat regulation in sheep. I. Energy metabolism in closely clipped sheep. *J. Agr. Sci.* **52**: 13–24.

Hales, J. R. S., and J. D. Findlay. 1968. The oxygen cost of thermally-induced and CO_2-induced hyperventilation in the ox. *Respirat. Physiol.* **4**(3): 353–362.

Hart, J. S. 1964. Insulative and metabolic adaptations to cold in vertebrates. *Symp. Soc. Exp. Biol.* **18**: 31–48.

Henshaw, J. 1968. The activities of the wintering caribou in northwestern Alaska in relation to weather and snow conditions. *Intern. J. Biometeorol. (Amsterdam)* **12**(1): 21–27.

Henshaw, R. E., L. S. Underwood, and T. M. Casey. 1972. Peripheral thermoregulation: foot temperature in two arctic canines. *Science* **175**(4025): 988–990.

Herrington, L. P. 1951. The role of the piliary system in mammals and its relation to the thermal environment. *Ann. N.Y. Acad. Sci.* **53**: 600–607.

Hess, E. A. 1963. Effect of low environmental temperatures on certain physiological responses of sheep. *Can. J. Animal Sci.* **43**(1): 39–46.

Hosley, N. W. 1956. Management of the white-tailed deer in its environment. In *The deer of North America,* ed. W. P. Taylor. Harrisburg, Pennsylvania: The Stackpole Co., pp. 187–259.

Irving, L., and J. Krog. 1955. Temperature of skin in the arctic as a regulator of heat. *J. Appl. Physiol.* **7**(4): 355–364.

Irving, L., L. J. Peyton, and M. Monson. 1956. Metabolism and insulation of swine as bare-skinned mammals. *J. Appl. Physiol.* **9**: 421–426.

Jansky, L., R. Bartunkova, J. Kockova, J. Mejsnar, and E. Zeisberger. 1969. Interspecies differences in cold adaptation and nonshivering thermogenesis. *Federation Proc.* **28**(3): 1053–1058.

Moen, A. N. 1966. Factors affecting the energy exchange of white-tailed deer, western Minnesota. Ph.D. dissertation, University of Minnesota, 121 pp.

Moen, A. N. 1968. The critical thermal environment: a new look at an old concept. *BioScience* **18**(11): 1041–1043.

Robbins, C. T. 1971. Energy balance of white-tailed deer in winter as affected by cover and diet. Special report, BioThermal Laboratory, Cornell University, 30 pp.

Russell, C. P. 1932. Seasonal migration of mule deer. *Ecol. Monographs* **2**(1): 1–46.

Severinghaus, C. W., and E. L. Cheatum. 1956. Life and times of the white-tailed deer. In *The deer of North America,* ed. W. P. Taylor. Harrisburg, Pennsylvania: The Stackpole Co., pp. 57–186.

Silver, H., J. B. Holter, N. F. Colovos, and H. H. Hayes. 1971. Effect of falling temperature on heat production of fasting white-tailed deer. *J. Wildlife Management* **35**(1): 37–46.

Stevens, D. S. 1972. Thermal energy exchange and the maintenance of homeothermy in white-tailed deer. Ph.D. dissertation, Cornell University, 231 pp.

Suggs, C. W. 1965. Some responses of humans to thermal radiation. *J. Appl. Physiol.* **20**(5): 1000–1005.

Taylor, C. R. 1969. The eland and the oryx. *Sci. Am.* **220**(1): 88–95 (Offprint No. 1131).

Wathen, P., J. W. Mitchell, and W. P. Porter. 1971. Theoretical and experimental studies of energy exchange from jackrabbit ears and cylindrically shaped appendages. *Biophys. J.* **11**: 1030–1047.

Webster, A. J. F. 1967. Continuous measurement of heart rate as an indicator of the energy expenditure of sheep. *Brit. J. Nutr.* **21**: 769–785.

SELECTED REFERENCES

Alexander, G., and D. Williams. 1968. Shivering and nonshivering thermogenesis during summit metabolism in young lambs. *J. Physiol.* **198**(2): 251–276.

Anderson, R. C., O. L. Loucks, and A. M. Swain. 1969. Herbaceous response to canopy cover, light intensity, and throughfall precipitation in coniferous forests. *Ecology* **50**(2): 255–263.

Franzmann, A. W., and D. M. Hebert. 1971. Variation of rectal temperature in bighorn sheep. *J. Wildlife Management* **35**(3): 488–494.

Irving, L. 1966. Adaptations to cold. *Sci. Am.* **214**(1): 94–101.

Irving, L., K. Schmidt-Nielsen, and N. S. B. Abrahamsen. 1957. On the melting points of animal fats in cold climates. *Physiol. Zool.* **30**: 93–105.

Kelsall, J. P. 1969. Structural adaptations of moose and deer for snow. *J. Mammal.* **50**(2): 302–310.

Martinka, C. J. 1967. Mortality of northern Montana pronghorns in a severe winter. *J. Wildlife Management* **31**(1): 159–164.

Porter, W. P., and D. M. Gates. 1969. Thermodynamic equilibria of animals with environment. *Ecol. Monographs* **39**: 245–270.

Pruitt, W. O., Jr. 1959. Snow as a factor in the winter ecology of barren ground caribou (*Rangifer arcticus*). *Arctic* **12**(3): 158–179.

Sakai, A., and K. Otsuka. 1970. Freezing resistance of alpine plants. *Ecology* **51**(4): 665–671.

Veghte, J. H. 1964. Thermal and metabolic responses of the gray jay to cold stress. *Physiol. Zool.* **37**(3): 316–328.

Whitton, G. C. 1962. The significance of the extremities of the ox (*Bos taurus*) in thermoregulation. *J. Agr. Sci.* **58**: 109–120.

THE ORGANIZATION OF
ENERGY AND MATTER IN
PLANT AND ANIMAL COMMUNITIES

Plant and animal communities are assemblages of individuals that live together in an ecologically organized manner. Individuals in a community have requirements that are met by the resources in the physical area that encompasses that community. As the community develops, interrelationships develop between individual plants, plant species, individual animals, animal species, and between plants and animals.

Many of the interrelationships between organisms are unique to the developed community. Some relationships are species dependent—in some cases obligatory. For example, the lichens that grow on the branches of trees in a dense coniferous forest are composed of both fungi and algae, which depend on each other for their existence, and the lichen as a whole depends on the conditions imposed on it by the dense coniferous overstory. The lichens may be attached to several different tree species, and several different species may be attached to a single tree.

The organization of communities can be described in many different ways. Taxonomic structure is a common way to describe it, especially through time inasmuch as different species are found in different successional stages. Different species have different morphologies, giving communities different appearances. The appearance of a community changes owing to phenological changes such as the falling of leaves, and the chemical composition of the different species varies with time also. A fundamental characteristic of an organized group of organisms is the flow of energy and matter through the group, beginning with solar energy and continuing through photosynthesis (primary production), ingestion (primary consumption), synthesis (secondary production), reingestion (secondary consumption), and decomposition resulting in the dissipation of energy.

15-1 TAXONOMIC RELATIONSHIPS

Plant communities contain members from different taxonomic groups. Oosting (1956) stresses the description of plant communities on the basis of their composition and structure. If two communities had identical taxonomic composition, they would have a maximum (1.0) *coefficient of similarity*. If they had no species in common, the coefficient of similarity would be zero. This kind of comparison shows that communities on similar soil and topography, with similar water resources, and in close physical proximity have high coefficients of similarity if they are of similar ecological age. Thus a maple-basswood forest in one location is quite similar in many respects to a maple-basswood forest in another area.

Animals, especially primary consumers, are closely tied to plant communities because of food resources. The coefficient of similarity can be applied to these animal populations. Some animals are found in rather specialized plant communities (spruce grouse, for example, are found only in rather mature coniferous forests), while others are much less selective in terms of plant-community characteristics. A white-tailed deer is an example of the latter type; it is found in a variety of plant communities ranging from grassland to coniferous forests.

15-2 MORPHOLOGICAL RELATIONSHIPS

Gross morphological characteristics of plants are well known to everyone. Leaves are the primary location for photosynthesis. The stems are supporting structures, as well as "pipelines" for the movement of water and absorbed nutrients to the leaves and for the movement of the products of photosynthesis from the leaves to the fruits and roots. The roots serve as anchors and are also metabolically active inasmuch as the growing tips absorb water and soil nutrients. Photosynthetic products, translocated to the roots, are the sources of energy for root metabolism. The reproductive structures take a variety of forms in plants, but all serve to concentrate enough energy in a small amount of matter (the seed) so that the life cycle can begin again under favorable conditions.

Different kinds of plant and animal communities have different appearances because of the life-form of their members. Tundra communities contain low-growing vegetation, without a great deal of taxonomic diversity. The prairie is another type of community with low-growing vegetation; the entire community is confined to a vertical height of a few feet. Variations in prairie communities relate to hydrologic and other physical factors. In general, the greater the amount of water available to a prairie community, the greater the vertical height of that community. This results in the following morphological classification of prairie types: tall-grass prairie, mixed prairie, and short-grass prairie. The distribution of prairie types from east to west in the prairie states generally follows the moisture gradients.

Deciduous forest communities have a distinctive morphology that includes layers of vegetation. The upper layer, made up of tree crowns, is the canopy. Beneath the canopy is the subcanopy, which is made up of shrubs and young trees (saplings) that have not yet reached the height of the canopy. Below the canopy and subcanopy is the understory, composed of herbaceous plants and tree seedlings that adapt to the conditions imposed on them by the canopy and subcanopy.

The vertical morphology of a deciduous forest community changes seasonally with the emergence and fall of leaves. In the spring, the canopy is a rigid, woody structure with a fairly low density, permitting sunlight to penetrate. The understory responds with a variety of fast-growing, short-lived flowering plants. These plants generally complete their life cycles before the canopy reaches maximum density. When the leaves are on the trees, the canopy is dense, shading the lower levels of the community. The understory is fairly quiescent at that time of the year.

A coniferous community has a more stable morphology. The loss of leaves (needles) takes place throughout the year, so seasonal differences in canopy characteristics are not marked. The density of mature coniferous canopies is usually great enough to prevent the growth of a diverse understory, resulting in a much simpler community morphology.

The size and distribution of trees in a forest community can be estimated by the use of standard methods. Square or circular plots are used as sample areas, and an estimate of the total number of plants in a community can be obtained by multiplying the number in a sample area by the quotient obtained by dividing the total area by the sample area. Using this method, several community characteristics can be described with the following terms:

Basal area: The area of a cross section of a tree trunk, generally taken at $4\frac{1}{2}$ feet above the ground.

Diameter, breast high (dbh): The diameter of a tree trunk at $4\frac{1}{2}$ feet.

Density, absolute: The number of trees of a given species in the sample plots.

Density, community: The total number of trees of each species per unit land area.

Density, relative: The number of individuals of a given species in relation to the total number of individuals of all species.

Dominance, absolute: The sum of the basal areas of individual trees of a given species in the sample plots.

Dominance, absolute mean: The average basal area of the individuals of a given species in each plot.

Dominance, community: The sum of the basal areas of individuals of a given species throughout the entire community.

Dominance, relative: The basal areas of a given species in relation to the basal area of all species, which can be expressed as a simple ratio or as a percentage.

Frequency, absolute: The number of plots in which a given species is found.

Frequency, relative: The number of plots containing a given species compared with the total number of plots sampled, which can be expressed as a simple ratio or as a percentage.

Frequency, relative community: The number of individuals of a given species in the sample plots as a percentage of the total number of individuals of all species.

The basal area can be found by measuring the circumference of a tree at a height of $4\frac{1}{2}$ ft with a tape that is marked for reading basal area directly (in square inches or square centimeters);[1] or the basal area can be calculated by measuring either the circumference or the diameter of the trunk at a height of $4\frac{1}{2}$ ft and using equations (15-1) and (15-2).

$$BA = \pi \left(\frac{C}{2\pi}\right)^2 \tag{15-1}$$

$$BA = \pi \left(\frac{D}{2}\right)^2 \tag{15-2}$$

The use of circular or rectangular plots for determining the number and distribution of trees is a straightforward mathematical technique, but it is frequently filled with statistical complications owing to the nonrandom distribution of trees in a forest. The location of the sample plots may be randomly determined or systematically determined, and the decision of which method to use should be based on the purpose of the sampling and on the general characteristics of the vegetation. Grieg-Smith (1964) discusses these considerations and many others in quantitative plant ecology in considerable detail.

There are several other methods for the determination of the distribution of trees in a forest. The inefficiency in using the large sample plots necessary in tree stands has resulted in the development of plotless sampling techniques. These are described further by Grieg-Smith (1964) and include the following:

1. *Closest individual method:* The distance from the sampling point to the nearest tree is measured.

2. *Nearest neighbor method:* The distance from the sampling point to the nearest tree is measured.

3. *Random pairs method:* A line from the sampling point to the nearest tree is first described, followed by a line perpendicular to the first line and through the sampling point. The distance from the tree nearest the sampling point to the tree closest to the nearest tree but on the other side of the second line (the one perpendicular to the first line) is measured.

4. *Point-centered quarter method:* The distance from the sampling point to the nearest tree in each quarter is measured, using a previously determined orientation of the quarters.

The calculation of density from the distances measured using these methods is based on a calculation of the mean area surrounding the trees described by the linear measurements. Grieg-Smith (1964, p. 49) states that the mean linear value obtained by using the first two methods is one-half the square root of the mean area. He discusses research on this that evaluates the applicability of the factor 2 (one-half) in the calculation, concluding that no single factor can be

---

[1]Basal-area tapes can be made by the student.

applied to allow for bias in the sample of distances. This is likely true for all four methods, although 0.8 is stated as the correction factor for the third method. The fourth method may be the most accurate; the mean of all distances measured is equal to the square root of the mean area. Thus the area surrounding each tree (the mean of all trees in the stand) can be calculated with equation (15-3).

$$\bar{A} = \bar{D}^2 \qquad (15\text{-}3)$$

where

$\bar{D}$ = average distance from points to trees

The density of trees per acre can be determined by equation (15-4).

$$\text{Density per acre} = \frac{43560}{\bar{A}} \qquad (15\text{-}4)$$

where

$\bar{A}$ = mean area (ft^2) surrounding the trees in the stand

The frequency of occurence and the dominant characteristics can be calculated as in the sample plot methods; the plotless sampling technique results in the same data but for plots of variable size.

There are very practical problems in using these approaches in the field. Trees are not all single bole structures, but often have sprouts at the base. This is especially true of some species in second-growth areas. *Tilia americana* (basswood) is an example of a tree that sprouts profusely. Some trees live together in groups called clones. *Populus* (aspen) is such a tree; a "mother" tree may have several offspring that have grown from shoots off the sprawling roots of the parent tree, resulting in a very bunched distribution. There often is a fairly symmetrical shape to the clone, with the parent tree in the center and progressively shorter, younger trees toward the periphery.

There can be similar problems in applying these techniques to herbaceous vegetation. Typically, large plots or plotless sampling are applied to the trees, smaller plots to saplings (1″ diameter at ground level to 4″ dbh), and still smaller plots to seedlings (less than 1″ diameter at ground level), shrubs, and herbaceous vegetation. The same type of sampling problems appear in each case, since many shrubs are clonal, sapling distribution is often related to the distribution of parent trees, and herbaceous vegetation is often clumped. This is especially true on the prairie where many species grow as "bunchgrass," or in clumps that may or may not form continuous mats of vegetation.

These sampling techniques, along with others described in the plant ecology references listed at the end of the chapter, are useful for describing community structure in terms of the parameters measured. Frequently, these parameters are not the pertinent factors in an animal-plant-environment relationship, however. Dominance based on the basal area of the bole may be of less significance than the dominant effect of canopy characteristics on the penetration of energy and the synthesis of matter into living tissue in the understory. There is a danger in using such general approaches because they may direct attention away from the functional relationship between organism and environment, resulting in an inability to perceive in an analytical way the transport of energy and matter through the system.

Animal communities have a morphology that can be described in a manner similar to that for plant communities. Animal communities exist within plant communities, of course, with individual animals usually having a considerable amount of mobility. Some species have a group mobility (migration), resulting in a seasonal movement of an entire group or population from one plant community to another.

One factor that plays a significant role in the distribution of energy and matter in the animal community is the effect of pheromones. These chemical substances, released from one animal, trigger specific reactions from another in the same species. An interesting characteristic of these pheromones is that such minute quantities markedly alter the behavior of members of the population. The expenditure of energy for their emission is miniscule compared with the expenditure of energy that may result from a response to them.

Analyses of the basic matter and energy characteristics of an animal might start with three basic questions: (1) What is its chemical composition? (2) What are its thermal energy characteristics? and (3) Where does the matter and energy come from?

The first question deals with the basic chemical structure of the animal. Elements—primarily carbon, hydrogen, oxygen, and nitrogen—are bound together with other elements found in lesser quantities to form compounds of unique biological capabilities when they are combined in cells, tissues, and organs. Energy is stored as tissue synthesis occurs. Animal products are also formed, including the fetus, milk, and eggs. Some tissue is formed at seasonal intervals, such as feathers, hair, and antlers. The amount of energy necessary for the synthesis of these materials can be quantified and, if considered in terms of time, becomes a dynamic measurement of the cost of life.

An animal's thermal energy characteristics play an important part in its physiological processes. These in turn affect behavior, and behavior can alter thermal characteristics as animals change their posture, activity, habitat, and so forth. Some animals maintain a fairly constant body temperature, whereas others have much more variation. Some are regulated at one level for part of the year and at a much lower level at other times (hibernation). All organisms live in a thermal-energy regime that results in a continual and dynamic exchange of heat energy with its environment.

Where this energy comes from can be answered in one word—food. The processes taking part in the ingestion, digestion, absorption, and utilization of food are very complex. An understanding of these processes is fundamental to an understanding of the productivity of an animal. The productivity of an individual animal is a basic part of the productivity of an animal community within an ecosystem. The productivity of primary consumers is directly related to the productivity of the plant community. The productivity of secondary consumers is directly related to that of primary consumers, which in turn is related to the productivity of the plant community. Thus there is an *n*-dimensional series of interrelationships in a plant and animal community. The challenge is that of analyzing significant parts of these relationships so that simple models can be developed that will lead to a greater understanding of the energy, matter, and time interactions that make the system what it is.

The "home range" of an animal generally varies according to the size of the animal. Small rodents such as *Microtus* and *Peromyscus* have home ranges of about an acre in size, whereas large mammals such as moose may range over several miles. There are many reasons for the relationship between body size and home-range size, including morphological ones—a small mammal simply cannot travel fast enough to cover a large home range; but one basic regulatory mechanism present is the distribution of energy and matter in relation to the requirements of the animal. Large mammals such as moose, elk, and deer must have a large supply of resources in order to satisfy their physiological requirements. Super-imposed on this energy and matter distribution is the effect of pheromones and other stimuli, which regulate the spacing of animals in a group. These stimuli may at times spread the animals out more than would be necessary to stay within the energy and matter limits of the resources. They may limit gregariousness that would cause an overuse of the resource supply, allowing the populations to be successful over a long period of time.

Animals that are gregarious during part or all of the year are usually quite mobile members of the animal community. Waterfowl congregate on the wintering grounds, and move as a group to feeding areas. Caribou move long distances in herds; elk migrate, congregating in wintering areas; and white-tailed deer often move to yarding areas in the winter and disperse in the spring. Snow conditions in these yards often confine the animals to limited areas, resulting in a rapid depletion of the food resource. The energy and matter balance is upset because the animals are confined, not only to specific yarding areas but to trails within the yards.

Physical characteristics of the habitat influence the distribution of animals within a community. In very homogenous habitats, territories may tend to be more regular in shape. In habitats with much variability in physical and biological features, territories conform to these features to some extent. Tree-nesting birds, for example, are found in plant communities of the right ecological age to include the right kinds of trees. Birds nesting in herbaceous vegetation are associated with plant communities of that type. These plant communities may be dependent on some physical feature—such as the distribution of cattails around a marsh, which provides nesting cover for red-winged blackbirds.

The morphology of the physical and biotic components of a habitat is often apparently very slightly different for closely related animals living in the habitat, but it may have more distinct ecological differences in both time and space. MacArthur (1958) has shown that ecological differences exist for five species of warblers that live in northeastern coniferous forests. These five species behave in manners that expose them to different kinds of food. Some feed in the crowns of trees, others in the lower reaches. Some feed on the periphery, and others in the interior of the crowns. The timing of changes in food requirements is such that peak requirements are slightly out of phase. Thus the members of the animal community utilize different aspects of the plant community and its physical and biological features. The *effective, functional* organization is different for different species and the roles of various community characteristics should only be evaluated in relation to the community members.

15-3 CHEMICAL RELATIONSHIPS

Since the distribution of energy and matter is a fundamental regulatory mechanism in the formation of plant and animal communities, it is useful to consider the distribution of chemicals (matter) that are the source of energy and matter for the metabolism of community members. The chemical characteristics of a plant community are related in part to the morphological characteristics of the community. Energy is stored in plant structures, with different amounts in different structures. Matter is distributed throughout the plant, with some plant parts serving specifically as storage organs.

The energy and matter cycle in plants occurs as a result of photosynthesis with the source of energy being the sun. The products of photosynthesis are used by the plant to synthesize its own tissue and are stored in the plant as a source of energy for subsequent plant growth (germinating seeds). These products, both in the leaves and seeds, are consumed by herbivores (primary consumers), who in turn synthesize their own body tissue with its particular energy and matter configuration.

The differences in the chemical morphology of plant parts result in different levels of usefulness to a primary consumer. Detailed data are lacking on the energy and matter distribution in wild plants, but values for corn illustrate this distribution pattern. A mature corn plant has the approximate energy and matter morphology shown in Figure 15-1. Note that the grain contains the highest protein and the highest digestible energy. This is because the grains are storage organs, whereas the other plant parts, such as the cobs and stem, are structures for mechanical support. They have rigid cell walls that provide structure, but they are not as nutritionally useful to a primary consumer as the grain is because the energy and matter is resistant to the action of digestive enzymes. The grain, however, contains sugar, starches, and proteins that are more readily available, except for the protective, highly lignified pericarp on the kernel.

The *presence* of nutrients such as protein must be carefully distinguished from the *availability* of nutrients in the plant parts. Digestible-protein data for cattle [data from Crampton and Harris (1969)] show that 46% of the protein in the aerial part of the plant is digestible: none of the protein in the cob is digestible; 28% of the protein in the husks is; and 76% of the protein in the grain is digestible. Thus the grain not only contains the highest percentage of crude protein but a distinctly greater amount of it is digestible by cattle. This illustrates that the

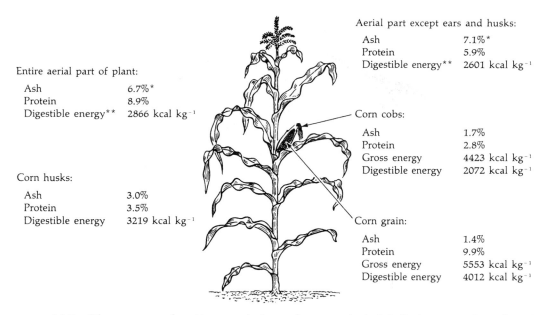

Entire aerial part of plant:

Ash   6.7%*
Protein   8.9%
Digestible energy**   2866 kcal kg^{-1}

Aerial part except ears and husks:

Ash   7.1%*
Protein   5.9%
Digestible energy**   2601 kcal kg^{-1}

Corn cobs:

Ash   1.7%
Protein   2.8%
Gross energy   4423 kcal kg^{-1}
Digestible energy   2072 kcal kg^{-1}

Corn husks:

Ash   3.0%
Protein   3.5%
Digestible energy   3219 kcal kg^{-1}

Corn grain:

Ash   1.4%
Protein   9.9%
Gross energy   5553 kcal kg^{-1}
Digestible energy   4012 kcal kg^{-1}

FIGURE 15-1.   The energy and matter morphology of a corn plant: * indicates percentages based on dry weight; ** indicates digestible energy for cattle. (Data from Crampton and Harris 1968.)

nutrient content of plants is of little significance by itself; the plant and animal must be considered together when evaluating the distribution of energy and matter in relation to productivity.

One other consideration will illustrate this point further. The protein content of corn grain is 10%, with 75% of the protein digestible by cattle. The absorption, metabolism, and synthesis of protein within the animal is a function of factors additional to digestibility. The *net* protein is of greatest significance because it represents the balance between the synthesis of protein in animal tissue and protein (nitrogen) extraction from the animal. Corn is deficient in the amino acid *lysine,* however, and as a result it is not a good single source of protein. Other forage containing lysine is necessary to make the amino acids in corn of greatest value to the animal. Since wild animals cannot very likely compensate for such kinds of deficiencies in their diets any more than domestic cattle can, it is clear that the analytical ecologist must pursue the animal-environment relationship to considerable depth if he is going to understand functional relationships in their entirety. Such subtle characteristics may have great ecological significance, especially when productivity is in a precarious balance.

The preceding discussion of chemical relationships showed how energy and matter is distributed within single plants. These combine to form a community distribution. Dense tree crowns form canopies that reduce photosynthetic activity in the subcanopies and at ground level. Some species have large numbers of suckers that are nourished by the parent tree. These individuals can survive in the subcanopy in part because they have a source of energy—the parent tree—that

TABLE 15-1 A DATA SHEET FOR MORPHOLOGICAL PHENOLOGY

Location: Observer:
  County: Date:

Phenology (P):

Stem emerging	= SE	Leaves withering	= LW	Flowers withering	= FW
Leaf buds	= LB	Leaves falling	= LF	Seeds forming	= SF
Leaves open	= LO	Floral buds	= FB	Seeds ripe	= SR
		Flowers open	= FO	Seeds disseminated	= SD

Genus	*Species*	*P*	*Notes*

extends beyond their own vertical height. Other species, particularly herbaceous ones in a deciduous forest, must rely on growth cycles that are out of phase with those of the canopy in order to absorb enough radiant energy to maintain the productivity necessary for life.

15-4 TEMPORAL RELATIONSHIPS

The energy and matter organization of plant and animal communities changes markedly with time. The morphological phenology changes as the plants progress from dormancy to leaf production, flower production, fruit and seed production, dispersal, and the abscission of leaves. Records of the morphological phenology can be kept initially on forms identifying various stages in the growth of plants (Table 15-1). These records are useful for reconstructing events in a plant community after an analytical model has been developed for the evaluation of the significance of temporal relationships in community organization.

Physiological changes involving photosynthetic processes in the leaf, translocation of the products of photosynthesis, development of fruits and seeds with the storage of photosynthetic products, and the dispersal of these products occur along with observed morphological changes. These physiological changes have a dramatic impact on the structure of the animal community. Several examples illustrate the importance of these changes. The fall of acorns, for example, results in a sudden abundance of food for ground-dwelling deer. The abundance of this food may have a significant effect on the amount of fat deposition in the deer prior to winter, which is important to winter survival. The development of corn in farmers' fields can have a dramatic effect on the food habits of racoon. The fall of leaves markedly alters the optical characteristics of the canopy, resulting in greater visibility for some kinds of predators and prey. These changes have an effect on the mechanical and thermal characteristics of the canopy as well; a sparse canopy intercepts less snow, has less effect on the reduction of wind, and is less effective as a barrier to the heat-sink effects of a cold night sky.

Changes in the chemical phenology of plants are as marked as changes in the morphological phenology. They are more subtle, however, and have not received as much attention from ecologists as have the more obvious morphological characteristics. Changes in the protein content of Kentucky bluegrass (*Poa pratensis*) as the plant goes from the immature stage to the mature stage range from 17.3% (dry-weight basis) to 3.3%, respectively (Figure 15-2). The continual decline in protein content is accompanied by a concomitant decline in digestability (by cattle), indicating that the nutritional usefulness of this forage plant actually deteriorates more than the crude-protein levels indicate.

Seasonal changes in the distribution and characteristics of energy and matter in plants owing to the processes of translocation and maturation occur also in alfalfa and timothy. In the immature stage, alfalfa has a protein content of 24.5%. This drops to 20.5% in prebloom, 19.3% in early bloom, 17.8% in midbloom, and 15.9% in full bloom [data from Crampton and Harris (1969)]. The data on timothy hay show the same trend since the protein content goes from 16.9% to 8.9% to 6.2% when cut in the immature, full bloom, and mature stages, respectively.

The rapid decline in protein content has considerable ecological significance. The timing of these changes in relation to the timing of the birth of deer fawns is of interest because fawns begin nibbling leaves shortly after birth. If the fawns are born at a time when both the protein content and the protein digestibility

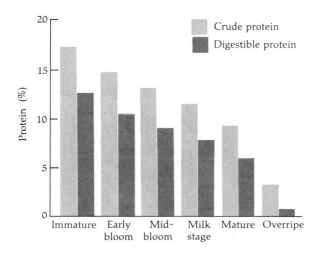

FIGURE 15-2. Protein content of Kentucky bluegrass at different stages of maturity. (Data from Crampton and Harris 1969, pp. 517–519.)

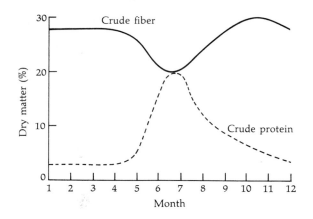
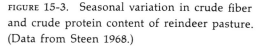

FIGURE 15-3. Seasonal variation in crude fiber and crude protein content of reindeer pasture. (Data from Steen 1968.)

is high, considerably more nutritive value can be derived from the plants. A month later the same forage species might have only half as much useful protein, resulting in a subtle deterioration in the range quality unless the fawn can compensate by selecting later-maturing species.

Changes in the chemical composition of plants occur continually, but at different rates at different physiological ages throughout the annual cycle. Steen (1968) shows the annual variation in the percentage of crude protein and crude fiber on reindeer pasture, illustrating the rapidity of the changes during the growing season (Figure 15-3). A similar graph by Kubota, Rieger, and Lazer (1970) for browse plants in Alaska is shown in Figure 15-4.

Chemical differences between species and between seasons for aspen, willow, blueberry, and grass on the Cache la Poudre Range in Colorado are reported by Dietz, Udall, and Yeager (1962). Protein, ash, and phosphorous percentages were higher in the spring than in the fall, and the fat, crude fiber, nitrogen-free extract, and calcium percentages were higher in the fall (Figure 15-5). Thus there is a shift from higher protein values during the growing season to higher fat and carbohydrate values in the fall when the plant is mature and the cell-wall fraction has increased.

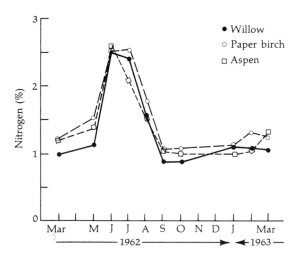

FIGURE 15-4. Seasonal changes in concentration of nitrogen in leaves and twigs of three Alaskan browse plants. (Data from Kubota, Rieger, and Lazer 1970, J. Wildlife Management.)

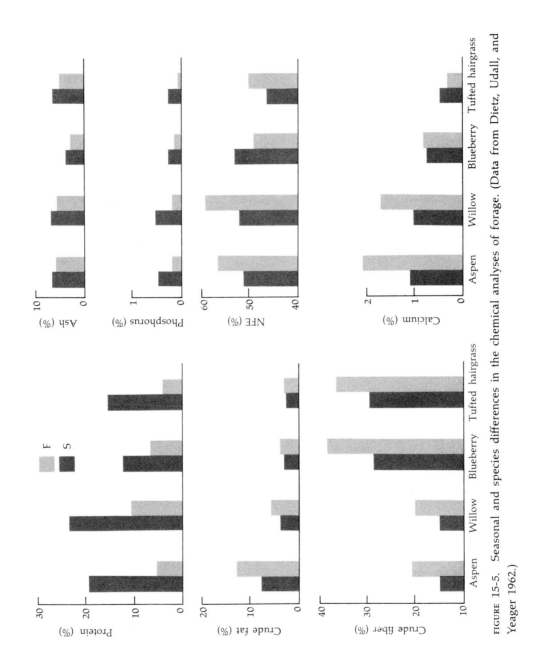

FIGURE 15-5. Seasonal and species differences in the chemical analyses of forage. (Data from Dietz, Udall, and Yeager 1962.)

Differences between seasons can be attributed to variations in the weather conditions that affect plant growth. These differences are greater for herbaceous plants than for woody species. The prairie, for example, can vary greatly from year to year because of differences in rainfall, solar radiation, and temperature.

It was pointed out earlier that the characteristics of the range cannot be evaluated in an ecologically realistic manner without an analysis of the animal characteristics as well. Ingestion, digestion, and nutrient metabolism were discussed in Part 3. Variation in the chemical characteristics of the range may also be reflected in the rumen content of animals on the range; the percentage of protein in the rumen of elk clearly coincide with the seasonal variation in the protein content of range plants (Figure 15-6).

Animals exhibit a chemical phenology as changes in body composition take place during growth and seasonally. The deposition of fat, for example, is a seasonal physiological function common to many animals. Rodents that enter hibernation have an accumulation of brown fat that is metabolized during hibernation. A lactating female deer has a lower fat content than a nonlactating animal, because the nutrient assimilation in the former is directed toward the synthesis of milk rather than fat. White-tailed deer reach their heaviest weights in the fall, with a decline beginning in November. The decline is greatest for males owing to their activity during breeding. Much of this weight loss can be attributed to

FIGURE 15-6.   Seasonal changes in the percentage of protein in elk rumens. Note the close similarity between this pattern and the protein content of the range shown in Figures 15-3 and 15-4. (Data from McBee 1964.)

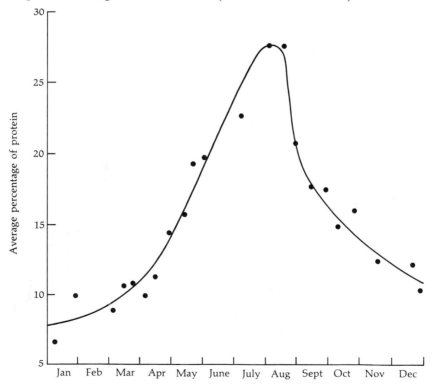

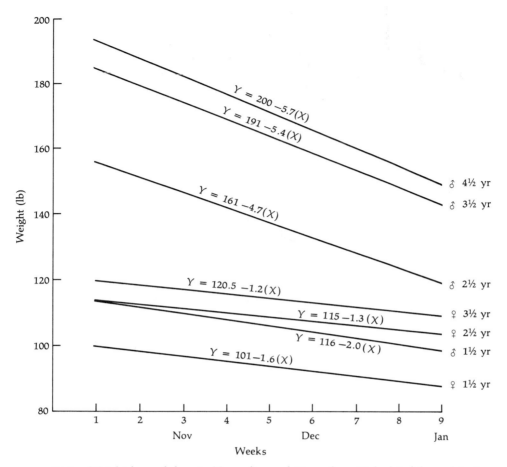

FIGURE 15-7. Weight loss of deer in November and December. (Calculated from data in Siegler 1968.)

the mobilization of body fat. The data in Figure 15-7 indicate a distinct difference in the weight losses between sexes and ages, with a more rapid loss for the older males. The older females hold their weight fairly well, perhaps even gaining during November and December if sufficient forage is available. Regression equations are in Appendix 5.

Minerals are mobilized on a seasonal basis for particular physiological functions. Antler growth is a seasonal phenomenon in white-tailed deer, normally occurring during the period from March to September. The deposition of calcium and phosphorus for antler growth occurs as a rich blood supply covers the antlers in a vascular tissue called "velvet." Some of the phosphorus comes from the ribs (Whelan, personal communication).

Many similar types of mobilization can be demonstrated for laboratory animals as the body shifts its physiological attention from one growing point to another. The occurrence of the redistribution of body components by free-ranging animals during the annual cycle is largely unknown. Its ecological significance in terms of productivity and even survival warrants further consideration.

Both plants and animals make provisions for survival during periods in which resources are unavailable or in short supply. This has been discussed with reference to the storage of the products of photosynthesis in seeds and fruits for later use as a source of energy for germination. The fat reserve of a hibernating animal functions in a similar manner inasmuch as it becomes a source of energy when ingestion processes are not active.

The requirements of animals for energy, protein, and other nutrients used in body-tissue maintenance and productivity are fairly regular over the annual cycle. The need for thermal energy to maintain homeothermy is a function of the thermal regime imposed on the animal by weather conditions and by its physiological condition. Weather conditions are regular on a long-term basis (the seasonal cycles) but irregular on a short-term basis, because of day-to-day weather fluctuations. The animal's physiological condition is fairly regular; seasonal shifts in metabolism and reproduction relate to such stimuli as daylength.

The bioenergetics of willow ptarmigan (*Lagopus lagopus*) over the yearly cycle illustrate the sequential occurrence of energy-demanding activities such as seasonal changes in weight, molt, egg laying, thermoregulation, and gross activity (West 1968). The spring molt begins two months before egg laying starts, and in females, the molt is interrupted during egg laying. Weight losses occur after egg laying and midway through the molt. West found that the birds in experimental pens were active during each of the 24 hours of daylight in the summer (College, Alaska), with a very small amount of activity in the long winter night. Additional energy requirements need to be considered for a bird in the field, including the cost of food gathering, migration, courtship behavior, incubating eggs, raising young, escaping from predators, responding to weather conditions, and others. West's article forms an excellent basis for the development of an analytical model describing the energetics of a free-ranging animal, and students are encouraged to continue within this framework for other species.

A similar bioenergetic model of a beaver (*Castor canadensis canadensis*) population has been described by Novakowski (1967). He shows that the energy supply cached for winter use when the colony was confined under the ice was not sufficient for the energy requirements of the colony, calculated on the basis of the number and weight of animals in each colony. This suggests that energy conservation is necessary, possibly including reduced activity, periods of dormancy, huddling, insulation by the lodge and fur, and fat deposition. The younger animals, both kits and yearlings, gained weight during the winter, whereas the older animals did not. An analysis of the energetics of beavers of different weights, with a consideration of the roles of different members of the population in the social structures, is basic to an understanding of the population dynamics and ecological organization of the beaver colony.

SUCCESSION. Plant communities that are undisturbed by the activities of man undergo natural successional changes. These changes are affected by physical factors, such as soil characteristics, rainfall, topography, and radiant energy, and biological factors that result from interactions between members of the plant community. This natural succession follows certain patterns that are predictable within the limits of variation in the physical factors from year to year.

Primary succession occurs as plants invade an area that has not supported life before in its present physical state. The bedrock exposed as a result of earthquakes, areas covered with volcanic ash, and so forth, are examples of potentially large-scale physical areas. On a smaller scale, overburden due to mining operations, exposure of the substrate owing to general removal, and local erosion over bedrock results in the exposure of physical substrates that have not supported life previously.

The plants that invade these areas are called pioneer plants. They include lichens that can grow on bare rock, which in turn contribute to a chemical and physical breakdown of the rock surface, preparing a substrate that is suitable for higher plants such as mosses. These modify further the rock-plant interface, resulting eventually in sufficient substrate that the higher seed plants can invade. Many higher plants invade newly exposed soil substrates directly without going through the sequence of lichen, moss, and higher plants. Floods, for example, redistribute soil that may be invaded directly by higher plants.

The disturbance of organisms in a developed community results in secondary succession. Fire, for example, can cause the destruction of a stand of mature trees. Soon after, herbaceous and woody plants invade the area, resulting in a new stand of young trees and an understory abundant in herbaceous plants. *Epilobium angustifolium* (fireweed) is a common herbaceous invader, and aspen (*Populus sp*) is a common woody invader, along with other trees and shrubs.

Plant succession occurs from an aquatic to a terrestrial substrate, as well as from virgin terra. A body of water, such as a pond, may support the growth of phytoplankton, and as these organisms die there is an accumulation of detritus on the pond's bottom that becomes a substrate for higher plants. At the same time, a mat of plants may be growing from the shore out into the water, resulting in a "floating bog" formation that encroaches on the open water. In time, the mat may cover the entire water surface unless it is destroyed by wind or some other natural force. As this occurs, rings of vegetation are also forming from the shore to the upland surrounding the water (Figure 15-8). Many examples of various stages of bog succession may be found, particularly in the northern part of the United States.

A succession of animals occurs concomitant with plant succession. Members of the grouse family illustrate this very well. Prairie grouse, such as the prairie chicken (*Tympanuchus cupido*), live on the prairie and do not tolerate dense canopies. If the shrub stage encroaches on the prairie, the habitat becomes more suitable for sharp-tailed grouse (*Pedioecetis phasianellus*). As a forest develops, particularly in the early stages of succession that include aspen trees, ruffed grouse (*Bonasa umbellus*) become more predominant. Finally, as the forest approaches a coniferous *climax* (the last stage in succession in any area, dependent on climatic condition), spruce grouse (*Canachites canadensis*) are natural members of the community (Figure 15-9). Similar trends in species distribution through successional stages can be observed for song birds, small rodents, rabbits and hares, and big game.

FIGURE 15-8. Successional stages from open water to the upland hardwood stand.

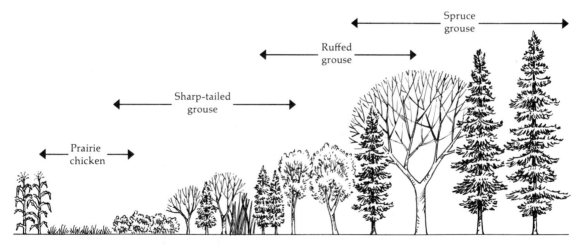

FIGURE 15-9. The relationship between prairie-to-forest succession and grouse species.

15-5 A PLANT-PRODUCTION MODEL

A GEOMETRIC MODEL. A geometric model of a plant is developed most easily if the plant is strikingly symmetrical. White pine (*Pinus strobus*) is such a plant. It is composed of a central axis with primary, secondary, tertiary, and quaternary whorls (Figure 15-10). The growth and forage production of twelve trees from 2 to 19 years old is shown in Table 15-2. Summation of these data results in a theoretical tree representing the average of measurements taken at identical growing points on different trees and variation in time for each tree.

A geometric model such as this need not be precise if the intent is to test the effect of variation in forage production on the productivity of the consumer. Everyone knows that forage production varies; the critical question for the ecologist is "How does variation in plant productivity affect animal productivity?" The value of a descriptive geometric model is in its mathematical form, which is compatible with computer analysis. Pictures of trees cannot be programmed; tables per se cannot be programmed. It is necessary to find some means to describe the growth over time in order for it to be synthesized with animal productivity in a dynamic way.

FIGURE 15-10. The geometry of a white pine.

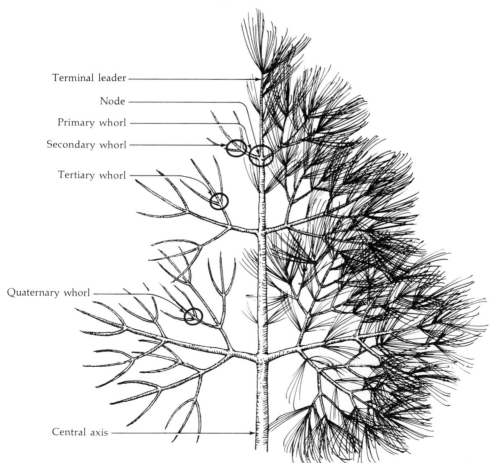

TABLE 15-2   GROWTH AND FORAGE PRODUCTION OF TWELVE WHITE PINE TREES FROM 2 TO 19 YEARS OF AGE

Growth in height	12.7 in yr^{-1}	32.3 cm yr^{-1}
Dry weight/length of current annual growth	0.44 g in^{-1}	0.173 g cm^{-1}
Total current annual growth in length	969 in yr^{-1}	2461 cm yr^{-1}
Total dry weight of current annual growth		425 g yr^{-1}
Dry weight of current annual forage production (under 6 ft)		350 g yr^{-1}

A tree, snow, and a primary consumer make a good combination for an analysis of the effect of interaction between chemical energy and forage, mechanical energy required for movement through snow by an animal, and the animal's energy metabolism. These relationships are depicted schematically in Figure 15-11; students are urged to expand on this model by using hypothetical components first, progressing towards greater realism as more data are assembled in successive models that describe the different functions.

The range ecologist and manager may be interested in only a specific aspect of the productivity of a tree or plant community. Deer and elk forage on the twigs of trees, usually the current annual growth (CAG), and it is the production of CAG that has the greatest short-term significance as a winter food supply. A mere estimation of range condition—excellent, good, or bad, for example—is not sufficient for a detailed analysis of animal-range relationships. A forage-production model composed of forage weight and chemical and nutritive characteristics of the forage for different consumers provides a much more realistic base for evaluating range quality.

The weight of CAG in relation to the number of CAG twigs of *Tilia americana* saplings is shown in Figure 15-12. This illustrates a technique for estimating the forage production on a given area, after which the nutritive characteristics can be related to forage production and, ultimately, to the net production of the consumer for maintenance and productive purposes.

The amount of CAG is dependent on many factors, of course, and it is desirable to determine the maximum production possible. The use of a number-weight equation for maximum production, followed by a series of forage reductions at set intervals permits the analysis of the effect of forage production on animal productivity.

A SEASONAL MODEL.   A descriptive model of seasonal variations in the distribution of energy and matter is useful for depicting the phenology of plants and

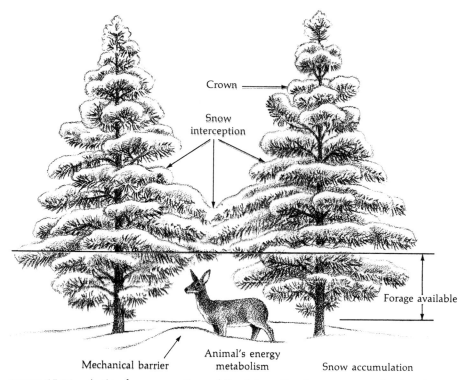

FIGURE 15-11. A simple representation of the interaction between an animal and snow, tree canopy, and forage.

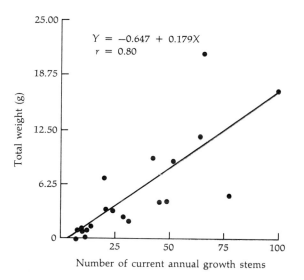

$Y = -0.647 + 0.179X$
$r = 0.80$

FIGURE 15-12. An illustration of a technique for predicting the weight of forage produced in relation to the number of current growth stems.

its relationship to primary consumers. The regression equation shown in Figure 15-12 could be a part of such a model. Another seasonal effect that could be considered is the growth, development, and fall of oak leaves and acorns. Leaf development must precede acorn development, since the leaves are sites of photosynthesis, which results in the formation of metabolites that can be translocated to the acorn for storage. The fall of leaves and acorns has a marked effect on the distribution of energy and matter. Thermal-energy relationships change owing to the differences in crown density. This was discussed in greater detail in Chapter 6. Chemical-energy distribution changes as the leaves and acorns fall. This results in an abundant supply of acorns for deer, and this may be important for fat deposition, which may be important for survival during the winter.

How is acorn production and fall utilized in a model that includes animal productivity? A working model can be developed first by using nondimensional numbers for acorn production, fall, and utilization by a primary consumer such as deer. A flow sheet is shown in Figure 15-13. Note the order in which the biological processes occur. It is not necessary to have complete information on each of these processes for useful dimensionless models for testing certain mechanisms. Select particular processes for the first model—acorn fall through tissue utilization in winter, for example. Use a linear regression equation to describe the rate of acorn fall (Y) over time (X). This takes the form of $Y = a + bX$. Suppose 100 units of acorns fell over 10 units of time. The specific equation is $Y = 0 + 10X$, where the limit of X equals 10. Suppose deer ate acorns at the rate of one half per unit time for a total of 50 units of time. The specific equation for that is $Y = 75 + \frac{1}{2}X$, where the limit of X is 50. This results in three known

FIGURE 15-13. A flow sheet for acorn production, fall, and utilization by deer.

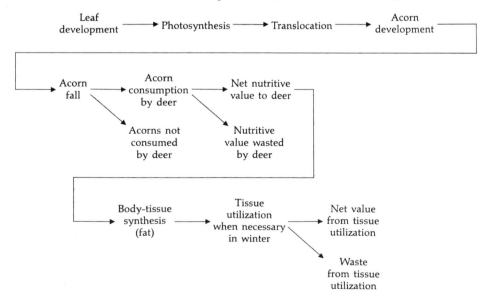

values for the flow sheet: 100 units of acorns produced, 25 units eaten, and 75 units not eaten. If the net nutritive efficiency is 30%, there would be 25 × .30 units available for tissue synthesis, or 7.5, with 17.5 units being nutritive waste. Fat mobilization at a later date might be 80% efficient, making .80 × 7.5 = 6.0 units available from tissue mobilization, with a waste of 1.5 units.

The total efficiency of this acorn-to-deer analysis is 6%, given the dimensionless units above. Keep in mind that all of the numbers used are arbitrarily selected to illustrate the mechanism of this kind of a calculation. Real numbers can be substituted to make the model more realistic. Many variations are possible within the simple framework used, including differences in acorn production, rate of fall, length of time the deer could consume acorns (this could depend on snow accumulation), and net nutritive value of acorns to the deer. Variations in any of these inputs will result in variations in the outputs. These variations could represent errors in measurements or natural variation in biological characteristics, or both.

This model is largely descriptive, especially if it is confined to acorn fall and tissue utilization, because there are no biologically dependent functions in the model. The model could be expanded to include dependent functions, such as photosynthesis = f (radiant energy, leaf surface area, leaf orientation, water absorption, transpiration . . .). The transition from a descriptive type of model to an analytical type can be exceedingly complex because so many more interactions, both direct and indirect, are included. Analytical models are conspicuously absent from the ecological literature, with a few exceptions.

One analytical model relating to plant productivity has recently appeared in *Science*. The model, developed by Dr. Edgar Lemon and his associates at the USDA Laboratory at Cornell University, illustrates the use of analytical data for the description of energy and matter interactions in a monotypic plant community.

A PHYSIOLOGICAL MODEL FOR PREDICTING NET PHOTOSYNTHESIS. The Soil-Plant-Atmosphere Model (SPAM) developed by Lemon, Stewart, and Shawcroft (1971) contains the basic mathematical format used to simulate a simple plant community—a corn field. The various submodels considered in SPAM are shown schematically in Figure 15-14. Note that the model is limited to the soil-surface boundary and the climate boundary surrounding the plant community. The gross structure of the plant community (crop submodel) and the fine structure at the site of photosynthesis (leaf submodel) are both considered. The former (crop submodel) has an effect on the distribution of energy and matter, including the geometric considerations of leaf angle and area distribution, light and wind distribution, and vertical diffusivity of energy and matter. The latter (leaf submodel) includes considerations of basic photosynthetic, metabolic, heat exchange and mass (water-vapor) transport from the leaves. Community behavior is predicted inasmuch as SPAM gives the vertical distribution of the activity of various community processes and vertical fluxes. Thus the vertical radiant-energy flux,

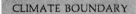

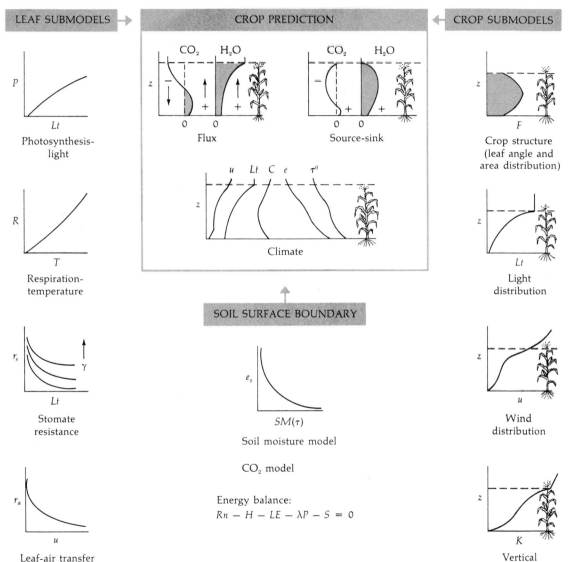

FIGURE 15-14.   Schematic summary of a mathematical soil-plant-atmosphere model (SPAM), giving required inputs, submodels, and representative daytime predictions of climate and community activity (that is, water vapor and carbon dioxide exchange). (From "The Sun's Work in a Cornfield," by E. Lemon, W. Stewart, and R. W. Shawcroft, *Science* **174:**371–378, Oct. 1971. Copyright 1971 by American Association for the Advancement of Science.)

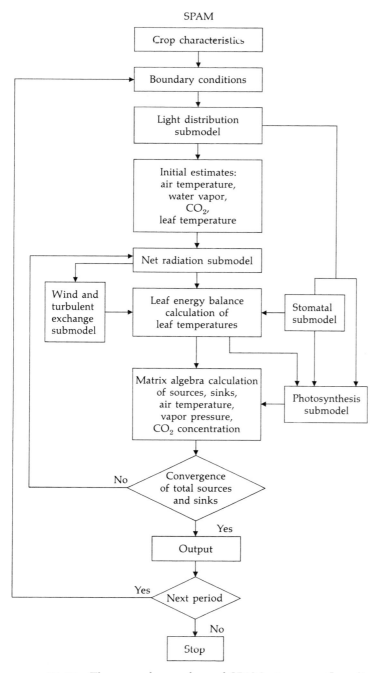

FIGURE 15-15. The general procedure of SPAM, given as a flow diagram. (From "The Sun's Work in a Cornfield," by E. Lemon, W. Stewart, and R. W. Shawcroft, *Science* **174**:371–378, Oct. 1971. Copyright 1971 by American Association for the Advancement of Science.)

air-temperature profile, wind profile, and profile describing water-vapor concentrations are used in predicting evaporation from the leaf surfaces at various heights. Similar kinds of predictions are made as other factors are considered in the interactions between energy and matter. A flow diagram with these relationships is shown in Figure 15-15.

The importance of these interactions is recognized by Lemon, Stewart, and Shawcroft: "It is obvious that solving the equation of any given part of SPAM is dependent on solving the equation of some other part." Once these interactions can be described, field tests can proceed to find the strengths and weaknesses of the model.

Lemon, Stewart, and Shawcroft have tested their model and find that some relationships were predicted quite accurately whereas others were much less so. The cause of discrepancies between measured values and predicted values must be determined by the analytical ecologist. The authors of SPAM suspect that classical fluid dynamic theory for turbulent boundary flow (such as wind in a plant canopy) is not applicable to tall vegetation that is porous and flexible. Work at the BioThermal Laboratory suggests that deer hair, too, has unique characteristics that affect the transfer of heat energy.

One consideration made possible with SPAM is that of carbon dioxide fertilization on a field scale. Green house fertilization with carbon dioxide has proved successful, but many more variables exist in the field that might reduce the efficiency of this practice. Simulations of different levels of carbon dioxide fertilization showed that high rates of release did not affect the concentration in the photosynthetically active part of the canopy. This is due to the vigorous diffusion processes in the atmosphere. Thus it appears that carbon dioxide fertilization is not a feasible solution for increasing yield. It was concluded, however, that wind plays such an important part in regulating the natural supply of carbon dioxide that the idea of designing crop communities to enhance air flow has merit.

Community structure was shown to have the greatest effect on productivity. This suggests that selective breeding might improve leaf angle, for example, resulting in a higher photosynthetic efficiency. The authors point out, however, that no one structure can be "ideal," but rather varies with climate, crop, latitude, and time of year.

It is interesting that a mathematical model based on extensive field experiments should result in such strong emphasis on community structure. Assuming that this is the case for natural plant communities, as well as a corn community, the student interested in plant ecology has a tremendous opportunity to analyse the structure of natural communities on the basis of energy, matter, and time interactions. This would result in a significant increase in our understanding of plant productivity in natural systems. The modeling and simulation approach requires the use of computer analyses, of course, along with several types of field instrumentation that may not have been used very frequently by plant ecologists in the past. Both gross and fine plant community structures have to be studied in relation to cellular physiology. These new dimensions only serve to add excitement to the field of ecology; indeed, many conclusions in the literature can be reevaluated with new insights that often lead to interesting new conclusions.

15-6   PERTURBATIONS

OVERGRAZING.   Plant communities on the range are dynamic biological units that respond not only to forces imposed on them by natural events but also to the additional pressures of grazing by both wild and domestic animals. The foraging activities of domestic and wild animals cause changes in the floral composition of a plant community. Stoddart and Smith (1955) review the terminology associated with changes in the species composition of the range community, including *climax decreasers, climax increasers,* and *invaders.* Species that become less abundant under heavy grazing are climax decreasers; they are likely to be the species most preferred by the grazing animals. Those species that become more abundant under heavy grazing are climax increasers; they are the species least preferred by the grazing animal. Invaders are species that are originally not present but which become established in the grazed area.

In general, overgrazing reduces the number of native (prior to grazing) plants on an area, with an increase in the number of weeds (unpalatable forbs) in the community. Shrubs can become abundant, too, often resulting in a plant community that is resistant to the effects of grazing because the energy and matter in the plants is largely unavailable to the primary consumers. It is a kind of dynamic nutrient filtration process that leaves only the non-nutrients in the system.

FIRE.   Fire is often thought of as a disaster, and indeed it is just that to the forester intent on harvesting mature trees for lumber. The ecological significance of fire is relative to the organisms under consideration, however. Thus a fire in a mature forest is bad for spruce grouse but good for ruffed grouse, bad for caribou but good for deer. The effect of fire is best evaluated within a framework of the successional stages desired. The prairie, a mixture of grasses and forbs, is maintained only when fire or some other perturbation prevents invasion by shrubs. Virgin prairie near Kensington, Minnesota (Douglas County, west-central part of the state), is being invaded by *Symphoricarpus occidentalis,* a low-growing shrub, because fire has been eliminated in one small area and because of overgrazing on an adjacent pasture. *Quercus macrocarpa* (Burr oak), a fire resistant tree, is found in some uncultivated areas along lakes and marshes, with sugar maple (*Acer saccharum*) and basswood found in areas protected from fire by the lakes. One stand of trees on a peninsula extending northward into a shallow lake contains about 80% sugar maple (Moen 1964). This sugar maple community would very likely not be present if the water had not protected it from fire, prior to settlement in that area.

Many other perturbations can occur, including floods, tornadoes, hurricanes, hail storms, and, over long periods of time, glaciation. All of these conditions may result in fairly long-term effects, and although they are interesting from an ecological point of view, they may not be the most important factors in the short-term population fluctuations that cause concern among sportsmen, legislators, and other members of the public whose interests and activities are geared to the present. Thus the resource analyst and manager must balance the theoretical long-term effects with the more immediate, short-term considerations, keeping in mind that in each case certain fundamental principles and physical laws affect the distribution of matter and energy in the ecosystem.

LITERATURE CITED IN CHAPTER 15

Crampton, E. W., and L. E. Harris. 1969. *Applied animal nutrition.* 2d ed. San Francisco: W. H. Freeman and Company, 753 pp.

Dietz, D. R., R. H. Udall, and L. E. Yeager. 1962. *Chemical composition and digestibility by mule deer of selected forage species, Cache la Poudre Range, Colorado.* Colorado Dept. Game and Fish, Technical Publication No. 14, 89 pp.

Grieg-Smith, P. 1964. *Quantitative plant ecology.* Washington, D.C.: Butterworth, 256 pp.

Kubota, J., S. Rieger, and V. A. Lazer. 1970. Mineral composition of herbage browsed by moose in Alaska. *J. Wildlife Management* **34**(3): 565–569.

Lemon, E., D. W. Stewart, and R. W. Shawcroft. 1971. The sun's work in a cornfield. *Science* **174**(4007): 371–378.

MacArthur, R. H. 1958. Population ecology of some warblers of northeastern coniferous forests. *Ecology* **39**(4): 599–619.

McBee, R. H. 1964. Rumen physiology and parasitology of the northern Yellowstone elk herd. Mimeographed. Bozeman: Montana State College, 28 pp.

McClaugherty, C. M., C. M. Haramis, and G. Trevail. 1972. Forage production of white pine: an analytical approach. Special report, BioThermal Laboratory, Cornell University, 24 pp.

Moen, A. N. 1964. A comparison of four sugar-maple–basswood stands on the prairie-forest margin. Unpublished report, 49 pp.

Novakowski, N. S. 1967. The winter bioenergetics of a beaver population in northern latitudes. *Can. J. Zool.* **45**: 1107–1118.

Oosting, H. J. 1956. *The study of plant communities.* 2d ed. San Francisco: W. H. Freeman and Company, 440 pp.

Siegler, H. R., ed. 1968. *The white-tailed deer of New Hampshire.* Concord: New Hampshire Fish and Game Dept., 256 pp.

Steen, E. 1968. Some aspects of the nutrition of semi-domestic reindeer. In *Comparative nutrition of wild animals,* ed. M. A. Crawford. Proceedings of Symposia of the Zoological Society of London, No. 21, pp. 117–128.

Stoddart, L. A., and A. D. Smith. 1955. *Range management.* New York: McGraw-Hill, 433 pp.

West, G. C. 1968. Bioenergetics of captive willow parmigan under natural conditions. *Ecology* **49**(6): 1035–1045.

SELECTED REFERENCES

Anderson, W. L., and P. L. Stewart. 1969. Relationships between inorganic ions and the distribution of pheasants in Illinois. *J. Wildlife Management* **33**(2): 254–270.

Beckwith, S. L. 1954. Ecological succession on abandoned farm lands and its relationships to wildlife management. *Ecol. Monographs* **24**(4): 349–376.

Beals, E. 1960. Forest bird communities in the Apostle Islands of Wisconsin. *Wilson Bull.* **72**(2): 156–181.

Black, C. A. 1968. *Soil-plant relationships.* New York: Wiley, 792 pp.

Bond, R. R. 1957. Ecological distribution of breeding birds in the upland forests of southern Wisconsin. *Ecol. Monographs* **27**: 351–384.

Bormann, F. H., and G. E. Likens. 1970. The nutrient cycles of an ecosystem. *Sci. Am.* **223**(4): 92–101 (Offprint No. 1202).

Bray, J. R., and J. T. Curtiss. 1957. An ordination of the upland forest communities of southern Wisconsin. *Ecol. Monographs* **27**: 325–349.

Caplenor, D. 1968. Forest composition of loessal and non-loessal soils in west-central Mississippi. *Ecology* **49**(2): 322–331.

Connell, J. H., D. B. Mertz, and W. W. Murdoch, eds. 1970. *Readings in ecology and ecological genetics.* New York: Harper & Row, 397 pp.

Curtis, J. T., and G. Cottam. 1962. *Plant Ecology Workbook.* Minneapolis: Burgess, 193 pp.

Daubenmire, R. F. 1936. The "big woods" of Minnesota, its structure and relation to climate, fire, and soils. *Ecol. Monographs* **6**: 233–268.

Daubenmire, R. 1968. *Plant communities: a textbook of plant synecology.* New York: Harper & Row, 300 pp.

Dix, R. L., and J. E. Butler. 1960. A phytosociological study of a small prairie in Wisconsin. *Ecology* **41**(2): 316–327.

Douglas, G. W., and T. M. Ballard. 1971. Effects of fire on alpine plant communities in the north cascades, Washington. *Ecology* **52**(6): 1058–1064.

Elton, C. S. 1966. *The pattern of animal communities.* New York: Wiley, 432 pp.

Giles, R. H., Jr. 1970. The ecology of a small forested watershed treated with the insecticide malathion-S35. *Wildlife Monographs* **24**: 1–81.

Golley, F. B. 1960. Energy dynamics of a food chain of an old-field community. *Ecol. Monographs* **30**: 187–206.

Goodrum, P. D., V. H. Reid, and C. E. Boyd. 1971. Acorn yields, characteristics, and management criteria of oaks for wildlife. *J. Wildlife Management* **35**(3): 520–532.

Gorham, E., and J. Sanger. 1967. Caloric values of organic matter in woodland, swamp and lake soils. *Ecology* **48**(3): 492–494.

Haberman, C. G., and E. D. Fleharty. 1971. Energy flow in *Spermophilus franklinii. J. Mammal.* **52**(4): 710–716.

Harris, S. W., and W. H. Marshall. 1963. Ecology of water-level manipulations on a northern marsh. *Ecology* **44**(2): 331–343.

Hodges, J. D. 1967. Patterns of photosynthesis under natural environmental conditions. *Ecology* **48**(2): 234–242.

Hughes, M. K. 1971. Seasonal calorific values from a deciduous woodland in England. *Ecology* **52**(5): 923–926.

Hurd, R. M. 1971. Annual tree-litter production by successional forest stands, Juneau, Alaska. *Ecology* **52**(5): 881–884.

Idso, S. B., and D. G. Baker. 1968. The naturally varying energy environment and its effects upon net photosynthesis. *Ecology* **49**(2): 311–316.

Keith, L. B., and D. C. Surrendi. 1971. Effects of fire on a snowshoe hare population. *J. Wildlife Management* **35**(1): 16–26.

Klein, D. R. 1970. Tundra ranges north of the Boreal Forest. *J. Range Management* **23**(1): 8–14.

Knight, D. H., and O. L. Loucks. 1969. A quantitative analysis of Wisconsin forest vegetation on the basis of plant function and gross morphology. *Ecology* **50**(2): 219–234.

Kozlovsky, D. G. 1968. A critical evaluation of the trophic level concept. I. Ecological efficiencies. *Ecology* **49**(1): 48–60.

Krefting, L. W., H. L. Hansen, and M. H. Stenlund. 1956. Stimulating regrowth of mountain maple for deer browse by herbicides, cutting and fire. *J. Wildlife Management* **20**(4): 434–441.

Kucera, C. L., R. C. Dahlman, and M. R. Koelling. 1967. Total net productivity and turnover on an energy basis for tall-grass prairie. *Ecology* **48**(4): 536–541.

Kydd, D. D. 1964. The effect of different systems of cattle grazing on the botanical composition of permanent downland pasture. *J. Ecol.* **52**(1): 139–149.

Larson, F. 1940. The role of the bison in maintaining the short grass plains. *Ecology* **21**: 113–121.

Leak, W. B. 1970. Successional change in northern hardwoods predicted by birth and death simulation. *Ecology* **51**(5): 794–801.

Lindeman, R. L. 1942. The trophic-dynamic aspect of ecology. *Ecology* **23**: 399–418.

Morowitz, H. J. 1968. *Energy flow in biology.* New York: Academic Press, 179 pp.

Orr, R. T. 1970. *Animals in migration.* New York: Macmillan, 303 pp.

Ovington, J. D., and D. B. Lawrence. 1967. Comparative chlorophyll and energy studies of prairie, savanna, oakwood, and maize field ecosystems. *Ecology* **48**(4): 515–524.

Palmblad, I. G. 1968. Competition in experimental populations of weeds with emphasis on the regulation of population size. *Ecology* **49**(1): 26–34.

Peek, J. M. 1970. Relation of canopy area and volume to production of three woody species. *Ecology* **51**(6): 1098–1101.

Phares, R. E. 1971. Growth of red oak (*Quercus rubra* L.) seedlings in relation to light and nutrients. *Ecology* **52**(4): 669–672.

Rosenzweig, M. L. 1966. Community structure in sympatric carnivora. *J. Mammal.* **47**(4): 602–612.

Schoener, T. W. 1968. Sizes of feeding territories among birds. *Ecology* **49**(1): 123–141.

Slatyer, R. O. 1967. *Plant-water relationships.* New York: Academic Press, 368 pp.

Sloan, C. E. 1970. Biotic and hydrologic variables in prairie potholes in North Dakota. *J. Range Management* **23**(4): 260–263.

Spedding, C. R. W. 1971. *Grassland ecology.* New York: Oxford University Press, 221 pp.

Tappeiner, J. C., II. 1971. Invasion and development of beaked hazel in red pine stands in northern Minnesota. *Ecology* **52**(3): 514–519.

Van Cleve, K. 1971. Energy- and weight-loss functions for decomposing foliage in birch and aspen forests in interior Alaska. *Ecology* **52**(4): 720–723.

Vezina, P. E. 1961. Variations in total solar radiation in three Norway spruce plantations. *Forest Sci.* **7**: 257–264.

Vezina, P. E., and G. Peck. 1964. Solar radiation beneath conifer canopies in relation to crown closure. *Forest Sci.* **10:** 443–451.

Vogl, R. 1969. One-hundred and thirty years of plant succession in a southeastern Wisconsin lowland. *Ecology* **50**(2): 248–255.

Weaver, J. E., and W. E. Bruner. 1954. Nature and place of transition from true prairie to mixed prairie. *Ecology* **35**(2): 117–126.

White, E. M. 1971. Some soil age-range vegetation relationships. *J. Range Management* **24**(5): 360–365.

Courtesy of Paul M. Kelsey
New York State Department of Environmental Conservation

PART

PRODUCTIVITY, POPULATIONS, AND DECISION-MAKING

The interaction of all of the factors affecting productivity results in a dynamic balance between the characteristics of plants, or producers, and the requirements of animals, or consumers. This dynamic balance is dependent on the distribution of energy and matter through time, and the whole system operates within the basic laws of energy and matter. Knowledge of the relative importance of the different factors is necessary for an understanding of the concept of carrying capacity, for insight into the causes of population dynamics, and for decision-making in the management of natural systems. This part of the book includes analyses of the biological relationships between an animal and its range, with a discussion of the role of economics and sociopolitical factors in the decision-making process.

A BIOLOGICAL BASIS
FOR THE CALCULATION
OF CARRYING CAPACITY

16-1 THE CONCEPTUAL DESIGN

How many boxes will fit into the one-inch-square box in Figure 16-1? The answer to that question is indeterminant because the sizes of the boxes to be fitted in are not known at this time. If a box to be fitted in is larger than the one illustrated, the number that will fit will be less than one. If boxes smaller than the one illustrated are selected, more than one will fit in. Further, nothing has been said about whether the boxes are the same size; they could all be different.

If the sizes of the boxes to be fitted in are known, it is possible to come up with an answer. Thus the box illustrated, exactly one-inch square, will hold 16 boxes exactly $\frac{1}{4}$-inch square. It will hold 13 boxes if 12 are $\frac{1}{4}$-inch square and the thirteenth is $\frac{1}{2}$-inch square. If the size of the holding box and the sizes of the inside boxes are determined, only one answer will satisfy the problem.

The example illustrates the kind of information needed to determine the number of animals that can be supported on a given area of land. Both the supply of resources on the range and the requirements of the animals must be known for an understanding of the relationship between the two. The numerical determination of carrying capacity is very complex since biological organisms, including both the animals and plants on the range, are dynamic assemblies of organic molecules that are highly organized into functional, living units whose "size" or requirements are changing continually.

There is an obvious need, then, for knowledge of the requirements of an animal for maintenance and productive purposes before a meaningful biological appraisal of carrying capacity can be made. The determination of the requirements of an

FIGURE 16-1. A box.

animal is a costly and time-consuming process consisting of feeding trials at different nutritional levels and the measurement of animal response, including weight, growth, and reproductive performance. These data are largely unavailable for wild ruminants. Data for domestic species can be used to make first approximations, and error analyses can be completed to find out how important variation in any one parameter is in the total animal-range relationship.

It is also necessary to know the quality of resources available to supply the requirements of the animal. These must be expressed in units that are biologically meaningful, such as the kilocalorie for energy and weight units for protein, with the relationship between animal and range analyzed through the use of *net* values.

16-2 PROTEIN REQUIREMENTS OF THE INDIVIDUAL ANIMAL

PROTEIN REQUIREMENTS FOR MAINTENANCE. Protein is necessary for the maintenance of basic life processes, including the synthesis of enzymes, the replacement of body tissue that is catabolized, and the replacement of tissue abraded from internal surfaces, such as the gastrointestinal tract, and from the skin.

Measurements of the protein requirements for these functions have been made for domestic cattle and sheep, and these experiments will form a base line for making first approximations for wild ruminants. Nitrogen requirements are calculated from protein requirements by dividing the latter by 6.25.

The nitrogen excreted in the urine that is of endogenous origin (EUN) is derived from the catabolism of body tissue, and this quantity is related to the metabolic weight of the animal according to Crampton and Harris (1969). They cite earlier work by Brody in assembling a table for the minimum daily requirements for protein, which can be expressed as endogenous urinary nitrogen in equation (16-1).

$$Q_{eun} = \frac{2 \times 70(W_{kg}^{0.75})}{1000} \qquad (16\text{-}1)$$

where

Q_{eun} = endogenous urinary nitrogen in g day^{-1}
 2 = the ratio of N in mg to kcal in the equation for basal metabolism = $(N_{mg}/kcal)$
W_{kg} = animal weight in kg

Estimates of endogenous urinary nitrogen excretion, with weight expressed as $W_{kg}^{0.75}$, are presented in an Agricultural Research Council publication (1965) for cattle and sheep. These estimates have been recalculated with weight expressed as $W_{kg}^{0.75}$, multiplied by 6.25, and shown as a protein requirement in Table 16-1. A comparison between the results using equation (16-1) and the ARC data shows that the endogenous urinary nitrogen calculated with equation (16-1) is slightly lower than the ARC data for the smaller animals within the weight range of deer and considerably higher for the larger animals.

TABLE 16-1 ESTIMATES OF ENDOGENOUS URINARY NITROGEN EXCRETION AND THE MINIMUM PROTEIN REQUIREMENT IN CATTLE AND SHEEP

			Calculated Protein‡ Requirement	
W_{kg}	$W_{kg}^{0.75}$	Endogenous Urinary Nitrogen (g per day per $W_{kg}^{0.75}$)*	Endogenous Urinary Nitrogen*	$[(2)(70)(W^{0.75})(6.25)]/1000$†
		Cattle		
50	18.80	.19	22.33	16.45
75	25.49	.17	27.08	22.30
100	31.62	.15	29.65	27.67
125	37.38	.14	32.71	32.71
150	42.86	.13	34.83	37.50
175	48.11	.12	36.09	42.10
200	53.18	.11 (200 kg+)	36.56	46.54
250	62.87		43.22	55.01
300	72.08		49.56	63.07
350	80.92		55.63	70.80
400	89.44		61.49	78.26
450	97.70		67.17	85.49
500	105.74		72.69	92.52
550	113.57		78.08	99.38
600	121.23		83.35	106.08
		Sheep		
2.5	1.99	.165	2.05	1.74
5	3.34	.16	3.34	2.93
10	5.62	.14	4.92	4.92
15	7.62	.13	6.19	6.67
20	9.46	.11	6.50	8.28
25	11.18	.09	6.29	9.78
30	12.82	.08 (30 kg+)	6.41	11.22
35	14.39		7.19	12.59
40	15.91		7.95	13.92
45	17.37		8.69	15.20
50	18.80		9.40	16.45
100	31.62		15.81	27.67

*Modified from Table 5.3, ARC 1965, p. 156.
†Crampton and Harris 1969.
‡Protein = N × 6.25.

The mechanical process of food passage through the gastrointestinal tract results in the abrasion of the epithelium lining the tract, resulting in a loss of protein. This loss, plus spent enzymes, bacterial residues, and other catabolized protein in the feces, is estimated to be 5 grams per kilogram of dry-matter intake per day for sheep and cattle on a forage diet, and 2.5 grams per kilogram of dry-matter intake for calves on a liquid diet (ARC 1965). Expressed mathematically for computational purposes,

$$Q_{mfn} = cF_{kg}/6.25 \qquad (16\text{-}2)$$

where

Q_{mfn} = metabolic fecal nitrogen in g day^{-1}
$c = 5$ for forage diets, 2.5 for milk diets, and $5 - (113.6 - 4.5$ $W_{kg})(2.5/100)$ for milk and forage diets for deer
F_{kg} = dry-matter intake in kg day^{-1}

PROTEIN REQUIREMENTS FOR PRODUCTION. The deposition of new tissue during growth represents a protein requirement that is directly related to the amount of gain of different kinds of body tissue. For growth, the nitrogen retention has been estimated to be 2.4% to 3.5% of the gain, with variation according to species and weight (ARC 1965). Higher nitrogen requirements per unit gain in weight are expected in the younger animals that are depositing more protein tissue than fat. Lower nitrogen requirements are characteristic of the older animals that are depositing more fatty tissue and less protein in each unit gain in weight. The mathematical expression for the nitrogen in the gain, using 2.5% nitrogen fraction, is shown in equation (16-3):

$$Q_{ng} = 2.5 \, \Delta W_{kg}/100 = .025 \, \Delta W_{kg} \qquad (16\text{-}3)$$

where

Q_{ng} = quantity of nitrogen required for daily gain in g day^{-1}
ΔW_{kg} = gain in weight in kg day^{-1}

The production of hair requires protein, but the amount required by wild ruminants has not been measured. An estimate of the loss of nitrogen in hair and scurf of cattle has been made by Blaxter and reported in the ARC publication (1965). It can be calculated with equation (16-4):

$$Q_{nh} = 0.02 \, W_{kg}^{0.75} \qquad (16\text{-}4)$$

where

Q_{nh} = quantity of nitrogen required for hair growth in g day^{-1}

This estimate is a small portion of the total nitrogen requirement. It is included here to draw attention to the fact that the growth of hair does involve a nitrogen "cost."

PROTEIN REQUIREMENTS FOR GESTATION. The fetus, placenta, uterus, and the fluids surrounding the fetus increase in weight as pregnancy progresses. The nitrogen retained also increases; the amount of nitrogen and its protein equivalent that is retained per day in fetal tissue is shown in Table 16-2 for cattle and sheep. The amount retained per day increases in a logarithmic manner (Figure 16-2). An estimation of the protein requirement for pregnancy of wild ruminants can be made as follows:

1. The protein requirements per day from Table 16-2 are expressed in a linear regression equation with a log transformation of Y, where Y = the protein required per day and X = time pregnant in days (t_d) (Table 16-3).

2. The protein requirements calculated with the equations in Table 16-3 are total daily requirements for pregnancy in cattle and sheep.

3. The calculated protein requirement per day during gestation is divided by the weight of the fetus at term, resulting in the expression of protein requirements per day per kilogram of fetus weight at term.

4. The gestation periods for cattle, sheep, and wild ruminants are different, so conversion factors that express the gestation periods on an equivalent physiological time scale are calculated (Table 16-4). The young of moose, elk, and bison are larger than lambs at birth, so the cattle data are used as a base for calculation. The young of other wild ruminants are more like lambs in size at birth, and sheep data (single lambs) are used as a base.

TABLE 16-2 RETENTION OF NITROGEN AND PROTEIN BY COWS AND EWES IN THE FETUS, PLACENTA, UTERUS, AND FLUIDS

Gestation Time		Retention of	Protein Equivalent
Month	Days	Nitrogen (g day^{-1})	(N × 6.25)
Cows: calf, 45 kg at birth			
5th–6th	185	1.7	10.63
7th	220	5.1	31.88
8th	250	12.0	75.00
9th	280	29.0	181.25
Ewes: single lamb, 5.9 kg at birth			
2nd	56	0.18	1.13
3rd	84	0.34	2.13
4th	112	1.45	9.06
5th	140	4.96	31.00
Ewes: twin lambs, 10.0 kg at birth			
2nd	56	0.24	1.50
3rd	84	0.96	6.00
4th	112	3.07	19.19
5th	140	7.40	46.25

SOURCE: Data from ARC 1965, pp. 163–165.

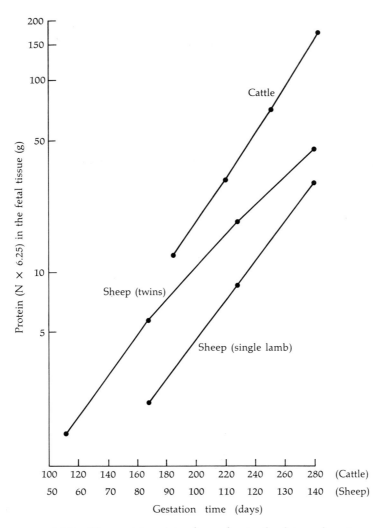

FIGURE 16-2. The protein retained per day in the fetus, placenta, uterus, and fluids in cattle and sheep. Note that the Y axis is a log scale.

TABLE 16-3 PROTEIN REQUIREMENTS FOR GESTATION (calculated from data in Table 16-2)

Species	Equation	Weight of Fetus at Term (kg)
Cattle	$Y_g = e^{(-3.1206 + 0.0298\, t_d)}$	45.0
Sheep (single lamb)	$Y_g = e^{(-2.3623 + 0.0407\, t_d)}$	5.9
Sheep (twins)	$Y_g = e^{(-1.7605 + 0.0409\, t_d)}$	10.0

TABLE 16-4 CONVERSION FACTORS FOR EXPRESSING EQUIVALENT TIMES IN THE GESTATION PERIODS OF DOMESTIC AND WILD RUMINANTS

Species	Gestation Period	Conversion Factor
Domestic cattle	280	—
Moose	245	0.87500
Elk	260	0.92857
Bison	290	1.03571
Domestic sheep	140	—
Mule deer	200	1.42857
White-tailed deer	200	1.42857
Pronghorn	240	1.71429
Bighorn sheep	150	1.07143
Mountain goat	180	1.28571
Caribou	220	1.57143

The protein requirement for pregnancy of any wild ruminant, expressed as the protein required at time t (days) in gestation per kilogram of fetus weight at birth, can then be calculated by equation (16-5):

$$Q_{pp} = [e^{a+b(t_d/c)}]/W_{kg} \tag{16-5}$$

where

Q_{pp} = quantity of protein required for pregnancy (grams per day per kg fetus weight at birth)

a and b = constants (see Table 16-5)

t_d = days pregnant

c = conversion factor for gestation periods (see Table 16-4)

W_{kg} = weight of fetus at term for cattle or sheep, depending on the base selected (see No. 4)

The numerical equations for calculating the protein requirements for pregnancy in wild ruminants are shown in Table 16-5. These protein requirements, expressed as grams per day per kilogram of fetus weight at term, are plotted through the entire gestation period in Figures 16-3 and 16-4. Absolute protein requirements are easily calculated by multiplying the grams per kilogram of fetus weight by the birth weight of the infant animal.

The equations for calculating the protein or nitrogen (dividing protein by 6.25) requirements can be stored in the memory of an electronic computing system, and the protein or nitrogen requirement for pregnancy can be calculated by entering the gestation time and the fetus weight at full term. An average birth weight for each species can also be stored, leaving the gestation time in days as the only variable to enter for a solution. This procedure is used in Chapter 17 for calculations of carrying capacity.

TABLE 16-5 PROTEIN REQUIREMENTS FOR GESTATION IN GRAMS PER DAY PER KILOGRAM OF FETAL WEIGHT AT TERM

Species using cattle requirements as a base

Cattle	$Y_g = [e^{(-3.1206 + 0.0298\, t_d/1)}]/45$
Elk	$Y_g = [e^{(-3.1206 + 0.0298\, t_d/.92857)}]/45$
Moose	$Y_g = [e^{(-3.1206 + 0.0298\, t_d/0.875)}]/45$
Bison	$Y_g = [e^{(-3.1206 + 0.0298\, t_d/1.03571)}]/45$

Species using sheep requirements as a base

Domestic sheep (singles)	$Y_g = [e^{(-2.3623 + 0.0407\, t_d/1)}]/5.9$
Deer	$Y_g = [e^{(-2.3623 + 0.0407\, t_d/1.42857)}]/5.9$
Pronghorn	$Y_g = [e^{(-2.3623 + 0.0407\, t_d/1.71429)}]/5.9$
Mountain goat	$Y_g = [e^{(-2.3623 + 0.0407\, t_d/1.28571)}]/5.9$
Bighorn sheep	$Y_g = [e^{(-2.3623 + 0.0407\, t_d/1.07143)}]/5.9$
Caribou	$Y_g = [e^{(-2.3623 + 0.0407\, t_d/1.57143)}]/5.9$
Domestic sheep (twins)	$Y_g = [e^{(-1.7605 + 0.0409\, t_d/1)}]/10$

FIGURE 16-3. Protein requirements of wild ruminants for gestation, using cattle data as a base for the calculations: *A*, moose; *B*, elk; *C*, cattle; *D*, bison.

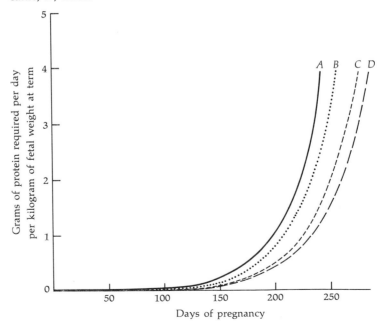

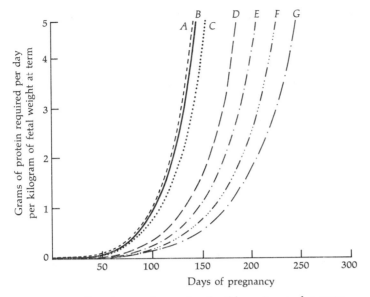

FIGURE 16-4. Protein requirements of wild ruminants for gestation, using sheep data as a base for the calculations: *A*, sheep (twins); *B*, sheep (singles); *C*, bighorn sheep; *D*, mountain goat; *E*, white-tailed and mule deer; *F*, caribou; *G*, pronghorn.

MILK PRODUCTION. The protein cost of milk production is an important consideration when determining the total protein requirements of an animal. The milk production of a wild ruminant has never been measured in a realistic way, however, so a method must be found to estimate it. Lactation has been studied extensively in dairy cattle, and some basic knowledge of the biological efficiency of this production process is available.

One necessary assumption for the calculation of milk production of a wild ruminant is that the nutritional requirements of the nursing ruminant are met by the milk and forage consumed. It can also be assumed that the milk production of a wild animal is in balance with the requirements of the nursing offspring. This is a reasonable assumption because the amount of milk produced is partly dependent on the demand, and it would be difficult to explain a significant imbalance in this mother-young relationship after many years of natural evolution.

The biological relationships included in the calculation of the milk production of a lactating female to meet both the protein and energy needs of the nursing young are shown in Figure 16-5. The concept is clear; the next step is the numerical representation of these relationships so that a mathematical expression can be formulated.

The protein and energy requirements of the young are necessary for the calculation of the milk production of a wild ruminant. Knowledge of rumen development is also necessary for the calculation of the percentage of their protein and energy requirements that are derived from milk.

Two criteria may be used for the development of the rumen. One, the proportional capacities of the rumen + reticulum and the omasum + abomasum change as the animal matures. The two divisions of the stomach are about equal when the fawn weighs 6–7 kg and is about one month old. This was discussed in detail in Chapter 8 for white-tailed deer. Two, the length of the papillae lining the rumen increases with rumen development. They are about 2 mm long at the age of one month, which is 50% of their length when the animal is 4 months old (see Table 8-3). These two changes indicate that the rumen is about half developed between one and two months of age when the fawn weighs 6–7 kg.

The relationship between rumen development and diet is a useful tool for calculating the amount of energy and protein that is derived from milk as growth occurs and the diet changes. The data in Chapter 8 on rumen development of deer can be used to express a rumen-development–nutrient-absorption ratio (Table 16-6). Since milk is very digestible and forage much less so, a coefficient expressing this ratio is applied to the rumen-development–nutrient-absorption curve. The value used is 10:6, where $10:6 \cong 97:58$, with 97 being the estimate of the digestibility of milk and 58 the digestibility of forage. This coefficient expresses the relative proportions of ingested milk and forage that are used to meet the energy and protein requirements. The application of this ratio to the rumen-development–nutrient-absorption data in Table 16-6 results in the percentage of rumen digestion shown in the table. The final step is the calculation of a regression equation representing the amount of milk required to meet the nutrient requirements supplied by milk. Regression equation (16-6) has been calculated for 100% milk utilization at birth (3 kg) to no utilization at 25.2 kg, or weaning.

$$\%MD = 113.6 - 4.5\ W_{kg} \qquad (16\text{-}6)$$

where

$\%MD$ = % of nutrients met by milk

W_{kg} = weight of the fawn

FIGURE 16-5. Steps in the calculation of milk production necessary to meet the needs of the growing fawn.

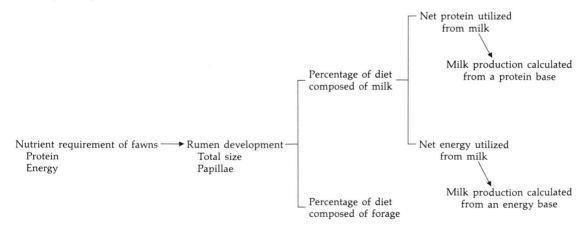

TABLE 16-6   RUMEN-DEVELOPMENT–NUTRIENT-ABSORPTION
RATIOS FOR WHITE-TAILED DEER

Weight of Fawn (kg)	% Nutrients Absorbed from Milk	% Rumen Digestion Composed of Milk
3	100	100
4	90	94
5	81	88
7	66	76
10	48	60
15	28	39
20	17	25
25	10	16
25	0	0

The protein requirement for lactation includes the protein that is in the milk and the additional protein requirement associated with the production of milk. The total requirement due to lactation can be expressed by equation (16-7):

$$Q_{nl} = \frac{(Q_{mp})(N\%)(I_{mp})}{100} \qquad (16\text{-}7)$$

where

Q_{nl} = grams of nitrogen required for lactation
Q_{mp} = quantity of milk produced in g day^{-1}
$N\%$ = percent nitrogen in milk = 1.76 [see Silver (1961)]
I_{mp} = metabolic increment for milk production

Weight increments of up to $\frac{1}{2}$ pound per day were recorded for white-tailed fawns receiving 2.1 grams of crude protein per day (calculated from data in Long et al. 1961). Using this information, the protein requirements can be estimated to range from 1.5 to 3 grams of crude protein per pound per day or 3.3 to 6.6 grams per kilogram per day during the nursing period. The milk production necessary to meet the protein needs of a white-tailed deer fawn can be calculated with equation (16-8):

$$Q_{mp} = \frac{(W_{kg})(MD)(Q_{pf}/6.25)}{(.0176)(.85)} \qquad (16\text{-}8)$$

where

Q_{mp} = quantity of milk produced (g) based on protein requirements
W_{kg} = weight of the fawn
MD = milk dependence = $(113.6 - 4.5W_{kg})/100$
Q_{pf} = quantity of protein required by the fawn in g kg^{-1} day^{-1}
6.25 = protein:nitrogen ratio for body tissue
.0176 = nitrogen fraction in deer milk
.85 = net protein coefficient for milk (Brody 1945)

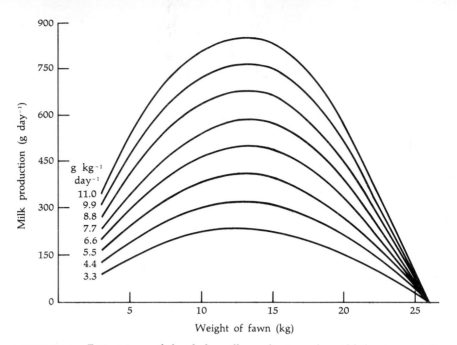

FIGURE 16-6. Estimations of the daily milk production of a wild doe to meet the protein needs of one fawn.

The results are shown in Figure 16-6 for eight estimates of the protein requirement of fawns. Note that the milk production rises rapidly at first, hits a peak, and then falls gradually to nothing when the young is weaned. A dairy farmer tries to maintain peak production in his cattle for as long as possible; the lactation curve would be flatter on top and elongated to the right.

The protein requirement for supplying milk for twin fawns may not be twice that for one. Twins are often smaller and there may also be a greater dependence on forage when there is competition for the mother's milk. The milk production for two fawns can be estimated to vary between 1.5 and 2 times the amount necessary for one fawn.

The protein requirement for lactation by the doe can be estimated by multiplying the nitrogen in the milk by 6.38 to convert it to a protein equivalent and then by a multiple between 1.25 and 1.50 (Figure 16-7). This multiple [based on data in Crampton and Harris (1969) for dairy cows] represents the cost of "overhead," or the protein costs to the doe over the protein in the milk alone.

SUMMARY OF THE PROTEIN REQUIREMENTS OF THE INDIVIDUAL. The amount of nitrogen metabolized by the body each day can be expressed as follows:

$$
\begin{Bmatrix} \text{Total} \\ \text{nitrogen} \\ \text{used daily} \end{Bmatrix} = \begin{Bmatrix} \text{The sum of} \\ \text{endogenous} \\ \text{urinary nitrogen} \end{Bmatrix} + \begin{Bmatrix} \text{Metabolic} \\ \text{fecal} \\ \text{nitrogen} \end{Bmatrix} + \begin{Bmatrix} \text{Nitrogen} \\ \text{for body} \\ \text{growth} \end{Bmatrix}
$$

$$
+ \begin{Bmatrix} \text{Nitrogen} \\ \text{for hair} \\ \text{growth} \end{Bmatrix} + \begin{Bmatrix} \text{Nitrogen} \\ \text{for} \\ \text{gestation} \end{Bmatrix} + \begin{Bmatrix} \text{Nitrogen} \\ \text{for milk} \\ \text{production} \end{Bmatrix}
$$

The equation can be rewritten using symbols:

$$Q_n = Q_{eun} + Q_{mfn} + Q_{ng} + Q_{nh} + Q_{np} + Q_{nl} \qquad (16\text{-}9)$$

where

Q_n = quantity of nitrogen required
Q_{eun} = endogenous urinary nitrogen
Q_{mfn} = metabolic fecal nitrogen
Q_{ng} = nitrogen in gain
Q_{nh} = loss of nitrogen in hair
Q_{np} = nitrogen required for pregnancy
Q_{nl} = nitrogen required for lactation

The amount of nitrogen used in each of the metabolic pathways in equation (16–9) is shown in numerical form in Table 16-7. The nitrogen used can be converted to protein requirements by multiplying the sum of Group A by 6.25, and the nitrogen requirement for milk production in Group B by 6.38.

The relative importance of these different nitrogen requirements is shown in Figure 16-8 for a fawn at different ages after weaning. A constant weight gain of 0.22 kg day^{-1} is used in the calculations. The nitrogen for hair growth is very small compared with the other requirements. Endogenous urinary nitrogen increases with body weight. The constant gain of 0.22 kg day^{-1} results in a constant requirement, of course. This rate of gain changes, but 0.22 kg seems to be a reasonable value for a fawn that is $3\frac{1}{2}$ to 5 months old. The requirement for metabolic fecal nitrogen is the highest of those shown. It is very diet-dependent, however, so its position in relation to the other nitrogen requirements may change.

The relative importance of the nitrogen requirements for different productive

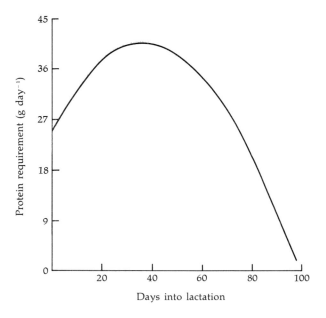

FIGURE 16-7. The protein requirement for lactation by a doe with one fawn.

TABLE 16-7 EQUATIONS USED IN THE CALCULATION OF THE PROTEIN REQUIREMENT

Group		Equation
A. $\dfrac{2 \times 70(W_{\text{kg}}^{0.75})}{1000}$	EUN	(16-1)
$+ \, cF_{\text{kg}}/6.25$	MFN	(16-2)
$+ \, .025 \, \Delta W_{\text{kg}}$	NG	(16-3)
$+ \, .02 W_{\text{kg}}^{0.75}$	NH	(16-4)
$+ \, e^{a \, + \, b(t_d/c)}/W_{\text{kg}}$	NP	(16-5)

(Σ Group A) 6.25 = protein requirements for all but milk production

B. $\dfrac{(Q_{\text{mg}})(\text{N}\%)(I_{\text{mp}})}{100}$	NMP	(16-7)

(NMP) 6.38 = protein requirements for milk production

Σ(Group A + Group B) = total protein requirement of the animal

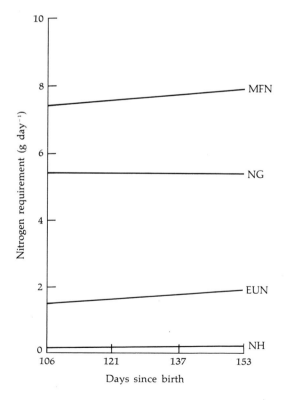

FIGURE 16-8. The nitrogen requirements for a fawn between 106 and 153 days old. The weight gain is 0.22 kg day^{-1}.

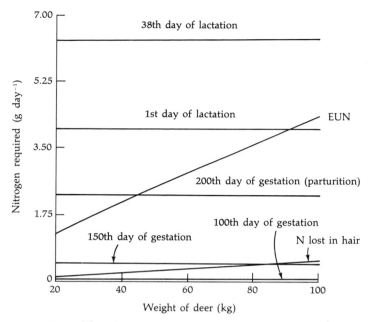

FIGURE 16-9.   The nitrogen requirements of deer of different
weights at three points in the gestation period and two points in
the lactation period.

processes of deer of different weights is shown in Figure 16-9. The requirement
for pregnancy at 200 days or full term is less than the requirement for lactation
just after parturition. The nitrogen requirements at peak lactation are greater than
for any other process. Deer weighing 60–80 kg on good range will usually have
two fawns each year, so the nitrogen requirement for lactation should be increased
by a factor of 1.5 to 2.0.

16-3   ENERGY REQUIREMENTS OF THE INDIVIDUAL ANIMAL

ENERGY REQUIREMENTS FOR MAINTENANCE AND ACTIVITY.   The transformation of
energy is necessary for sustaining life processes. The amount of energy required
for basal metabolism is expressed in equation (7-2). The observed similarity
between species of widely different weights results in a very useful biological
rule since it permits the establishment of a base line for all homeothermic species.
Experiments on different wild species have been in fairly close agreement with
the predicted rate of basal metabolism (see Chapter 7). Deviations can be attributed to the many uncontrolled variables in the metabolism tests and to differences
between individual animals of the same or different species. Most species have
daily metabolic cycles, and some have seasonal metabolic cycles also.

The amount of energy expended by free-ranging animals is unknown because
at this time there is no feasible method for measuring it in the field. The energy
expenditure of the free-ranging animal can be estimated from data for domestic
species that have been studied in the laboratory or in pastures. This is useful

because it permits one to test the effect of possible variation in the energy relationships on the total animal-range relationships.

Anything that an animal does "costs" something in terms of energy. The energy requirements for different activities can be calculated, and the sum of these is the total daily energy requirement. The total can be expressed as follows:

$$\begin{Bmatrix} \text{Total} \\ \text{daily energy} \\ \text{requirement } (Q_{me}) \end{Bmatrix} = \begin{Bmatrix} \text{Basal metabolic} \\ \text{energy} \\ \text{expenditure } (Q_{mb}) \end{Bmatrix} + \begin{Bmatrix} \text{Activity} \\ \text{expenditure} \\ (Q_{ma}) \end{Bmatrix}$$

$$+ \begin{Bmatrix} \text{Production} \\ \text{expenditure} \\ (Q_{mp}) \end{Bmatrix} + \begin{Bmatrix} \text{Additional cost} \\ \text{to maintain} \\ \text{homeothermy } (Q_{mh}) \end{Bmatrix}$$

Factors included in the activity increment include standing, running, walking, foraging, playing, breeding, ruminating, and bedding. Production increments include the energy necessary for the deposition of additional body tissue, such as muscles, bones, fat, and hair, and for the production of fetal tissue during pregnancy and milk during lactation. The additional cost to maintain homeothermy is a part of the total energy requirement only when the sum of the heat production resulting from the first three items $(Q_{mb} + Q_{ma} + Q_{mp})$ is less than the total heat loss of the animal. This was discussed in Chapter 13.

The bedding posture is one of the standard conditions during a basal metabolism test, so the increment due to this activity (I_a), expressed as a multiple of Q_{mb}, is 1.0. Metabolism tests are often continued for several hours, and the experimental animals do stand up during the measurement periods. The amount of energy expended in standing is about 9% of that of basal (Crampton and Harris 1969). The energy cost of standing can then be expressed with an activity increment of 1.1 ($100 \div 9 = 1.1$). Thus a standing animal in thermoneutral conditions and in a postabsorptive state will have an energy expenditure that can be predicted with equation (16-10).

$$Q_{es} = (70) \ W_{\text{kg}}^{0.75}(1.1) \tag{16-10}$$

where

Q_{es} = energy expenditure of an animal in standing posture in thermoneutral conditions in kcal day^{-1}

W_{kg} = body weight in kg

The energy cost of walking adds to an animal's energy requirement. Clapperton (1961) studied the energy expenditures of sheep walking on a level surface and on gradients in a treadmill. Two levels of nutrition and two speeds were used. His results and the results of measurements on several other species that were summarized by Brody (1945) and by Blaxter (1967) are shown in Table 16-8. The energy cost of lifting the body on a vertical gradient for some of the species included in Table 16-8 is over 10 times greater than that for walking on the level. The energy-cost values for ascent that were determined by Clapperton (1961) show an increase with speed but not with gradient.

TABLE 16-8 ENERGY COST OF WALKING ON A
LEVEL SURFACE AND OF ASCENT

Species	Level Walking (kcal kg^{-1} km^{-1})	Ascent (kcal kg^{-1} km^{-1})
Sheep	0.59 $\pm$ 0.05[A]	6.45 $\pm$ 0.47[A]
Cattle	0.452[B]	
Cow	0.48[F]	
Horse	0.385[B]	
Horse	0.40[F]	6.83[F]
Human	0.544[B]	
Human	0.54[C]	6.92[E]
Dog	0.58[D]	

[A] Clapperton 1961.
[B] Brody 1945.
[C] Clapperton citing Smith, Carnegie Inst. Publ. No. 309, 1922.
[D] Clapperton citing Lusk, *The Science of Nutrition*, 1931.
[E] Clapperton citing Lusk.
[F] Baxter 1967.

If basal metabolism and walking are considered together, the energy cost can be expressed mathematically as follows:

$$Q_{ew} = (70)\ W_{kg}^{0.75} + (E_{wl})(W_{kg})(V)(24) + \frac{(E_{wv})(W_{kg})(V)(H)(24)}{100} \qquad (16\text{-}11)$$

where

Q_{ew} = energy expended during walking (kcal day^{-1})
W_{kg} = weight in kg
E_{wl} = energy cost of walking on level = 0.59 kcal kg^{-1} km^{-1}
E_{wv} = energy cost of lifting the body weight vertically = 6.45 kcal kg^{-1} km^{-1}
V = rate of speed in km hr^{-1}
H = vertical height ascended expressed as percentage of km on level

Applying the data for sheep or other species measured under experimental conditions to wild ruminants on free range may result in error. The similarity in the energy cost of different activities for the species listed in Table 16-8 is striking. The amount of energy involved can be compared with the total basal energy requirement of the animal by dividing the part of the equation for walking by the basal energy expenditure as follows:

$$(Q_{ew}/Q_{mb}) =$$
$$\frac{70W_{kg}^{0.75} + (E_{wl})(W_{kg})(V)(24) + [(E_{wv})(W_{kg})(V)(H)(24)]/100}{(70)W_{kg}^{0.75}} \qquad (16\text{-}12)$$

The use of values for E_{wl} and E_{wv} that are midway between the possible extremes, including the variation shown for sheep, results in E_{wl} = .59 $\pm$ .13 and

$E_{wv} = 6.45 \pm 0.47$. If a 100-kg animal walks one kilometer a day and ascends 100 meters (or 0.1 km) the amount of energy used in walking compared with the amount required for basal energy processes is 2.7% for walking on the level, 2.9% for vertical ascent, and 5.6% for the two combined (Table 16-9).

These percentages are small, and they are even smaller if other activity and production processes are considered since the total energy expenditure per day is increased. Further, the estimations of distances walked and ascended are probably overestimations of the real situation unless there is a long distance to water, a herd is migrating, or there is some other cause for long-distance traveling. Thus, the error due to estimations from data on domestic species is very small.

No information on the energy cost for ruminants descending a gradient is available. Studies on humans indicate that it is small (ARC 1965), and the authors of the ARC publication consider it sufficient to equate descent with walking on the level.

There is an energy cost for browsing or grazing that can be attributed to the prehension and mastication of the forage material (Young 1966). Young also points out that psychic factors are involved since there is a change in heart rate when food is first given to penned sheep receiving their daily ration at regular times. A wild ruminant would very likely have a lesser response to the onset of feeding because the time of feeding is regulated by the animal itself.

Graham (1964) measured the energy expenditure of a 50-kg sheep while it was grazing in a respiration chamber, with fresh sod brought in to duplicate grazing conditions (Table 16-10). The experiment is quite artificial in many ways, but it does result in a first approximation from which additional calculations can be made. Graham also considered the difference in cost between grazing on good range and on poor range by including the energy cost of walking while foraging

TABLE 16-9 THE ENERGY COST OF WALKING COMPARED WITH BASAL METABOLISM (2214 kcal) OF A 100-kg ANIMAL

Activity	Energy Expenditure (kcal)	%BM	Energy Expenditure (kcal)	%BM	Energy Expenditure (kcal)	%BM
Walking on level	Distance = 1 km		Distance = 2 km		Distance = 3 km	
Upper limit	62	2.8	134	6.1	201	9.1
Midpoint	59	2.7	118	5.3	177	8.0
Lower limit	56	2.5	102	4.6	153	6.9
Vertical ascent	Height = 100 m		Height = 200 m		Height = 300 m	
Upper limit	69	3.1	138	6.2	208	9.4
Midpoint	65	2.9	129	5.8	194	8.8
Lower limit	60	2.7	120	5.4	178	8.0
Walking on gradient						
Upper limit	136	6.1	205	9.3	275	12.4
Midpoint	124	5.6	188	8.5	253	11.4
Lower limit	111	5.0	171	7.7	229	10.3

TABLE 16-10 COMPARISONS OF THE ENERGY COSTS OF VARIOUS ACTIVITIES FOR GRAZING SHEEP

Comparison	Number of Sheep	Body Weight (kg)	Number of Estimates	Energy Cost (kcal hr^{-1} kg^{-1} of body weight)	
				Range	Mean with Standard Error
Standing and grazing with standing	4	30–110	23	0.29–0.79	0.54 ± 0.05
Standing and eating with standing or lying	4	40	23	0.24–0.98	0.54 ± 0.05
Standing with lying	1	110	11	0.29–0.42	0.34 ± 0.02
Lying and ruminating with lying	3	30–110	21	0.08–0.52	0.24 ± 0.03

SOURCE: Adapted from Graham 1964.

for food. The additional energy spent while foraging on poor range may not add to the total daily requirements of wild ruminants, however, since an animal that is rapidly filled on good range may spend more time in nonforaging activity, such as investigation or play.

The energy cost of running has been measured in reindeer (*Rangifer tarandus*) by Hammel (1962). The activity increment (I_{ma}) in the metabolic rate equation is 8.0 (from Table 16-15). Crampton and Harris (1969) indicate that sustained work is from three to eight times as costly as standing. If a maximum increment for running of eight times Q_{mb} is used for a first approximation, the energy cost of running can be considered in relation to the total daily expenditure. The amount of time that elk spend running is less than 1% (Struhsaker 1967); it is very likely true that wild ruminants spend very little of their time running unless they are frequently disturbed by man or predators.

The validity of the use of such broad estimates for the energy requirement of a free-ranging animal can be determined with much more confidence after the time element has been realistically included in the calculations. The energy cost of running may have the greatest variability per unit time, but an animal that runs for 30 minutes a day is running only about 2% of the total time. A comparison of the energy expenditure during that time with the basal metabolic requirement of a 100-kg animal can be made with equation (16-13).

$$Q_{er} = \left[\frac{(70)(W_{kg}^{0.75})(I_{ar})(t_h)/24}{(70)(W_{kg}^{0.75})} \right] 100 \qquad (16\text{-}13)$$

where

Q_{er} = % daily energy for running
I_{ar} = activity increment for running
t_h = hours spent in running

At a maximum value of $I_{ar} = 8$, the energy expenditure of a 60-kg animal for running one-half hour a day is about 17% of its daily basal energy expenditure. If other activity and production processes are included in the total daily energy expenditure, the percentage of the total that is attributed to running decreases.

The energy cost of maintaining homeothermy is not a part of the energy requirements of an animal until the heat loss due to prevailing weather conditions exceeds its heat production at that time. The amount of heat energy produced during the exothermic chemical reactions of basic life processes—from muscular activity, the heat of fermentation in the rumen, and the heat of nutrient metabolism in all body tissue—exceeds the heat loss in many situations. Thus the problem often facing a homeothermic animal is the dissipation of heat rather than the conservation of heat. The principles underlying the exchange of heat were discussed in Chapter 13 and the responses an animal can make to changes in the thermal regime in Chapter 14.

Another energy requirement of an animal that needs to be mentioned, but for which no data are available, is the energy cost of a parasite or pathogen load. Any nutrient or body tissue that is absorbed by a parasite represents an energy drain on the host, but the energy cost has not been quantified through research. Further, the metabolic experiments conducted have in most cases been on animals carrying some kind of a parasite load, so the basic energy requirements of the animal include these additional requirements. Parasites and pathogens probably have a greater effect by upsetting the metabolic process of the host than they do as an extra energy demand, but it is well to consider the idea since the first law of the conservation of energy does apply to the host-parasite relationship.

ENERGY REQUIREMENTS FOR PRODUCTION.   Productive processes of wild ruminants include growth of body tissue, growth of the fetus, and the production of milk. The energy cost of these processes has been studied in both domestic animals and wild ruminants under experimental conditions.

The intake of total digestible nutrients (TDN) for different rates of gain in white-tailed fawns was determined by Cowan at The Pennsylvania State University (personal communication), and these values have been converted to caloric intake by multiplying the TDN by 2000 (Figure 16-10). The metabolizable energy can be estimated to be 80% of digestible energy [82% in dairy cattle (Crampton and Harris 1969)]. The net energy available remains after the heat of nutrient metabolism has been considered.

ENERGY REQUIREMENTS FOR GESTATION.   The additional energy requirement due to gestation remains small from conception through the first two-thirds of the gestation period. The last one-third of pregnancy is marked by accelerated growth of the fetus, and the energy requirement increases. The metabolizable energy required for pregnancy in cattle has been computed from data of Jakobsen [see ARC (1965)]. The increase is logarithmic, so a linear regression equation for the log of the energy requirement in relation to gestation time (t_d) can be calculated.

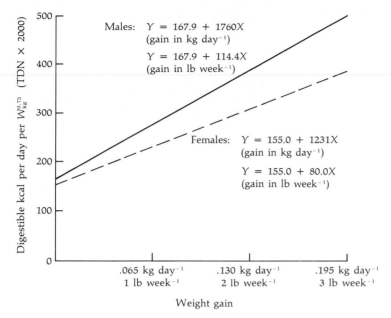

FIGURE 16-10. Energy in the feed for weight gain in white-tailed deer fawns. (Calculated from data from Cowan, personal communication.)

Equation (16-14) expresses this relationship:

$$Q_{ep} = [e^{(2.8935 + .0174\, t_d)}]/45 \qquad (16\text{-}14)$$

where

Q_{ep} = energy requirement for pregnancy per kg fetus weight at term for cattle
t_d = gestation time in days
45 = calf weight in kg at birth

Note that the energy required is divided by the average calf weight at term. The final expression of this relationship is energy per kilogram of fetus weight at term, and the application of the value obtained to wild ruminants results in a first approximation for use in testing the relative importance of the energy requirement for gestation. This is done for deer with equation (16-15), which includes a conversion factor that makes the gestation time of deer equivalent to that of cattle. Similar curves for other wild ruminants can be calculated with the appropriate conversion factors for gestation time (see Table 16-4). They are shown in Figure 16-11.

$$Q_{ep} = [e^{2.8935 + (.0174\, t_d/0.71429)}]/45 \qquad (16\text{-}15)$$

where

Q_{ep} = energy requirement for pregnancy of deer per kg fetus weight at term
0.71429 = 200/280 = (gestation period of deer)/(gestation period of cattle)

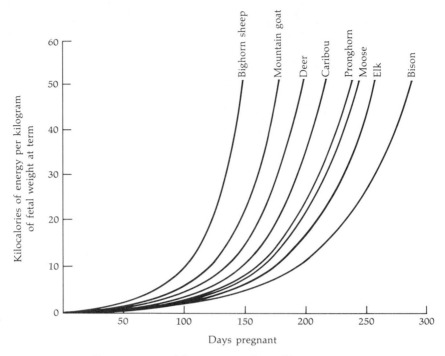

FIGURE 16-11.   Energy required for gestation by wild ruminants.

MILK PRODUCTION.   The energy required for milk production by dairy cattle has been estimated to be over 1.6 times the energy contained in the milk (Crampton and Harris 1969). This is in addition to maintenance. The energy cost of lactation is related to the amount of milk produced, however, so milk production and the energy cost of lactation by wild ruminants can be calculated from known biological relationships, based on the premise that the amount of milk produced is sufficient to meet the requirements of the nursing fawn. The biological relationships involved in estimating milk production were discussed in Section 16-2 and summarized in Figure 16-5.

Nutrient requirements change during growth, and as the rumen develops the percentage of the diet that is composed of forage changes accordingly, along with the percentages of the energy and protein requirements that are met by milk and forage. When the net protein and net energy utilized from milk has been determined, the amount of milk necessary to meet these levels of utilization can be determined.

The first item of information needed is an estimation of the energy requirements of fawns. The energy metabolism of black-tailed deer fawns has been measured by Nordan, Cowan, and Wood (1970), and it is equal to $(2.1) (70)W_{kg}^{0.75}$ (see Table 7-5). This is in line with data on young domestic ruminants, and the range of values for I_{ma} of 2.0 to 3.5 should cover the normal requirements of the growing fawn.

The next step necessary in the calculation of milk production based on energy needs is the expression of rumen development and its relationship to the digestion and absorption of nutrients from milk and forage. This was discussed in Section 16-2 and illustrated in Table 16-6.

The final step is the calculation of milk production necessary to meet the energy requirements of fawns. The equation is:

$$Q_{mp} = [(I_{ma})(I_{mp})(70)(W_{kg}^{0.75})][(RD)(1/E_{net})]/GE_m \qquad (16\text{-}16)$$

where

Q_{mp} = milk production based on energy requirements
I_{ma} = energy increment for activity of the fawn
I_{mp} = energy increment for production by the fawn
RD = rumen development = $(113.6 - 4.5W_{kg})/100$
E_{net} = net energy coefficient for milk = 0.8
GE_m = energy in milk = 0.7 kcal g^{-1}

Estimates of the milk production necessary to meet a range of energy needs of the growing fawn are shown in Figure 16-12. The shape of the lactation curve is about as expected, with an increase in milk production from parturition up to the age at which the rumen capacity is more than 50% of the total stomach size and a decrease until the young animal is weaned. The absolute amount of energy required by the fawn increases with increasing age, the relative amount decreases with increasing age, and the amount derived from milk increases up to a fawn weight of about 10 kg and then begins to decrease.

The final step is the determination of the energy requirement of the lactating female in order to produce the milk necessary to support the fawns. This consideration is analogous to overhead in a business operation; the total cost of the

FIGURE 16-12. Estimations of the daily milk production of a wild doe to meet the energy needs of one fawn.

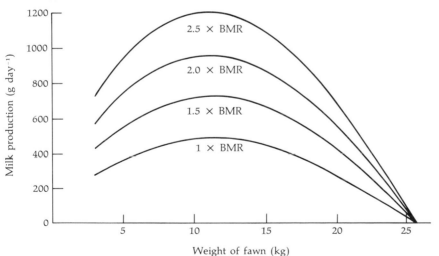

final product includes the cost of materials plus the costs associated with production. Using the increment of 1.6 times the energy contained in the milk produced (Crampton and Harris 1969), the total daily energy requirement can be calculated by adding the requirements for basal metabolism and activity to the energy requirement for milk production.

SUMMARY OF THE ENERGY REQUIREMENTS OF THE INDIVIDUAL.   The energy requirement of an individual animal is dependent on its basal metabolic characteristics, its activity, and the amount of production occurring. The total daily energy requirement is composed of the energy requirements for each of these biological processes. The energy cost equation is:

$$\left\{\begin{array}{c}\text{Total daily}\\ \text{energy required}\end{array}\right\} = \left\{\begin{array}{c}\text{The sum of the energy}\\ \text{required for bedding}\end{array}\right\} + \left\{\text{Ruminating}\right\}$$

$$+ \left\{\text{Standing}\right\} + \left\{\text{Feeding}\right\} + \left\{\text{Walking}\right\} + \left\{\text{Running}\right\}$$

$$+ \left\{\text{Breeding}\right\} + \left\{\text{Social activity}\right\} + \left\{\text{Production energy}\right\}$$

The energy cost of each of these activities is summarized in Table 16-11, with a comparison of the rate of energy expenditure for each activity compared with

TABLE 16-11   ENERGY EXPENDITURE PER HOUR BY A 100-kg ANIMAL IN DIFFERENT ACTIVITIES

Activity	Rate per Hour	Cost above Basal Metabolism (kcal hr^{-1})	Basal Metabolism + Activity Cost as Multiple of Basal Metabolism
Basal metabolism	$[(70)(W_{kg}^{0.75})]/24$	0 (BM = 92)	1.0
Standing[1]	$[(70)(W_{kg}^{0.75})(1.1)]/24$	9	1.1
Running[2]	$[(70)(W_{kg}^{0.75})(8)]/24$	646	8.0
Walking 1 km on level[3]	$(0.59)(W_{kg})(D_{km})$	59	1.64*
Vertical ascent of 0.1 km[3]	$(6.45)(W_{kg})(H_{km})$	65	1.71*
Walking 1 km, 10% gradient	(Sum of rates for walking and vertical ascent)	124	2.35*
Foraging[4]	$(0.54)(W_{kg})$	54	1.59*
Playing	$[(70)(W_{kg}^{0.75})(3)]/24$	185	3.0
Ruminating[4]	$(0.24)(W_{kg})$	24	1.26*

[1] Crampton and Harris 1969.
[2] Estimated from Hammel 1962 and Crampton and Harris 1969.
[3] Clapperton 1961.
[4] Graham 1964.
* These values are dependent on body weight and cannot be applied directly to all weights. New multiples must be determined since the values will decrease at higher weights because of the combined effect of the basal metabolic component and the activity cost.

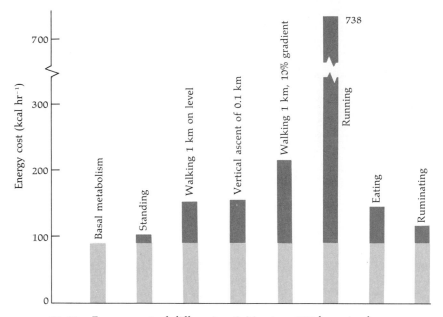

FIGURE 16-13. Energy cost of different activities to a 100-kg animal.

the basal metabolic rate of $Q_{mb} = 70 \; W_{\mathrm{kg}}^{0.75}$ for a 100-kg animal (Figure 16-13).

When the daily proportion of time spent in each of these activities is considered, the total daily requirement can be calculated. The results for five activity regimes of a 60-kg animal are shown in Table 16-12. Note that the daily energy expenditure, expressed as a multiple of the basal rate, varies from 1.23 to 1.98 for the five different activity regimes. This result provides an insight into the amount of normal variation expected for free-ranging animals. The maximum value (1.98) is found for an extremely active animal—far in excess of the expected amount of activity of a deer in its natural habitat.

The activity patterns of a 39-kg female white-tailed deer monitored by telemetry in the summer (Jeter and Marchinton 1964) have been used in the calculation of energy expenditure calculated (Table 16-13). Three 24-hour observation periods were used in the calculations, including the two extremes and an intermediate activity pattern. The multiples of BMR range from 1.24 to 1.45, indicating that different observed activity regimes for 24-hour periods do not cause very great differences in the total energy expenditure.

Montgomery (1963) observed nocturnal behavior of white-tailed deer in central Pennsylvania, and estimates of the percentage of time spent in different activities have been made for the calculation of seasonal differences in the energy expenditure for nighttime activity (Table 16-14). There is a general trend from higher activity levels in the summer to lower in winter. Values for the actual energy expenditure compared with the basal metabolic rate $Q_{mb} = (70)(W_{\mathrm{kg}}^{0.75})$ range from 1.59 to 1.70. The seasonal differences in energy expenditure parallel seasonal differences in basal and fasting metabolic rates reported by Silver (see Tables 7-2 and 7-3).

TABLE 16-12 ENERGY EXPENDITURE (Q_e = kcal day^{-1}) OF A 60-kg DEER IN FIVE DIFFERENT

Activity	Hours	Q_e	%Q_e	Hours	Q_e	%Q_e
Standing	0.50	35	1.8	0.75	52	2.4
Running	0.00	0	0.0	0.25	126	5.9
Walking	1.00	172	9.2	2.00	306	14.3
Foraging	4.00	387	20.8	6.00	581	27.1
Playing	0.50	94	5.0	0.75	142	6.6
Bedding and ruminating	18.00	1175	63.1	14.25	939	43.8
Totals	24.00	1863	99.9	24.00	2146	100.1
Multiple of BMR		1.23			1.42	

Seasonal differences in the energy expenditure of elk in different reproduction conditions, calculated from behavior data reported by Struhsaker (1967), are shown in Table 16-15. The amount of time spent in different activities clearly reflects the reproductive status of the individual within the herd. A spike bull in velvet is more sedentary than one with no velvet; they bed 50% and 21% of the time, respectively. The two-and-one-half-year-old bull spends 21% of the time bedded, and the three-and-one-half-year old solitary bull beds 39% of the time, but 46% of the time is spent standing, with little running or walking. In general, an aggressive but subdominant bull is considerably more active than older bulls, and the energy expenditure is clearly related to the reproductive activity regime.

TABLE 16-13 ENERGY EXPENDITURE OF A 38.6-kg FEMALE WHITE-TAILED DEER DURING THREE 24-HOUR MEASUREMENT PERIODS

Gross Activity Pattern	June 29–30			July 8–9			July 18–19		
Moving	38%			60%			83%		
Still	62%			40%			17%		
Activity	% Time	Q_e	%Q_e	% Time	Q_e	%Q_e	% Time	Q_e	%Q_e
Basal Metabolism	100	1084	—	100	1084	—	100	1084	—
Bedding	31	336	25	20	217	14	9	98	6
Standing	31	370	28	20	239	16	9	108	6
Feeding	29	143	11	45	225	15	62	311	20
Ruminating	4	9	<1	6	13	1	9	19	1
Walking	38	487	36	60	829	54	83	1046	67
Totals	*	1345		*	1523		*	1582	
Multiple of BMR		1.24			1.40			1.46	

SOURCE: Based on behavior data reported by Jeter and Marchinton, 1964, using telemetry, and analyzed by Stevens 1970.
* Some activities are concurrent; total exceeds 100.

ACTIVITY REGIMES

Hours	Q_e	$\%Q_e$	Hours	Q_e	$\%Q_e$	Hours	Q_e	$\%Q_e$
1.00	69	2.8	1.25	86	3.2	1.50	104	3.5
0.50	252	10.4	0.75	377	13.9	1.00	503	16.8
3.00	440	18.1	4.00	573	21.2	5.00	707	23.6
8.00	774	31.9	10.00	968	35.7	12.00	1161	38.8
1.00	189	7.8	1.25	236	8.7	1.50	283	9.5
10.50	703	29.0	6.75	468	17.3	3.00	232	7.8
24.00	2427	100.0	24.00	2708	100.0	24.00	2990	100.0
	1.61			1.79			1.98	

The difference in the multiple of BMR for a spike bull in velvet and after the velvet has been shed is considerable—1.44 to 1.82. Older bulls expend 1.74 times the basal rate, with more time spent in standing and breeding and less in feeding and moving about. The percentage of time spent in breeding activity and the relative cost of breeding for elk of different ages and reproductive status are shown in Figure 16-14. The energy cost of activity during the breeding season is related to the social position of the bull in the herd!

TABLE 16-14 ENERGY EXPENDITURE FOR NOCTURNAL ACTIVITY OF A 75-kg WHITE-TAILED DEER DURING DIFFERENT SEASONS

Gross Activity Pattern	Summer			Fall			Winter		
Bedded	16%			18%			25%		
Other activity	84%			82%			75%		
Activity	% Time	Q_e	$\% Q_e$	% Time	Q_e	$\% Q_e$	% Time	Q_e	$\% Q_e$
Basal metabolism	100	1784	—	100	1784	—	100	1784	—
Bedding	16	280	9	18	318	11	25	442	16
Standing	16	308	10	18	349	12	25	487	17
Feeding	69	667	22	64	626	21	50	490	17
Ruminating	16	68	2	18	77	3	25	107	4
Walking	69			64			50		
On level		(1529)			(1437)			(1124)	
Vertical ascent		(179)			(179)			(179)	
Total for walking		1708	56		1616	54		1303	46
Totals	*	3031		*	2986		*	2829	
Multiple of BMR		1.70			1.67			1.59	

SOURCE: Activity pattern based on data in Montgomery 1963; analyzed by Stevens 1970.
*Some activities are concurrent; total exceeds 100%.

TABLE 16-15 THE ENERGY EXPENDITURE OF ELK DURING THE RUT

Activity	160-kg Cow			145-kg Spike in Velvet			170-kg Spike (no velvet)		
	% Time	Q_e	% Q_e	% Time	Q_e	% Q_e	% Time	Q_e	% Q_e
Basal metabolism	100	3149	—	100	2925	—	100	3296	—
Bedding†	34	1083	21	50	1460	35	21	683	11
Standing†	41	1428	28	28	887	21	42	1529	26
Eating†	46	947	19	38	714	17	63	1379	23
Ruminating	26	241	5	29	244	6	16	155	3
Walking†	24	1310	26	21	790	19	35	2035	34
Running†	0	0	0	<1	13	<1	0	0	0
Breeding†	1	53	1	1	92	2	2	210	4
Totals	*	5062		*	4200		*	5991	
Multiple of BMR		1.61			1.44			1.82	

†Percentages for different activities calculated by Stevens 1970, from data in Struhsaker 1967.
*Some activities are concurrent; total exceeds 100%.

The energy expenditure of pronghorn of different weights has been calculated using activity data based on observations by Prenzlow, Gilbert, and Glover (1968). Resting activity consumed 46% of the animal's time, with feeding and other activities 54%. Differences between the energy cost of activity for pronghorn of 30, 45, and 60 kg are slight, with the multiples of BMR equal to 1.40, 1.42, and 1.45, respectively (Table 16-16). The middle activity regime shown in Table 16-12

FIGURE 16-14. The percentage of time spent in breeding activity and the percentage of the total daily energy expenditure for breeding activity by elk of different ages and reproductive status.

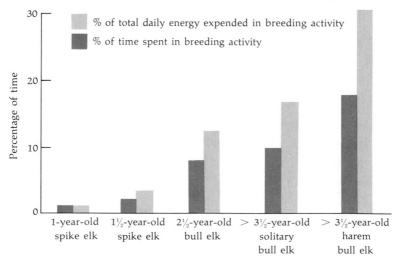

215-kg 2½-year-old Bull			327-kg < 3½-year-old Solitary Bull			327-kg < 3½-year-old Harem Bull		
% Time	Q_e	% Q_e	% Time	Q_e	% Q_e	% Time	Q_e	% Q_e
100	3930	—	100	5383	—	100	5383	—
21	835	11	39	2121	23	19	1029	11
44	1886	26	29	1736	18	46	2694	29
48	1329	18	27	1146	12	17	615	7
21	254	3	28	520	6	28	446	5
27	1893	26	21	2173	23	18	1681	18
<1	205	3	<1	150	2	<1	41	<1
8	927	13	10	1566	17	18	2836	30
*	7329		*	9412		*	9342	
	1.86			1.75			1.74	

TABLE 16-16 THE EFFECT OF WEIGHT ON THE DAILY ENERGY EXPENDITURE OF PRONGHORN

	Weight of Animal (kg)								
Gross Activity Pattern	30			45			60		
Resting	54			54			54		
Nonresting activity	46			46			46		
Activity	% Time	Q_e	% Q_e	% Time	Q_e	% Q_e	% Time	Q_e	% Q_e
Basal metabolism	100	897	—	100	1216	—	100	1509	—
Bedding	23	206	16	23	280	16	23	347	16
Standing	23	227	18	23	308	18	23	382	17
Eating	54	210	17	54	315	18	54	420	19
Ruminating	23	40	3	23	60	3	23	79	4
Walking	53	499	40	53	668	38	53	846	39
Running	1	72	6	1	97	6	1	121	6
Totals	*	1254		*	1728		*	2195	
Multiple of BMR	1.40			1.42			1.45		

SOURCE: Activity times based on data in Prenzlow, Gilbert, and Glover 1968; analyzed by Stevens 1970.
*Some activities are concurrent; total exceeds 100.

has been used to calculate the energy requirements of deer weighing 30, 60, and 90 kg with similar results. This illustrates that the weight effect is not an overriding consideration in the calculation of the energy requirements if activity levels are held constant.

The energy requirements of 60-kg deer at different levels of production are shown in Figure 16-15. The energy expenditure for activity is 1.42 times basal, increasing to 1.53 at the end of gestation with one fawn, 1.64 with two fawns, and then rising to 1.86 and 2.30 at the peak of lactation with one and two fawns, respectively. A gain of 0.15 kg day^{-1} results in an energy expenditure of 1.82.

The energy requirements of free-ranging animals have been estimated in the preceding calculations by using data from different sources in the literature. An interesting fact that emerges is that energy expenditure of a 60-kg deer, expressed as a multiple of basal metabolism, does not exceed 2, except at the peak of lactation when it is 2.3. This indicates that lactation is a costly process and that variations in weight, activity, and pregnancy all have a lesser effect. This does not mean that the latter are not important cost items in the energy budget; their importance must be evaluated in relation to the energy available on the range and the efficiency of the animal in using it.

Biologists have recognized differences in the energy and protein requirements of animals for years, but little effort has been made to analyze the importance

FIGURE 16-15.   Energy cost of activity and production by a 60-kg deer.

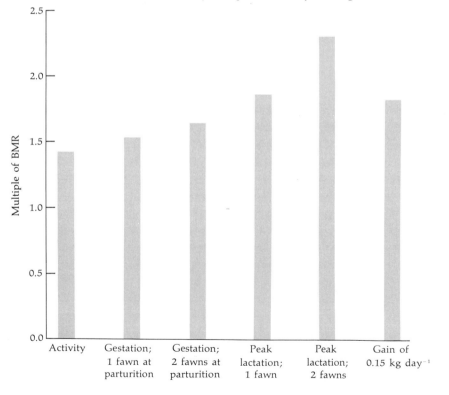

of these requirements for free-ranging animals. The calculations in this chapter illustrate what can be done to make first approximations. The results indicate that animals of different weights and at different activity levels may not vary widely in some energy requirements. Since range characteristics change also, it is necessary to relate animal requirements with range conditions throughout the annual cycle in order to determine the times at which more critical balances between the two exist. The importance of these biological characteristics, both inputs and outputs, is analyzed in Chapter 17.

LITERATURE CITED IN CHAPTER 16

Agricultural Research Council. 1965. *The nutrient requirement of farm livestock.* No. 2. *Ruminants.* London: Agricultural Research Council, 264 pp.

Blaxter, K. L. 1967. *The energy metabolism of ruminants.* London: Hutchinson, 332 pp.

Brody, S. 1945. *Bioenergetics and growth.* New York: Reinhold, 1023 pp.

Clapperton, J. L. 1961. The energy expenditure of sheep in walking on the level and on gradients. *Proc. Nutr. Soc.* **20**: xxxi–xxxii.

Crampton, E. W., and L. E. Harris. 1969. *Applied animal nutrition.* 2d ed. San Francisco: W. H. Freeman and Company, 753 pp.

Dietz, D. R., R. H. Udall, and L. E. Yeager. 1962. *Chemical composition and digestibility by mule deer of selected forage species. Cache la Poudre Range, Colorado.* Colorado Game and Fish Dept., Technical Publication No. 14, 89 pp.

Graham, N. McC. 1964. Energy costs of feeding activities and energy expenditure of grazing sheep. *Australian J. Agr. Res.* **15**(6): 969–973.

Hammel, H. T. 1962. *Thermal and metabolic measurements of a reindeer at rest and in exercise.* Technical Report. Fort Wainwright, Alaska: Arctic Aeromedical Laboratory.

Jeter, L. K., and R. L. Marchinton, 1964. Preliminary report on telemetric study of deer movements and behavior on the Eglin Field reservation in Northwestern Florida. *Proc. 18th Ann. Conf. S. E. Assoc. Game Fish Comm.* pp. 140–152.

Long, T. A., R. L. Cowan, C. W. Wolfe, and R. W. Swift. 1961. Feeding the white-tailed deer fawn. *J. Wildlife Management* **25**(1): 94–95.

Montgomery, G. G. 1963. Nocturnal movements and activity rhythms of white-tailed deer. *J. Wildlife Management* **27**(3): 422–427.

Nordan, H. C., I. McT. Cowan, and A. J. Wood. 1970. The feed intake and heat production of the young black-tailed deer (*Odocoileus hemionus columbianus*). *Can. J. Zool.* **48**(2): 275–282.

Prenzlow, E. J., D. L. Gilbert, and F. A. Glover. 1968. *Some behavior patterns of the pronghorn.* Special Report No. 17. Colorado Dept. of Game, Fish, and Parks, 16 pp.

Silver, H. 1961. Deer milk compared with substitute milk for fawns. *J. Wildlife Management* **25**(1): 66–70.

Stevens, D. S. 1970. Activity patterns and energy expenditure of wild ruminants. Special report. BioThermal Laboratory, Cornell University, 19 pp.

Struhsaker, T. T. 1967. Behavior of Elk (*Cervus canadensis*) during the rut. *Z. Tierpsychol.* **24**: 80–114.

Young, B. A. 1966. Energy expenditure and respiratory activity of sheep during feeding. *Australian J. Agr. Res.* **17**(3): 355–362.

SELECTED REFERENCES

Basile, J. V., and S. S. Hutchings. 1966. Twig diameter-length-weight relations of bitter-brush. *J. Range Management* **19**(1): 34–38.

Cook, C. W. 1966. Factors affecting utilization of mountain slopes by cattle. *J. Range Management* **19**(4): 200–204.

Coop, I. E., and M. K. Hill. 1962. The energy requirements of sheep for maintenance and gain. II. Grazing sheep. *J. Agr. Sci.* **58**: 187–199.

Enlen, J. M. 1966. The role of time and energy in food preferences. *Am. Naturalist* **100**: 611–617.

Hansen, R. M., and D. N. Ueckert. 1970. Dietary similarity of some primary consumers. *Ecology* **51**(4): 640–648.

Hungerford, C. R. 1970. Response of Kaibab mule deer to management of summer range. *J. Wildlife Management* **34**(4): 852–862.

Julander, O., W. L. Robinette, and D. A. Jones. 1961. Relation of summer range condition to mule deer herd productivity. *J. Wildlife Management* **25**(1): 54–60.

Malone, C. R., and B. G. Blaylock. 1970. Length- and weight-diameter relations of service-berry twigs. *J. Wildlife Management* **34**(2): 456–460.

Nellis, C. H. 1968. Productivity of mule deer on the National Bison Range, Montana. *J. Wildlife Management* **32**(2): 344–349.

Nestler, R. B., W. W. Bailey, L. M. Llewellyn, and M. J. Rensberger. 1944. Winter protein requirements of bobwhite quail. *J. Wildlife Management* **8**(3): 218–222.

Nestler, R. B., W. W. Bailey, M. J. Rensberger, and M. Benner. 1944. Protein requirements of breeding bobwhite quail. *J. Wildlife Management* **8**(4): 284–289.

Schuster, J. L. 1965. Estimating browse from twig and stem measurements. *J. Range Management* **18**: 220–222.

Scotter, G. W. 1967. The winter diet of barren-ground caribou in northern Canada. *Can. Field Nat.* **81**: 33–39.

Segelquist, C. A., and W. E. Green. 1968. Deer food yields in four Ozark Forest types. *J. Wildlife Management* **32**(2): 330–337.

Thetford, F. O., R. D. Pieper, and A. B. Nelson. 1971. Botanical and chemical composition of cattle and sheep diets on pinyon-juniper grassland range. *J. Range Management* **24**(6): 425–431.

MATHEMATICAL ANALYSES
OF FACTORS AFFECTING
CARRYING CAPACITY

When the biological components of the animal-range relationship have been identified, they can be assembled into a sequence of calculations that represent the biological functions involved. The number of components in the calculation of carrying capacity—including both the animal requirements and the characteristics of the range supply—is large, and the use of electronic computing equipment greatly facilitates the analyses. The model that is developed becomes an electronic analog of the system being analyzed, designed in a manner that permits the user to vary biological characteristics within the model, simulating changes that might take place in the natural environment.

17-1 THE CARRYING-CAPACITY MODEL

A basic consideration in the calculation of carrying capacity is that both the requirements of the animal and the range supply must be known before the calculation can be completed. This was illustrated very simply in Figure 16-1 in which the large box represents the range supply and the number of smaller boxes that fit inside this box depends on the sizes of the smaller boxes. Requirements for energy and protein (the small boxes) were calculated and range characteristics (the large box) were discussed.

The assembling of biological information for computer analyses needs to follow a logical pattern that represents biological relationships. A flow sheet showing the relationships between the items of information considered in the current analysis is shown in Figure 17-1. A more detailed verbal description of the model

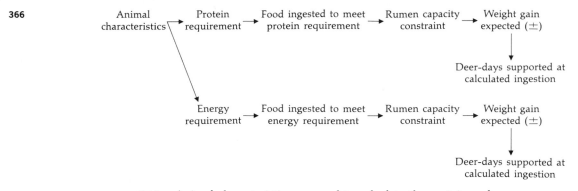

FIGURE 17-1. Animal characteristics are used to calculate the protein and energy requirements and the amount of food necessary to meet those requirements. The expected weight gain is calculated, and the number of deer-days supported by the range is determined on both a protein and an energy base.

is shown in Figure 17-2. The complexity of a flow sheet increases with the size of the model; the most detailed flow sheet used in the present model includes seven sheets of 17″ × 11″ paper. The details in such a flow sheet are continuously undergoing changes.

The schematic displays of the factors considered are followed by the assembling of the mathematical equations for energy and protein requirements and the quantity and quality of the forages on the range into a working mathematical design that represents the interrelationships of these biological factors. The order in which the calculations are made is important because of the characteristics of protein and energy metabolism. The relationship between protein and energy metabolism is essentially a one-way street since protein can be converted to energy but energy cannot be converted to protein. Catabolism of body fat is useful only as a source of energy, whereas protein-containing tissue can be mobilized to supply nitrogen for specific production purposes or as an energy source.

This distinction between protein and energy metabolism is an over-simplification of the simultaneous metabolic processes taking place in body tissue. Energy is necessary for protein metabolism, both in the rumen where the microflora are active and in the body tissue of the ruminant animal. Indeed, Crampton and Harris (1969) suggest that it is the energy needs of the animal that are met first and other nutritive needs are likely to be satisfied also if the diet is balanced.

The distinction between the mobilization of protein-containing tissue for either nitrogen or energy and the mobilization of fat reserves for energy alone is an important one, however. There are times at which the animal can be in a positive energy balance but a negative nitrogen balance. The opposite can also occur. Thus the weight changes (ΔW; ± gain) calculated in the carrying-capacity analyses do not necessarily coincide on both a protein and an energy base.

A paradoxical conclusion can be reached in the assembling of such models. Little basic information is known about wild ruminants, but analyses of their characteristics based on biological principles and known facts about domestic animals results in a large body of information of value in the carrying-capacity analyses. Thus although little is known, much can be learned from these analyses.

17·2   THE PROGRAM FORMAT

INPUTS FOR ANIMAL REQUIREMENTS.   The items of information used to calculate animal requirements throughout the entire year include the following:

1. Age in years, expressed as the lowest whole number. The number 1 is entered for a $1\frac{1}{2}$-year-old animal.

2. Age in days, above the year entry. The number 183 is entered for a $1\frac{1}{2}$-year-old animal.

3. Body weight in kg.

4. Rumen-fill coefficient.

5. Number of fawns *in utero* or nursing.

6. Protein requirements of nursing fawns in g per kg fawn weight per day.

7. Energy requirements for activity expressed as a multiple of BMR.

FIGURE 17-2.   An expanded flow sheet showing the sequence of calculations in the carrying-capacity analyses.

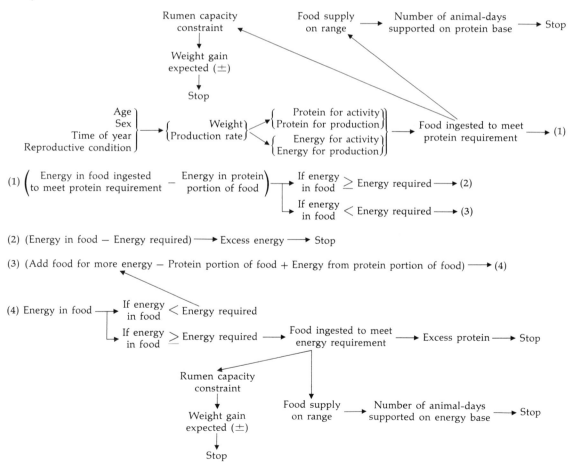

These items are variable inputs into the program. The constants in the equations for calculating protein and energy requirements are written in the computer program; they need not be changed for deer with different characteristics. Some of these constants may vary from animal to animal, but the variation may be small enough to overlook. It may also be that so little is known about the variability of these characteristics of deer of different ages or weights that it is best to use constants first, testing for the importance of these factors in the entire calculation by running a variability analysis or an error analysis.

Another characteristic of inputs such as those listed is that a single number may be used to represent a complex biological process. The energy requirements for activity, for example, may be entered into the analysis as a single value, but the single value may have been determined from a subroutine that is external to the main program. This arrangement is often desirable because the effect of different factors on energy requirements can be treated separately from the main program, yet the main program can be run using a number of possible values that could cover the whole range of energy requirements for activity in the wild. The effect of variation in energy requirements can be analyzed in the main program, with more realistic values determined after further refinement in the subroutine used to determine the energy requirements for activity.

The rumen-fill coefficient has also been treated as an external subroutine that results in a single input into the main program. This is necessary because so little is known about both the appetite of a wild ruminant and the passage rates of different foods. The amount of food in the rumen is a function of the appetite level, the passage rates of diet components, and the physical capacity of the rumen. A lack of this information necessitates the use of a range of somewhat arbitrary values for rumen fill. Resulting outputs expressing the amount of food necessary to meet energy or protein requirements as a fraction or multiple of rumen capacity are useful, however, since the importance of rumen capacity can be displayed on dimensionless graphs and the role of rumen capacity can be evaluated. This is done in later figures in which the importance of rumen capacity for deer of different weights is shown.

The number of fawns *in utero* or nursing is an important input for determining the requirements for pregnancy and lactation. Twin fawns may not necessarily place twice the demand on a doe that a single fawn does, although the outputs displayed in Chapter 18 have been made on that assumption.

INPUTS DESCRIBING THE RANGE SUPPLY.   The following characteristics of the range are included in the present model:

1. Quantity in kg of each species of forage available to the deer.
2. Percentage of each species in the ingested diet.
3. Percentage of crude protein on a dry-weight basis.
4. Net protein coefficient (NPC—the fraction of the crude protein that is available for maintenance and production).

5. Gross energy of each forage in kcal kg^{-1}.

6. Net energy coefficient (the fraction of the gross energy that is available for maintenance and production).

7. Percentage of water in the field-condition forage.

These seven characteristics apply to each of the forages in the diets used in the calculations discussed in this chapter. Six forages are included in each of the analyses. Deer eat more than six forages, of course, but analyses of rumen contents seldom show more than six to be present in significant quantities. Thus the use of only six forages in the diet is adequate for analyses of the relative importance of some of the factors that affect carrying capacity.

Each of the characteristics of the range can be varied to test their effect on the total animal-range relationship. The quantity available to deer, for example, can be stratified so that the number of deer-days possible at each one-foot interval can be determined. The effects of snow depth can be analyzed by entering the total quantity available to the deer from 0–6 feet, 1–6 feet, 2–6 feet, and so on. The reduction in the food supply due to snow depths can also be combined with changes in the energy requirements due to the effects of walking through snow.

The percentage of each food in the whole diet can be varied to represent either real diets determined in the field or simulations that can vary from one extreme to another. The effect of passage rate can also be included here. Diet changes due to weather effects, depletion of preferred foods, seasonal shifts in the foods available, and so forth, can be simulated through this input.

The percentage of crude protein varies seasonally. This variation, combined with the variation in net-protein coefficients for different animal requirements, permits an analysis of the effect of food quality. The higher protein content and greater digestibility of spring growth is certainly an improvement in range quality, but its net benefit to the population can be determined only after the animal requirements are considered. Protein requirements for pregnancy and lactation, for example, increase rapidly at the same time that the protein supply on the range increases.

The calculations for both net-protein and net-energy utilization are complicated by the fact that the net coefficients for each are different for maintenance, for production, and at different feeding levels. Thus the net coefficients should not be expressed as single values; however, a lack of information about protein and energy metabolism for different purposes necessitates a simplified approach at this time.

DECISIONS. One of the characteristics of a computer is its ability to make decisions. This consists of the comparison of two numbers—whether n_1 is greater than, equal to, or less than n_2—with the computations proceeding to different subroutines in the program according to the directions following each decision. This capability is used in the carrying-capacity model for determining the reproductive characteristics of the animals.

The decisions on reproductive condition are based on a deer calendar stored in the program. The age of the deer is entered in years and days, with the parturition date set in the present model as June 1. This date can be varied; a more comprehensive program is currently being planned that includes a distribution of parturition dates for a population. With a June 1 beginning date, a fawn is at age 0 + 30 on June 30, and 0 + 100 in September at weaning time. Breeding takes place at 0 + 165 for the fawn, 1 + 165 for a yearling, and so on. The significant dates for deer and moose are given in Chapter 10.

The decision-making capabilities of the computer are utilized by comparing the entered age of the animal with the significant dates in the deer calendar. For example, a deer's age of 0 + 90 is compared with the significant date for weaning—0 + 100. Since the fawn has not yet been weaned, the computer continues to calculate the milk ingested as a part of the total food intake. An age of 1 + 180 is compared with the significant date for breeding (1 + 165), and computations are then made for protein and energy requirements for pregnancy. Parturition occurs after day 365, so for any yearling or adult female over 1 year and between days 1 and 100 a lactation requirement will be calculated, unless it is not pregnant or lactating because of unsuccessful conception or fawn mortality. In that case, the number of fawns entered as an input is zero so the requirements for pregnancy or lactation will also be zero.

Another decision is made in the carrying-capacity program in the calculation of metabolic fecal nitrogen (MFN). All of the nitrogen requirements except MFN are calculated first. This is followed by the calculation of the amount of food necessary to meet these requirements. The requirement for the addition of MFN at that level of food intake is then calculated. A nitrogen requirement for MFN results in a new total nitrogen requirement, and this is followed by a second calculation of food intake. This increased food intake will result in a second calculation of MFN, and the cycle will be repeated again, resulting each time in a new calculation of food intake. When the last food intake calculated is less than one gram greater than the previous one, the cycle stops through the use of a decision-making routine in the program. If the protein component of the food is too low to allow for successful convergent iteration, this subroutine is bypassed and the total nitrogen requirement is calculated on the basis of maximum rumen fill.

17-3   CONSTRAINTS IN THE ANIMAL-RANGE RELATIONSHIP

Factors that affect the extent of a relationship between an animal and its environment are called constraints. They may reduce the rate of a biological process, and if the reduction is sufficient to limit all other processes, then the particular constraint is analogous to a limiting factor. It is important to remember, however, that many constraints are always present, whether or not they are limiting factors at a particular moment in time.

Several constraints are included in carrying-capacity analyses, either as an integral part of the main program or as external subroutines. The rumen-capacity constraint, analyzed through the use of a rumen-fill coefficient, is an example of a constraint that is an integral part of the main program. When the rumen is filled, the portion of the requirements for energy and protein that have not been met by ingested food must be met by mobilizing body reserves. Thus the amount of food ingested in relation to the rumen-fill constraint is a determinant of the sign (+ or −) for weight gain.

Some constraints are easily handled by external subroutines, or even quick glances at outputs. Summer densities of deer, for example, are never very high, so an abundance of forage at the peak of the growing season may support many more deer than are found on an area because of social constraints operating among deer. Thus psychological characteristics of the animals themselves would limit the density rather than the food supply.

Some factors or interactions that act as constraints are outputs from the computer runs. The relationships between MFN, crude protein, and NPC, for example, are determined through the computer runs, with some outputs indicating that a critical situation exists.

17-4 PROGRAM OUTPUTS

Two categories of program outputs result from the use of the carrying-capacity model under discussion. One is the listing of protein and energy requirements for the particular animal described by the input characteristics. The outputs in this category include the nitrogen requirements for:

1. Endogenous urinary nitrogen
2. Hair growth
3. Pregnancy
4. Milk production
5. Metabolic fecal nitrogen

Energy requirements for (1) basal metabolism and (2) activity are also expressed as kcal day^{-1}.

The second category of outputs includes different relationships between animal and range:

1. Amount of dry-weight forage ingested to meet protein and energy requirements at maintenance (zero gain).
2. Quantity ingested of each forage species to fill the rumen.
3. The rate of gain on both a protein and an energy base at rumen capacity.
4. The number of deer-days that each forage species will support.

The amount of forage ingested is shown both in kilograms and as a multiple of the physical rumen capacity and the rumen fill. The latter output indicates the importance of appetite, rumen size, the passage rate of food, and the quality of the food in relation to the size of the animal. The changes in body weight can be used to continuously change the input weight of the deer being analyzed, or a series of weights can be analyzed to test the relationships between body weight and the various outputs.

17-5 THE DYNAMIC CHARACTERISTICS OF THE ANIMAL-RANGE RELATIONSHIPS

The complexity of the dynamic relationships between an animal and its environment is so great that the human mind cannot fully comprehend it. Many analyses, covering a range of set conditions, provide insight into the relative importance of different factors. Thus carrying capacity is more a concept than a straightforward, definable, biological relationship. In this respect it is similar to the concept of homeothermy in which the dynamic balance between heat production and heat loss is so complex for a free-ranging animal that thermal energy relationships can be analyzed only to determine the relative importance of different factors.

The use of this biological model for analyzing factors that are important to carrying capacity provides several very interesting insights into their relative importance. Analysis of weight changes in relation to body weight, maintenance-gain comparisons, differences in forage consumption, physiological efficiency, and the calculation of deer-days on different ranges indicate that very definite differences exist among the deer, with weight being a particularly important consideration. The analyses have been made using both field data and arbitrary but representative data for hypothetical situations.

The use of arbitrary data can be an advantage rather than a disadvantage in computer modeling because it permits the analysis of the relative importance of different factors without time-consuming field collections. If a factor is found to be unimportant in the total analysis, then field measurements are unnecessary.

The outputs expressing relationships between factors at different times during the year are handled in the same way. Selected time periods that illustrate certain relationships that appear to be of significance are included in this text. Work is continuing on revised and updated models for the calculation of carrying capacity; the picture will approach completion as more factors are analyzed in the model.[1]

17-6 WEIGHT CHANGES

ABSOLUTE VALUES. Calculations of weight changes of deer on both a protein and an energy base have been made for several diets throughout the year. The predicted weight loss for deer on a winter diet, calculated on an energy base, is shown in Figure 17-3. Note that the actual weight loss predicted for a 20-kg deer is less than the weight loss predicted for a 100-kg deer. The weight loss

---

[1]Charles T. Robbins, "Biological basis for the determination of carrying capacity" (Ph.D. diss., Cornell University, in preparation).

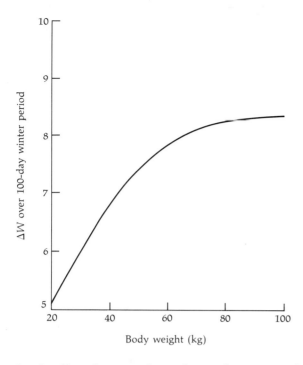
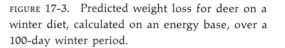

FIGURE 17-3. Predicted weight loss for deer on a winter diet, calculated on an energy base, over a 100-day winter period.

levels off as the animals get larger, however, indicating that deer over 60 kg do not benefit much more from being larger.

PERCENTAGE OF BODY WEIGHT. The expression of the predicted weight loss as a percentage of the initial body weight shows very clearly that smaller deer are at a distinct disadvantage (Figure 17-4). It is predicted that a 20-kg deer, given a winter diet for 100 days, will lose 25% of its initial body weight at the beginning of winter, whereas a 100-kg animal will lose less than 10% of its initial weight. This indicates that the fat reserve of a small deer allows for a much narrower

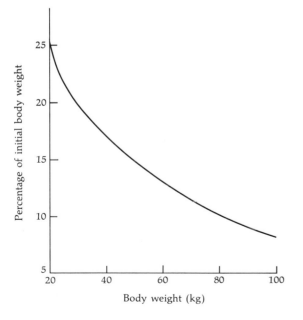

FIGURE 17-4. Predicted weight loss expressed as a percentage of initial body weight over a 100-day winter period, calculated on an energy base.

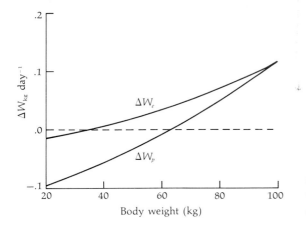

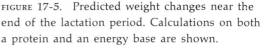

FIGURE 17-5. Predicted weight changes near the end of the lactation period. Calculations on both a protein and an energy base are shown.

margin of safety; therefore small deer are in a more precarious balance during the winter period.

The weight changes shown in Figures 17-3 and 17-4 have been calculated with all other parameters in the program held constant. All deer were on the same winter diet, for example. This may not be the case in the field since small deer are usually the subdominant animals in a population, and the forage available to them may not be as high in quality as the forage ingested by the large, dominant deer. This is frequently observed at feeding stations; the smaller deer eat last and must be content with the leftovers. If the quality of the small deer's diet were reduced in the calculations to simulate the effects of these behavioral factors, the disadvantage shown for the small deer would be accentuated.

LACTATION EFFECTS.   Analyses completed to date indicate that lactation is a costly biological process with high weight losses predicted. This is in agreement with data on domestic cattle; weight losses usually occur during peak lactation. The accuracy of these predictions for deer has not yet been analyzed sufficiently because of the lack of information on the summer diets of free-ranging deer. Weight losses have been observed in captive animals, however. A doe on a low plane of nutrition at The Pennsylvania State University weighed 81 pounds at parturition, dropped to a low of 55 pounds 65 days later, and reached 72 pounds by weaning time (36 days later) while on grain. Her fawn reached 33 pounds by weaning (Ondik, personal communication). Verme (1970) stated that minimum weights of female moose were recorded during lactation.

The predicted weight changes for lactating deer near the end of the lactation period (92 days of lactation) are shown in Figure 17-5. Note that both the energy-base and the protein-base calculations are related to body weight in a nonlinear fashion, with the smaller deer showing the greatest losses. Since the milk production drops off rapidly in the last few days of the lactation period, the positive gains predicted for the large does near the end of the lactation period represent an earlier start in the fall weight-gaining period.

17-7 FORAGE INGESTED

RELATIONSHIP TO BODY WEIGHT.   The amount of forage ingested by a small deer to meet its maintenance and production requirements for protein is less than the amount needed by a larger deer (Figure 17-6). The data shown are for a spring diet, or just before parturition. The relative amount of forage necessary to meet the maintenance and production requirements of deer on a spring diet is greater, however, for small deer than for large deer. The quantity of food that must be ingested to meet the needs of a deer, expressed as a percentage of its body weight, is more than 10% for a 20-kg deer and 4% for a 100-kg deer (Figure 17-7). These analyses were completed for pregnant does, each carrying one fawn. The relatively greater efficiency of the larger animal is again clear.

MAINTENANCE-GAIN COMPARISONS.   The metabolic efficiency of a large animal is greater than that of a small animal. This suggests that there should be relatively less forage ingested to meet the maintenance needs (0 gain) of a large deer than those of a small one. Predicted ingestion rates show this to be the case (Figure 17-8). However, large deer need to ingest more forage than small deer. The curvature of the line showing field-weight forage ingested in relation to body weight is not very obvious because the greater efficiency of the larger deer is masked by the increase in the absolute quantities of food ingested.

The importance of the higher efficiency of large deer compared with that of small deer is shown clearly in Figure 17-9, in which the amount of forage ingested to meet maintenance and production needs is expressed as a percentage of body weight. For the spring diet used in this calculation, maintenance needs are met if a 20-kg deer ingests an amount of forage equal to 3.6% of its body weight, whereas a 100-kg deer can meet its maintenance needs by ingesting forage equal to 1.4% of its body weight. The nonlinear relationship between the field-weight

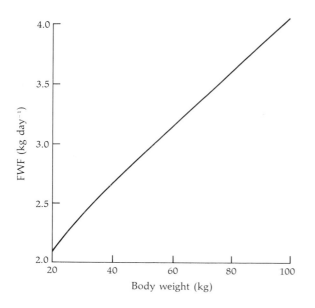

FIGURE 17-6.   Field-weight forage ingested (spring diet) to meet the protein requirements of a female deer at the end of pregnancy (one fetus).

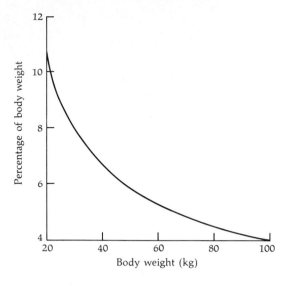

FIGURE 17-7. Field-weight forage ingested (spring diet) in relation to body weight to meet the protein requirements of a female deer at the end of pregnancy (one fetus).

forage ingested, expressed as a percentage of body weight, and body weight is due to the greater metabolic efficiency of the large deer.

The amount ingested to meet the calculated weight gain expressed as a percentage of body weight is constant for all body weights because the rumen capacity was calculated as a constant percentage of body weight. The amount of ingested forage available for production purposes is clearly greater for the large deer, providing a greater likelihood of weight gains for the large animal. This may be very important during the winter when the range quality is reduced and the animal is forced to draw on its body reserves.

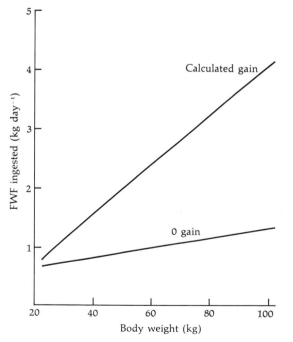

FIGURE 17-8. Field-weight forage ingested to meet maintenance (0 gain) and production needs of a deer on a spring diet.

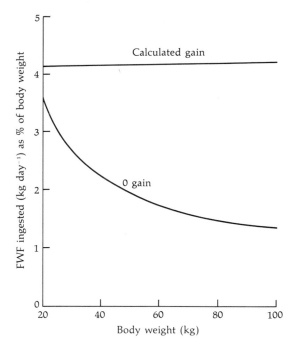

FIGURE 17-9. Field-weight forage ingested, expressed as a percentage of body weight, to meet maintenance and production needs of a deer on a spring diet.

FORAGE CONSUMPTION DURING GESTATION. The increase in the amount of forage necessary at different stages of pregnancy illustrates how the quantity ingested is related to animal requirements (Figure 17-10). There is only a slight increase in the quantity ingested during the first 150 days of pregnancy, but during the last 50 days there is a marked increase. The data in Figure 17-10 are for deer on a constant diet. Normally, there is a shift in a deer's diet during the last part

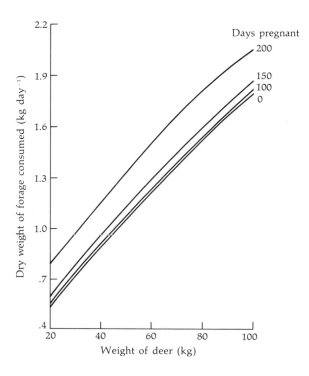

FIGURE 17-10. Predicted dry-weight forage necessary to meet the protein requirements of a pregnant doe on the same diet throughout the gestation period.

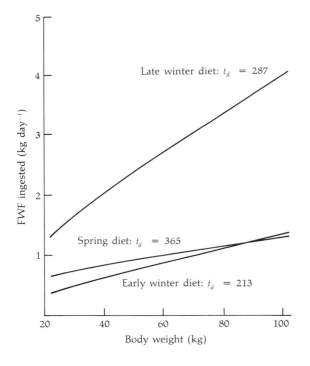

FIGURE 17-11. Predicted field-weight forage necessary to meet protein requirements at three different times in the gestation period: 48 days—early winter diet, time in days (t_d) on the deer-calender = 213; 122 days—late winter diet, t_d = 287; 200 days—spring diet, t_d = 365.

of the gestation period as spring growth begins and deer disperse from the winter concentration areas. The higher quality of forage available in spring results in a reduction in the quantity of forage ingested to meet the protein requirements for gestation (Figure 17-11). This reduction occurs even though the deer has a higher total protein requirement during the last quarter of the gestation period; increased forage quality compensates for the increase in the protein requirement.

FORAGE INGESTED AS A FRACTION OF THE PHYSICAL RUMEN CAPACITY. The absolute quantities of food ingested to meet protein and energy requirements have not been calculated with enough accuracy in the model being discussed to state definitely that these predicted quantities are sufficient. The calculations could be made more precise by considering such factors as the efficiency of protein and energy metabolism for deer of different weights and for different metabolic processes and the recycling of nitrogen. Little is known about these biological functions in deer and other wild ruminants. Research in progress indicates that the recycling of nitrogen can be of definite advantage to white-tailed deer on a low protein diet (Robbins et al., in preparation). Klein and Schonheyder (1970) suggest it may be important in other cervidae.

The amount of forage ingested is important inasmuch as the rumen and reticulum have a finite capacity that can act as a physical constraint. First approximations of the amount of food that should be ingested have been compared with the physical size of the rumen. A base-line expression of 7% of body weight has been used for estimating rumen size. Since the absolute values of both the amount of forage ingested and the rumen size are uncertain, an alternative is to express the amount of forage ingested as a fraction of the physical rumen capacity on a scale from 0 to 1.0.

A comparison of the fraction of the rumen filled for three calculated diets and one series of field measurements is shown in Figure 17-12. The three calculated rumen fills are based on simulated winter diets, with comparisons of both protein- and energy-base calculations. The field data were collected at the Seneca Army Depot near Ithaca, New York, with measurements made on 52 animals that were field dressed at the check station and the rumens collected for later volumetric measurements.

The two lines are curvilinear, with the smallest animal having a greater fraction of its rumen filled in each case. This is a further indication of the physiological advantage that a larger animal has.

If a constant weight loss of 0.05 kg (50 grams) per day is introduced into the calculations, the fraction of the physical rumen capacity that will be utilized to meet the needs of the deer becomes lower for smaller deer and higher for larger ones (Figure 17-13). This appears at first glance to be an advantage to the small deer, but this is an illusion since a constant weight loss of 0.05 kg for all deer ranging in weight from 20 to 100 kg represents a much faster depletion of body reserves for the small deer. Thus mobilization of the fat reserve that results in similar weight losses for deer of different weights is of greater benefit to a small animal in terms of rumen fill, but the time span over which this benefit can continue will be much less. The small deer will deplete its reserve much earlier in the winter than a large deer.

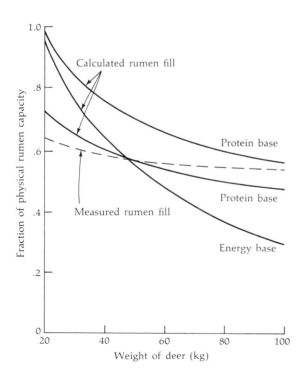

FIGURE 17-12. Rumen fill in relation to body weight for three calculated diets and one series of field measurements at the Seneca Army Depot.

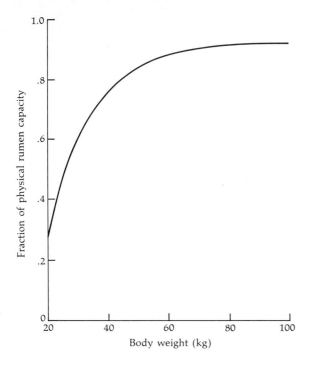

FIGURE 17-13. Fraction of the physical rumen capacity used to meet protein requirements at a daily weight loss of 0.05 kg (50 gms) for deer of all weights.

17-8 PHYSIOLOGICAL EFFICIENCY

METABOLIC FECAL NITROGEN AND NET PROTEIN RELATIONSHIPS. The ingestion of forage results in the production of enzymes for digestion and in the abrasion of the gastrointestinal tract as a result of the physical passage of food material. The amount of nitrogenous material of metabolic origin in the feces is partially dependent on the level of intake. As more food is ingested, more enzymes are produced, and there is more abrasion from the gastrointestinal tract. An analysis of the relationship between the amount of MFN and the NPC indicates a nonlinear relationship between the two (Figure 17-14). As the protein quality goes down, the amount of MFN increases rapidly, with the point of inflection being at an NPC of about 0.40. The animal may be forced to go into a negative nitrogen balance or find an alternative that will have a compensatory effect.

METABOLIC FECAL NITROGEN, FORAGE INGESTED, AND NET PROTEIN COEFFICIENT RELATIONSHIPS. The importance of the forage quality in terms of its *net value* to the animal is shown in Figure 17-15. An NPC of 0.50 compared with one of 0.75 results in a greater MFN requirement and an increase in the quantity of forage necessary to meet this requirement. As the NPC increases, the amount of MFN decreases along with the quantity of forage ingested. Note that the difference between an NPC of 0.75 and one of 1.00 is quite small compared with the difference between an NPC of 0.50 and one of 0.75. It is clear that the net value of forage can reach a low point at which the rumen is not large enough to hold the quantity of ingested material necessary to meet the MFN requirement. The animal will be forced either

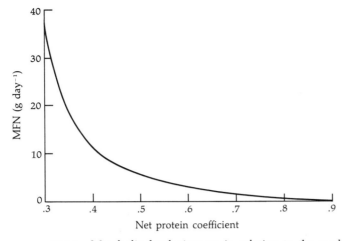

FIGURE 17-14. Metabolic fecal nitrogen in relation to the quality of the food as expressed by net protein coefficients.

to reduce its intake and go into a negative nitrogen balance or to change its diet. When the latter is impossible on depleted winter ranges, weight losses are inevitable. If the body reserves are too low to meet the nitrogen requirements for an extended period of time, death will result.

METABOLIC FECAL NITROGEN, BODY WEIGHT, AND NET PROTEIN COEFFICIENT RELATIONSHIPS. The absolute amount of MFN of a large deer is greater than that of a small deer (Figure 17-16). The effect of differences in the NPC is obvious. There is a slight curvature to the lines expressing this relationship, which indicates that

FIGURE 17-15. Metabolic fecal nitrogen in relation to the amount of field-weight forage ingested at three net protein coefficients. Each line includes deer ranging in weight from 20 to 100 kg.

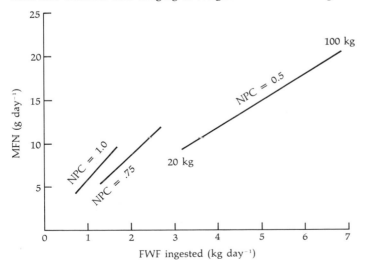

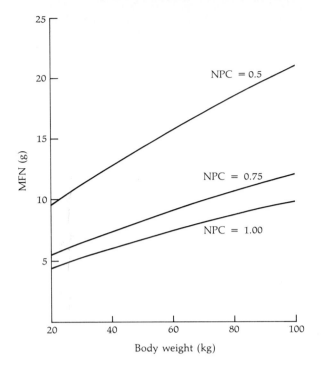

FIGURE 17-16. Metabolic fecal nitrogen in relation to body weight for three net protein coefficients.

large deer have *relatively* lower MFN than small deer. This is shown more clearly in Figure 17-17, in which MFN is expressed as a ratio, MFN (g): body weight (W_{kg}).

BODY WEIGHT, NET PROTEIN COEFFICIENT, AND PHYSICAL RUMEN CAPACITY RELATIONSHIPS. It is interesting to compare body weight, NPC, and the fraction of the physical rumen capacity used to meet protein requirements (Figure 17-18). A small

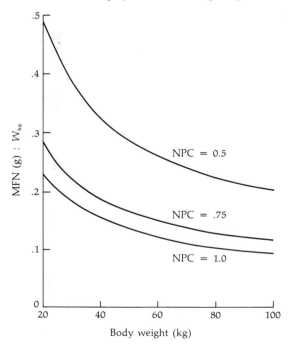

FIGURE 17-17. The ratio of metabolic fecal nitrogen to body weight (W_{kg}) illustrates that the smallest animal has a relatively higher metabolic fecal nitrogen output than the largest one. Data for three net protein coefficients are illustrated.

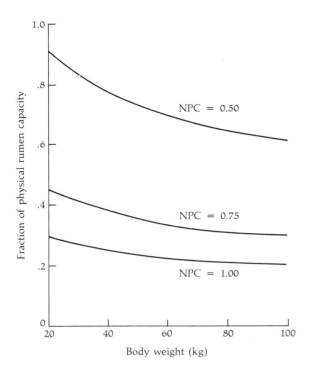

FIGURE 17-18. Fraction of the physical rumen capacity used to meet protein requirements for different net protein coefficients.

deer uses a greater fraction of its rumen capacity to meet protein requirements than does a large deer. The difference between deer of different weights is greatest at low NPCs. Body weight makes little difference for an NPC of 1.00, but such an NPC is biologically unrealistic because no animal is 100% efficient in the utilization of the protein in forage. An NPC of 0.75 results in some differences between the 20- and 100-kg deer, but the disadvantage for the smaller deer is still quite small. As the NPC is reduced to 0.50, the smaller deer are at a distinct disadvantage, with less space in the rumen for forage that can be utilized to meet production needs. Note also the relative spacings of the three NPC curves in Figure 17-18; the effect of reduction of the NPC from 0.75 to 0.50 is much greater than the effect of reduction from 1.00 to 0.75. This again illustrates the diminishing returns for the animal in relation to net protein values.

17-9 THE EXPRESSION OF CARRYING CAPACITY IN DEER-DAYS

Carrying capacity can be expressed in deer-days by dividing the amount of forage ingested per deer each day into the quantity available on the range. These calculations can be made on a protein base, by using the protein requirements of an animal to determine the amount of food that needs to be ingested, or on an energy base, by using its energy requirements to determine the ingestion. The quantity of forage available on the range is expressed in the model in terms of each species of forage, so that the number of deer-days that each forage will support—given the percentage of the diet composed of that forage—can be calculated.

If deer ate forages in strict proportion to their abundance on the range, all

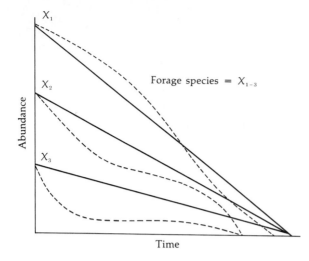

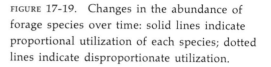
Forage species $= X_{1-3}$

FIGURE 17-19.   Changes in the abundance of forage species over time: solid lines indicate proportional utilization of each species; dotted lines indicate disproportionate utilization.

forages would be used up at the same time. This does not happen, however; some forages are depleted more quickly than others (Figure 17-19). Short-term (daily, if necessary!) shifts in diet can be analyzed with the model. Little is known about the amount of each forage consumed in relation to forage abundance in the field, however, so such precision is hardly necessary at this point in the analysis.

Forage abundance has been measured in three different stands in the vicinity of Ithaca, New York. Measurements were made at one-foot intervals from zero to six feet high. These data may be used to test the effect of differences in vertical stratification in relation to snow depths, as well as determining the total number of deer-days supported by each stand based on the requirements of the deer. In the calculations, the percentage of each forage in the diet has been taken as the percentage of each forage on the range; any preference for certain forages would lower the available number of deer-days. The actual numbers expressed in Figures 17-20–17-23 are somewhat arbitrary, but the relationships between the relative importance of such things as body weight, snow depth, and successional stages is clear.

EFFECTS OF BODY SIZE.   The amount of food ingested to meet the requirements of a small animal is less than the amount ingested to meet the requirements of a large animal. The metabolic efficiency of a large animal is greater, however, so less food per unit of body weight is necessary to meet the requirements of a large animal. This is expressed in the exponent 0.75 in equation (7-2) for basal metabolism.

A comparison of weights from 20 to 100 kg indicates that the number of deer supported declines in a curvilinear fashion with increasing weight (Figure 17-20). The greater efficiency per unit weight of larger deer and its effect on the number of deer-days supported is striking if a range of possible deer weights are analyzed. More small deer can be supported, but the curvature of the line clearly indicates that the larger deer are relatively more efficient in utilizing the food supply. The relationship between deer-days and deer weight shown in Figure 17-20 would be the same for any diet, unless there are compensatory differences in nutrient utilization by deer of different weights.

The curvature in the line showing the relationships between body weight and deer-days on an energy base is due entirely to the effect of the 0.75 exponent in the metabolic rate equation. It is interesting to see the importance of the relative efficiency of deer of different weights; the effect of the 0.75 exponent is much more dramatic in an ecological context such as the expression of carrying capacity than it is in the expression of heat production as shown in Chapter 7.

The energy requirements or "ecological metabolic rates" of free-ranging animals are higher than basal metabolic rates, of course. The energy requirements of deer of different weights vary because of behavior differences, differences in reproductive condition, and many other factors. Some of these variations are compensatory. Small fawns, for example, are at a disadvantage in snow because their legs are shorter than those of large deer. Small fawns, however, are less likely to be pregnant so their energy requirements do not include the metabolic cost of pregnancy. Smaller, subdominant animals may have a lower energy requirement because they do not have to maintain a high social position in the herd. The energy requirement may be lower because they need not be as alert as older deer; they can rely on older deer to signal approaching danger. These are interesting considerations, but it is doubtful if these factors compensate enough to equalize the effect of weight on the metabolic efficiency of small and large deer.

Some variations are additive in their effect on energy requirements inasmuch as smaller deer cannot reach the forage that larger deer can. As subdominant

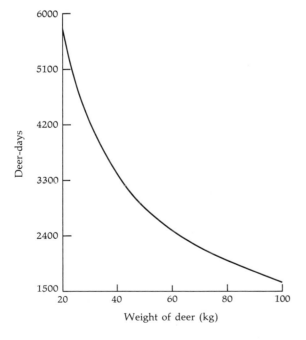

FIGURE 17-20. The number of deer-days supported in relation to deer weight in a mixed upland hardwood stand in winter.

members of a population, they are "last in line" for whatever food is available. The quality of this food is likely to be lower than that of food selected first by larger deer.

EFFECTS OF SNOW DEPTH.   Calculation of the number of deer-days that each of three forest types near Cornell University will support shows the effect of differences in the vertical distribution of food in these stands. The reduction in deer-days in McGowan's Woods amounts to about 50% after the first foot (0–1) of forage is removed from the calculation. The removal of another foot of forage, from 1 to 2 feet, results in another reduction of approximately 50% (Figure 17-21). These reductions could be a result of the effect of snow, which would make this forage unavailable.

Measurements of the quantity of forage available to deer in an invasion zone between a stand of mixed hardwoods and conifers and an abandoned field in Connecticut Hill Game Management area south of Ithaca show that the effect of a foot of snow is much less important in that habitat (Figure 17-22). The reduction in deer-days is less than 10% when the first foot of forage becomes unavailable in that stand, with a further reduction of less than 20% when the second foot of forage is covered with snow. Most of the forage in this invasion zone is between 2 and 6 feet.

The effect of one foot of snow is quite different in a second-growth hardwood stand in the Connecticut Hill Area (Figure 17-23). The number of 30-kg deer-days is reduced from 300 to 8 when 1 foot of snow is on the ground. This is a result

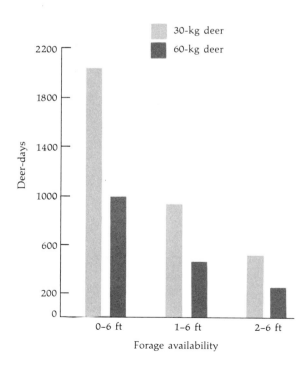

FIGURE 17-21.   The number of deer-days per square mile in a second-growth hardwood stand with a comparison of snow depths of 0, 1, and 2 ft (calculated on a protein base).

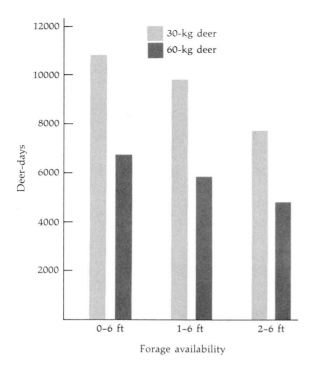

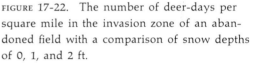

FIGURE 17-22. The number of deer-days per square mile in the invasion zone of an abandoned field with a comparison of snow depths of 0, 1, and 2 ft.

of a very uneven vertical distribution of forage. About 97% of the forage is seedlings located in the first vertical foot of the stand.

An interesting analysis of the effect of error in estimating net-energy coefficients in relation to the effect of a foot of snow on carrying capacity can be made. If the net-energy coefficient for each forage is varied ± 10%, the number of deer-days supported by the stand shown in Figure 17-23 varies from 270 to 330 for a 30-kg deer foraging at heights between 0 and 6 feet. This is a small variation compared with the effect of a foot of snow that reduced the carrying capacity from 300 to 8! In this particular case the effect of a foot of snow is far greater than the effect of experimental error due to estimation of the net-energy coefficient. This clearly indicates that the importance of errors or variation in one parameter cannot be determined until the importance of that parameter has been analyzed in relation to other factors.

The effect of 1 or 2 feet of snow on the total animal-environment relationship extends beyond a simple reduction in the food supply. Snow depths of 16 to 24 inches cause deer, especially smaller deer, to expend additional energy for walking. The reduction in the food supply plus the added energy requirements act together to reduce the carrying capacity. Further, snow may be a mechanical barrier causing deer to remain on well-traveled paths. The restricted movement further reduces the quantity of forage available to the deer, and they are forced to depend more and more on their body reserves.

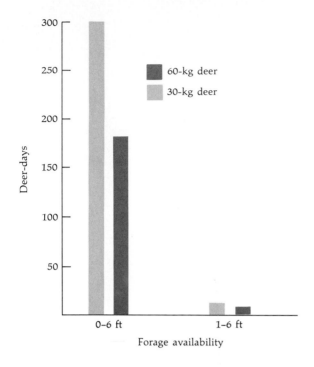

FIGURE 17-23. The number of deer-days per square mile in a second-growth hardwood stand in the Conneticut Hill Game Management Area south of Ithaca, New York. The effect of snow depths of 0 and 1 ft are shown.

FORAGE PRODUCTION AT DIFFERENT STAGES IN SUCCESSION. The number of deer-days expressed in Figures 17-21, 17-22, and 17-23 indicates the importance of successional stages in providing forage for deer. The invasion zone supports many more deer per square mile than either of the other stands. The second-growth hardwood stand on Connecticut Hill, which includes many seedlings but little else in the understory, supports an insignificant number of deer.

17-10 SUMMARY

Analyses of the interactions between a deer and its range clearly indicate that there are many relative and compensatory factors to consider. Because of such considerations, the idea of carrying capacity is best approached as a concept rather than a simple, definable entity. It may be that significant factors will be isolated, which can be used to quantify carrying capacity in a very practical manner. The identification of one or more significant factors is best made after a thorough analysis of animal-range interactions rather than by selecting a particular relationship and hoping that it proves to be suitable for practical use.

LITERATURE CITED IN CHAPTER 17

Crampton, E. W., and L. E. Harris. 1969. *Applied animal nutrition.* 2d ed. San Francisco: W. H. Freeman and Company, 753 pp.

Klein, D. R., and F. Schonheyder. 1970. Variation in ruminal nitrogen levels among some cervidae. *Can. J. Zool.* **48:** 1437–1442.

Robbins, C. T., R. L. Prior, A. N. Moen, and W. J. Visek. Nitrogen metabolism of white-tailed deer (in preparation).

Verme, L. J. 1970. Some characteristics of captive Michigan moose. *J. Mammal.* **51**(2): 403–405.

SELECTED REFERENCES

Bailey, J. A. 1968. Rate of food passage by caged cottontails. *J. Mammal.* **49**(2): 340–342.

Bell, R. H. V. 1971. A grazing ecosystem in the Serengeti. *Sci. Am.* **225**(1): 86–93.

Duke, G. E., G. A. Petrides, and R. K. Ringer. 1968. Chromium-51 in food metabolizability and passage rate studies with the ring-necked pheasant. *Poultry Sci.* **47**(4): 1356–1364.

Dodds, D. G. 1960. Food competition and range relationships of moose and snowshoe hare in Newfoundland. *J. Wildlife Management* **24**(1): 52–60.

Godwin, K. O. 1968. Abnormalities in the electrocardiograms of young sheep and lambs grazing natural pastures low in selenium. *Nature* **217**(5135): 1275–1276.

Mautz, W. W., and G. A. Petrides. 1971. Food passage rate in the white-tailed deer. *J. Wildlife Management* **35**(4): 723–731.

Van Soest, P. J., and L. H. P. Jones. 1968. Effect of silica in forages upon digestibility. *J. Dairy Sci.* **51**(10): 1644–1648.

N-DIMENSIONAL
POPULATION STRUCTURES

A population is a collection of organisms of the same species located within a prescribed area. This suggests that a population has the characteristics of *number* and *area,* from which the *density* or number per unit area can be calculated. This is a useful calculation for determining how many organisms are present at a point in time, and the number present at one time can be compared with the number present at another time. Since the numbers are usually not stable for any length of time, biologists and mathematicians may begin to calculate the rate of change and other characteristics of the change in numbers.

The analytical ecologist is interested in *cause and effect* relationships. Observed responses by organisms, both plant and animal, must be caused by something. These responses may range from mortality, at one extreme, to maximum productivity at the other. Within the productivity gradient there is room for an infinite number of combinations of responses, and these responses constitute the observed population dynamics through time.

The many possible observed responses are due to the many different characteristics of the members of a population. Some of these characteristics have been discussed in preceding chapters; variations in weight have been related to a number of different biological functions, for example. The significance of many of these variations between individual animals cannot be determined without a detailed "systems analysis" of the individual on one hand and of the individual in relation to its environment, including both physical and biological factors, on the other. Biological factors include internal, intraspecies, and interspecies relationships, implying that the variation associated with one individual may be a function both of itself and of the entire community in which it lives. This community is a very complex entity that cannot be described analytically without an understanding of how its component parts function.

Various characteristics of a population will be examined in this chapter, with an indication of how they might be used in predicting population dynamics given in Chapter 19. Both of these chapters contain examples of the kinds of considerations that can be made, but without a full complement of considerations for the many different kinds of wild organisms that could be analyzed. Many of the characteristics of a population are based on data in a descriptive article by Taber and Dasmann (1957) on the dynamics of a mule-deer (*Odocoileus hemionus columbianus*) population in shrubland habitat in California (Table 18-1). Students are urged to apply the same principles to species in which they are interested in their own localities.

18-1 SEX AND AGE RATIOS

The sex ratio is the expression of the number of males in relation to the number of females present. Sex-ratio data from Taber and Dasmann for a mule-deer population inhabiting 100 square miles of shrubland are shown in Table 18-2. The unbalanced ratio in favor of females is in part a result of selective hunting of the male deer; this is reflected in the drop from 66 to 47 males per 100 females between July and December.

Expressing sex ratios without considering age ratios results in a masking of certain factors since different age classes respond differently to mortality factors. There were more male fawns born than female fawns (Table 18-3), but differential mortality apparently resulted in a reversal of that balance by December, inasmuch as there were fewer male than female fawns at that time. The ratio of male to female deer dropped further in the one-year age group, and then dropped drastically in the two-year age group after the hunting season.

The cost of life is related to the weight of an organism. The productivity of an organism is dependent on the ability of that organism to meet the cost of life

TABLE 18-1 DYNAMICS OF A DEER POPULATION INHABITING 100 SQUARE MILES OF SHRUBLAND IN CALIFORNIA

Season	Males				Females				Total
	3+ yr	2 yr	1 yr	Fawns	3+ yr	2 yr	1 yr	Fawns	
Late May (fawn-drop)	616	482	556	2042	1935	706	858	1810	9005
July herd-composition count	616	482	556	1692	1906	706	858	1586	8402
December herd-composition count	467	182	536	836	1791	669	838	1039	6358

SOURCE: Data from Taber and Dasmann 1957.

TABLE 18-2 SEX RATIOS OF A MULE-DEER POPULATION IN CALIFORNIA

Season	Males	Females	Males per 100 Females
Late May	3696	5309	70
July	3346	5056	66
December	2021	4337	47

SOURCE: Calculated from data in Taber and Dasmann 1957.

both for itself and for its offspring. Thus it is logical to calculate the weight of deer in different age groups to determine the cost of life for each age group as a whole and on a per animal basis.

The weights of mule deer in the California shrubland population are shown in Table 18-4, calculated using an equation from Wood, Cowan, and Nordan (1962; see Figure 9-8). This procedure results in only an approximation of the weight, of course, but it illustrates the effect of analyzing a population by weight rather than by number. The average weights of 1- to 4+-year-old males increased from late May to July to December, but the total weight of the male portion of the

TABLE 18-3 SEX RATIOS FOR EACH AGE CLASS IN A MULE-DEER POPULATION IN CALIFORNIA

Age	M	F	M/100F
Late May			
3+	616	1935	32
2	482	706	68
1	556	858	65
Fawns	2042	1810	113
July			
3+	616	1906	32
2	482	706	68
1	556	858	65
Fawns	1692	1586	107
December			
3+	467	1791	26
2	182	669	27
1	536	838	64
Fawns	836	1039	80

SOURCE: Calculated from data in Taber and Dasmann 1957.

TABLE 18-4 CONVERSION OF THE AGE STRUCTURE OF A MULE-DEER (Odocoilcus hemionus hemionus) POPULATION TO A WEIGHT STRUCTURE

Age Classes

Season	Males					Females				Total
	4+	3	2	1	Fawns	3+	2	1	Fawns	
NUMBER OF DEER AT DIFFERENT AGES*										
Late May (fawn drop)	434	182	482	556	2,042	1,935	706	858	1,810	9,005
July	434	182	482	556	1,692	1,906	706	858	1,586	8,402
December	350	117	182	536	836	1,791	669	838	1,039	6,358
WEIGHT OF DEER IN EACH AGE CLASS*										
Late May										
Individual weights (lbs)	210	195	169	124	9	195	169	124	9	
Individual weights (kgs)	95	89	77	56	4	89	77	56	4	
Total weight (lbs)	91,140	35,490	81,458	68,944	18,378	377,325	119,314	106,392	16,290	914,731
Total weight (kgs)	41,230	16,198	37,114	31,692	8,168	172,215	54,362	48,048	7,240	416,267
July										
Individual weights (lbs)	244	207	200	154	20	198	174	132	20	
Individual weights (kgs)	111	94	91	70	9	90	79	60	9	
Total weight (lbs)	105,896	37,674	96,400	85,624	33,840	377,388	122,844	113,256	31,720	1,004,642
Total weight (kgs)	48,174	17,108	43,862	38,920	15,228	171,540	55,774	51,480	14,274	456,360
December										
Individual weights (lbs)	271	255	233	191	92	205	185	152	92	
Individual weights (kgs)	123	116	106	87	42	93	84	69	42	
Total weight (lbs)	94,850	29,835	42,406	102,376	76,912	367,155	123,765	127,376	95,588	1,060,263
Total weight (kgs)	43,050	13,572	19,292	46,632	35,112	166,563	56,196	57,822	43,638	481,877

SOURCE: Age data from Taber and Dasmann 1957.
*Male weights calculated by using an equation from Wood, Cowan, and Nordan 1962; female weights calculated with their equation for minimum male weights.

population decreased because of a loss of number. The females increased slightly in average weight and decreased slightly in total weight. The fawns increased the most in weight (from 4 to 42 kg) but decreased about 60% in number. Numerically, the fawn portion of the population went down, but the amount of animal tissue in deer fawns increased over four times for males and six times for females.

All of the animal tissue in the population, including both adults and young, is supported metabolically in a nonlinear fashion in relation to weight. A unit weight of larger adult deer can be supported at less cost to the range than a unit weight of smaller deer. This was discussed in Chapter 17 in relation to carrying capacity. Using the fractional exponent 0.75, the metabolic weights are shown in Table 18-5. Multiplication of the metabolic weight by 70 gives the energetic cost of life under basal conditions, with the ratios expressing this cost showing an increase in efficiency with increases in body weight.

Free-ranging animals do not live at a basal rate, but rather at an "ecological metabolic rate" that is a reflection of the maintenance and production factors present at a given point in time. Thus it is necessary to calculate the cost of the biological functions specific to each group of animals in order to come up with a total energy cost. The total energy cost can be expressed as a multiple of BMR by dividing the total by $(70)W_{kg}^{0.75}$.

TABLE 18-5 AVERAGE WEIGHT, AVERAGE METABOLIC WEIGHT, AND BASAL METABOLISM OF A MULE-DEER POPULATION

Age	Males W_{kg}	$W_{kg}^{0.75}$	$BMR\ day^{-1}$	Females W_{kg}	$W_{kg}^{0.75}$	$BMR\ day^{-1}$
Late May						
4+	95	30.4	2128			
3	89	29.0	2030	89	29.0	2030
2	77	26.0	1820	77	26.0	1820
1	56	20.5	1433	56	20.5	1433
Fawns	4	2.8	196	4	2.8	196
July						
4+	111	34.2	2394			
3	94	30.2	2114	90	29.2	2044
2	91	29.5	2065	79	26.5	1855
1	70	24.2	1694	60	21.6	1512
Fawns	9	5.2	364	9	5.2	364
December						
4+	123	36.9	2583			
3	116	35.3	2471	93	29.9	2093
2	106	33.0	2310	84	27.7	1939
1	87	28.5	1995	69	23.9	1673
Fawns	42	16.5	1155	42	16.5	1155

TABLE 18-6 THE ENERGY COST OF LACTATING ADULTS AND YEARLINGS IN JULY

	Adults	*Yearlings*
Number	2612	858
Average weight (kg)	87	57
Number with 2 fawns	1698	0
Number with 1 fawn	914	858
Energy cost per doe with 2 fawns*	4586 kcal day^{-1}	—
Energy cost per doe with 1 fawn†	3709 kcal day^{-1}	2701 kcal day^{-1}
Energy cost for does with 2 fawns	7,787,028 kcal day^{-1}	—
Energy cost for does with 1 fawn	3,390,026 kcal day^{-1}	2,317,458 kcal day^{-1}
Total	11,177,054 kcal day^{-1}	2,317,458 kcal day^{-1}
Average cost per deer	4279 kcal day^{-1}	2701 kcal day^{-1}

Note: Reproductive rate is 1.65 fawns per adult doe and 1.00 fawns per yearling doe.
*$Q_e = (2.30)(70)W_{kg}^{0.75}$
†$Q_e = (1.86)(70)W_{kg}^{0.75}$

Estimates of the cost of homeothermy were discussed in Part 5, and of growth, activity, gestation, and milk production for deer earlier in Part 6. The cost of activity ranged from 1.23 to 1.98 × BMR for several different activity regimes. The cost of activity for a 60-kg deer was 1.42 × BMR in one case (see Figure 16-12), increasing to 1.86 at the peak of lactation with one fawn and to 2.30 with two fawns. Applying these data to an analysis of the energy cost of a female population results in the costs shown in Table 18-6. The cost of lactation in different age classes includes consideration of the reproductive rates of yearlings and adults. Taber and Dasmann state that the former produce about 1.0 fawns per year and the latter 1.65.

Another population dimension that varies between geographical areas and between ages of deer is the conception date. The data in Figure 18-1 show that more adult deer breed at an earlier date in the northern part of New York than in the southern part. Conception normally occurs the second week of November in the northern part, and the third week of November in the southern part of the state. Thus the high energy and protein costs at the termination of pregnancy would be expected about a week earlier for the deer in the northern part, but wintery weather conditions are likely to be present a week or two longer there. This essentially alters the pregnancy-period–winter-period relationship by two to three weeks and may have significance in terms of productivity. Further, the earlier fawns in the northern part have a greater chance of being born in cold, wet weather. The effects of wet weather may be greater than those of cold; data are lacking on the survival of fawns in the wild that are exposed to different weather conditions, but data on sheep show that wet weather can cause higher mortality.

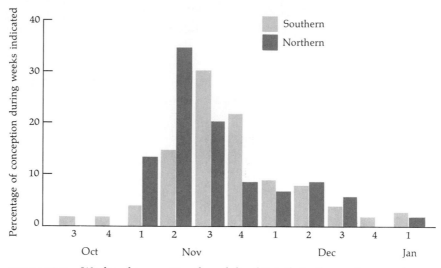

FIGURE 18-1.   Weeks of conception for adult whitetails in the northern and southern part of New York State. (Data from Cheatum and Morton 1946.)

Differences in the breeding dates of white-tailed adults and fawns (Figure 18-2) are of interest in terms of productivity because the fawn bred in the middle of December will drop a fawn during the first week in July. If the second fawn were 3 kg at birth and gained 0.2 kg day^{-1}, it would weigh 6 kg (13 pounds) less than the first fawn in the fall if growth rates of both fawns were equal. Given these assumptions, the predicted weights on November 15 would be about 65 pounds for the offspring of a late-bred fawn, and 78 pounds for the offspring of the earlier bred adult. The critical weight necessary for fawn breeding to occur in western New York seems to be about 65 pounds (Hesselton, personal communication)

FIGURE 18-2.   Weeks of conception for white-tailed adults and fawns in the southern part of New York State. (Data from Cheatum and Morton 1946.)

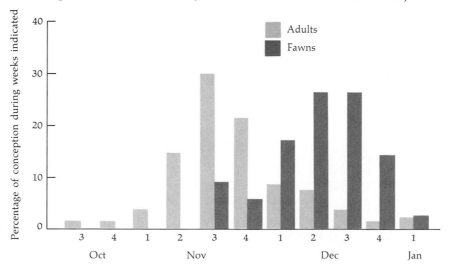

so the fawns born later might consistently be in a precarious balance between breeding and nonbreeding. These are oversimplifications of the real situation, of course, but they do illustrate the possible role of these dimensions in the productivity of particular members of a deer herd.

They also raise the question of compensatory growth. Assuming equal birth weight and growth rates, the distribution of fawn weights would be identical to the distribution of conception dates. Since there is variability in both birth weight and growth rate owing to range quality and other factors, it is logical to consider whether that variability has a beneficial effect consistently or if it is more or less random.

The variability of birth date in relation to conception date may be compensated for by the physiological stage of development at birth. White-tailed fawns born earlier than expected at the BioThermal Laboratory are less developed physiologically; they are "premies" and it takes one to two weeks for them to develop to the same point at which a later-born fawn might be at birth. These considerations are discussed further in Chapter 19.

18-2 PARASITISM

The dynamic relationships between parasite and host have not been analyzed as a system except in a few instances such as the work of Whitlock et al., which was discussed in earlier chapters. There are several basic relationships between parasite and host populations that can be expressed, however, at least in dimensionless form.

The number of parasites present in a host is partially dependent on age (Figure 18-3). Newborn animals are not infected, but as time passes the number of parasites increases up to a maximum. The shape of the parasite-number curve is very likely that of the typical sigmoid-growth curve.

The *number* of parasites present may not be strictly related to the *effect* of the parasites. The effect might be suspected to be greater on young adults that have

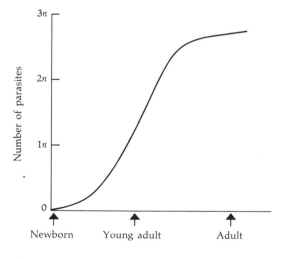

FIGURE 18-3. Nondimensional display of the relationship between the number of parasites and the age of the host.

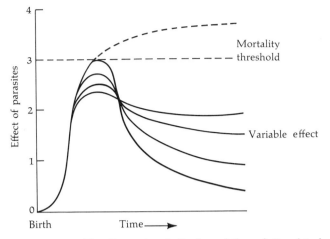

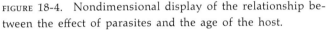

FIGURE 18-4. Nondimensional display of the relationship between the effect of parasites and the age of the host.

not built up a physiological resistance to the parasite(s) than on older adults that
have reached a kind of equilibrium between parasite and host (Figure 18-4). The
effect of parasite productivity is difficult to measure because of its subtle characteristics, as well as its interaction with other factors, such as nutrition, weather,
and intraspecific relationships due to density-related factors, and the temporal
physiological rhythms of the host itself. Somewhere within the effect of variables
on productivity a "normal" physiological load due to parasites might be recognized
since they are a part of the life of every animal, except those raised in sterile
environments.

18-3 GEOMETRY AND POSTURE

The geometric characteristics of animals in a population affect the animals'
relationships to certain physical factors in the habitat. The depth of snow, for
example, is important in determining its effectiveness as a barrier to travel. The
heights of the bellies of deer of different weights are given in Figure 18-5.

The energy cost of traveling through snow of different depths is very likely
not linearly related to snow depths. There may be little effect from snow depths
up to about two-thirds of the length of the legs. However, the energy cost increases
in snow depths that equal the length of the legs, and it increases sharply if the
belly is resting on the snow, forcing the deer to leap from one spot to another.
The energy cost varies for different types of locomotion and for different snow
characteristics. Loose fluffy snow is less a barrier than more dense snow, and a
hard wind-pack may support the weight of the deer, presenting no barrier. Very
deep, soft snow may be almost a complete barrier. One of the most costly
combinations may be that of a running deer on a snow pack that supports the
animal's weight only part of the time. If such a situation exists, dogs can become
a problem in some areas.

Another geometric factor of possible importance in a population is the height

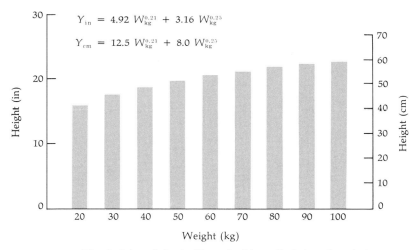

$$Y_{in} = 4.92\ W_{kg}^{0.21} + 3.16\ W_{kg}^{0.25}$$

$$Y_{cm} = 12.5\ W_{kg}^{0.21} + 8.0\ W_{kg}^{0.25}$$

FIGURE 18-5. The height of the bellies of white-tailed deer in relation to deer weight.

to which a deer can reach. This may be especially critical in areas in which overbrowsing has occurred, since a large deer may be able to reach a supply of food that is unavailable to smaller deer. The estimated heights reached by deer of different weights are shown in Figure 18-6, calculated from data used in determining the relationships between surface area and weight that were discussed in Chapter 13. These geometric considerations have an effect on the deer's ability

FIGURE 18-6. The height of forage that white-tailed deer can reach while standing on their hind legs.

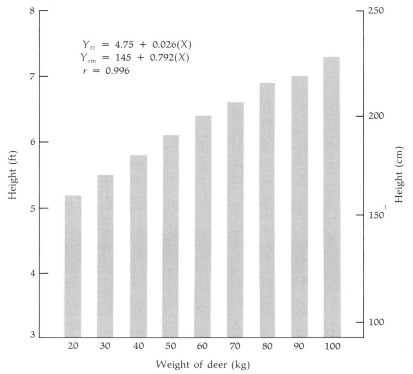

$$Y_{ft} = 4.75 + 0.026(X)$$
$$Y_{cm} = 145 + 0.792(X)$$
$$r = 0.996$$

to select and ingest forage. This is important in determining how well the requirements for protein and energy are met by ingestion and how much of these requirements must be met by urea recycling, mobilization of the fat reserve, and other physiological compensations during depressed nutritive conditions.

18-4 POPULATION REQUIREMENTS FOR PROTEIN

The protein requirement of a herd of animals is the sum of the protein requirements of individuals in the herd. The average requirement per animal can be determined as a weighted mean. An arithmetic mean is not appropriate since the total protein requirement is nonlinear in relation to weight. The time dimension must also be considered, since the rate of gain changes, females are pregnant or lactating for a part of the year, males have seasonal changes in activity, and so forth. Further, the total requirements change as the weight structure of the population changes.

The mean-protein requirement of a population can be determined for a given time by multiplying the sum of the protein requirements for specific body functions (see Table 16-7) by the percentage of the population in each weight class under consideration. This can be illustrated by equation (18-1).

$$Q_p = [(Q_{pw1})(\%W_1) + (Q_{pw2})(\%W_2) + \cdots (Q_{pwn})(\%W_n)]/100 \quad (18\text{-}1)$$

where

$$Q_p = \text{mean protein requirement}$$
$$Q_{pw1,2,\dots,n} = \text{protein requirement for each weight class}$$
$$\%W_{1,2,\dots,n} = \text{percentage of animals in each weight class}$$

The number of combinations of conditions is large because activity, weight gains, pregnancy, and lactation can all vary.

18-5 POPULATION REQUIREMENTS FOR ENERGY

The energy requirement of a herd of animals is the sum of the energy requirements of the individuals in the herd. As with protein, the mean requirement of each individual is not a simple arithmetic mean, since energy requirements are nonlinear with respect to weight. The effects of growth, pregnancy, and lactation on the total energy requirement must also be considered. The average energy requirement of individuals in a population can be calculated by using equation (18-2). Each Q_e is the sum of the energy costs listed in Table 16-11.

$$Q_e = [(Q_{ew1})(\%W_1) + (Q_{ew2})(\%W_2) + \cdots (Q_{ewn})(\%W_n)]/100 \quad (18\text{-}2)$$

18-6 OTHER POPULATION DIMENSIONS

The population characteristics mentioned thus far in this chapter illustrate the kinds of considerations that can be made in dividing a population into functional biological groups. Many others have been mentioned in earlier chapters without expressing them directly as population characteristics. All of the dependent relationships in which weight was shown as an independent variable are population dimensions whose distributions are dependent on the weight distribution within a deer population. Many thermal factors were shown in relation to weight, and several nutritive considerations were also shown to be related to weight.

Many of these characteristics are very subtle. The bedding posture of a deer or the amount of head extension of a grouse are both difficult to detect in the field, yet both can be demonstrated to be heat-conservation responses that may be important if the animal is in a critical thermal environment. On the other hand, some members of a population may employ these heat-conservation responses although other members may not need to, resulting in no differences among the animals because of their capabilities to compensate. These examples illustrate that populations have internal labile characteristics that may not appear to be of consequence except in analyses of factors affecting productivity.

Whenever one of these parameters has a negative effect on productivity, another factor (or more) has to compensate or else there will, in fact, be a reduction in productivity. Thus the different age classes in a population are in a constant state of change inasmuch as factors and forces affect biological functions in both negative and positive ways.

Population ecologists have recognized the effects of the existence of many of these factors. Chapman, Henny, and Wight (1969) discuss the vulnerability of Canada geese to hunting in relation to age, sex, harvest areas, time periods, and all-day hunting. They also discuss mortality in relation to age, sex, and hunting regulations. These are valuable considerations, but the potential for population growth lies not so much with the number and characteristics of animals removed from a population as it does with the productivity of the remaining members. Wildlife biologists have been saying this for some time. Hunting is not as important as hatching and rearing in determining a fall population of game birds, for example; but there has been a dearth of information on the real, fundamental causes of high production, with too much emphasis on the causes of mortality and on the characteristics of the dead.

One reason for the lack of information and understanding of factors affecting productivity is that many biological investigations are short-term and limited in scope—too short and too small to observe changes over many generations. There are exceptions, such as the ruffed-grouse project at the Cloquet Forest in Minnesota, under the direction of Dr. W. H. Marshall, University of Minnesota. Gullion (1970) discusses the factors influencing grouse populations there, pointing out that although predation is the ultimate fate of more than 80% of the grouse in the Cloquet Forest, it is seldom a limiting factor per se. This indicates the n-dimensional characteristics of the population, and he concludes that ". . . periodic fluctuations are not so much the result of any 'die-off' or accelerated losses among living grouse as they are the failure to recruit young grouse [i.e., productivity] each season to replace birds lost through 'normal' attrition." He also suggests that annual reproductive success has been largely determined before nesting begins, indicating that it is the physiological condition of the bird prior to nesting that determines its productivity and, subsequently, the current season's production. He also points out that the red-phase grouse do not live as long as the grey-phase, especially in unfavorable snow conditions. The color itself may be of some importance in predation, but Gullion suggests that data for the entire continent indicate that the red phase may be less tolerant to cold than the grey phase. He also discusses nutritive factors and social behavior, both of which affect productivity.

The roles of the many population dimensions that might be considered are difficult to identify, but they can be conceptualized, given the right philosophical approach to the prediction of population dynamics. Some suggestions are discussed in Chapter 19.

LITERATURE CITED IN CHAPTER 18

Chapman, J. A., C. J. Henny, and H. M. Wight. 1969. The status, population dynamics, and harvest of the dusky Canada goose. *Wildlife Monographs* 18:1–48.

Cheatum, E. L., and G. H. Morton. 1946. Breeding season of white-tailed deer in New York. *J. Wildlife Management* 10(3): 249–263.

Gullion, G. W. 1970. Factors influencing ruffed grouse populations. *Trans. North Am. Wildlife Nat. Resources Conf.* 35:93–105.

Taber, R. D., and R. F. Dasmann. 1957. The dynamics of three natural populations of the deer, *Odocoileus hemionus columbianus*. *Ecology* 38(2): 233–246.

Wood, A. J., I. McT. Cowan, and H. C. Nordan. 1962. Periodicity of growth in ungulates as shown by deer of the genus *Odocoileus*. *Can. J. Zool.* 40: 593–603.

IDEAS FOR CONSIDERATION

Examine the characteristics of individual members of different populations, identifying as many differences as possible. Then relate these different characteristics to productivity, predicting the production of each individual based on its own characteristics and its relationship to other members of the population. This can be done from the literature and in the field.

SELECTED REFERENCES

Atkins, T. D., and R. L. Linder. 1967. Effects of dieldrin on reproduction of penned hen pheasants. *J. Wildlife Management* 31(4): 746–753.

Beale, D. M., and A. D. Smith. 1970. Forage use, water consumption, and productivity of pronghorn antelope in western Utah. *J. Wildlife Management* 34(3): 570–582.

Bergerud, A. T. 1971. The population dynamics of Newfoundland caribou. *Wildlife Monographs* 25: 1–55.

Charlesworth, B. 1971. Selection in density regulated populations. *Ecology* 52(3): 469–474.

Christian, J. J., and D. E. Davis. 1964. Endocrines, behavior, and population. *Science* 146: 1550–1560.

Errington, P. L. 1945. Some contributions of a fifteen-year local study of the northern bobwhite to a knowledge of population phenomena. *Ecol. Monographs* 15: 1–34.

Freeman, M. M. R. 1971. Population characteristics of musk-oxen in the Jones Sound Region of the Northwest Territories. *J. Wildlife Management* **35**(1): 103–108.

Hairston, N. G., D. W. Tinkle, and H. M. Wilbur. 1970. Natural selection and the parameters of population growth. *J. Wildlife Management* **34**(4): 681–690.

Hazen, W. E. 1964. *Readings in population and community ecology.* Philadelphia: Saunders, 388 pp.

Henny, C. J., W. S. Overton, and H. M. Wight. 1970. Determining parameters for populations by using structural models. *J. Wildlife Management* **34**(4): 690–703.

Jordan, P. A., D. B. Botkin, and M. L. Wolfe. 1971. Biomass dynamics in a moose population. *Ecology* **52**(1): 147–152.

Kemp, G. A., and L. B. Keith. 1970. Dynamics and regulation of red squirrel (*Tamiasciurus hudsonicus*) populations. *Ecology* **51**(5): 763–779.

Meslow, E. C., and L. B. Keith. 1968. Demographic parameters of a snowshoe hare population. *J. Wildlife Management* **32**(4): 812–834.

Mitchell, G. J. 1967. Minimum breeding age of female pronghorn antelope. *J. Mammal.* **48**: 489–490.

Newson, R., and A. DeVos. 1964. Population structure and body weights of snowshoe hares on Manitoulin Island, Ontario. *Can. J. Zool.* **42**: 975–986.

Robinette, W. L., J. S. Gashwiler, J. B. Low, and D. A. Jones. 1957. Differential mortality by sex and age among mule deer. *J. Wildlife Management* **21**(1): 1–16.

St. Amant, J. L. S. 1970. The detection of regulation in animal populations. *Ecology* **51**(5): 823–828.

Schultz, V., and V. Flyger. 1965. Relationship of sex and age to strontium-90 accumulation in white-tailed deer mandibles. *J. Wildlife Management* **29**(1): 39–43.

Spencer, A. W., and H. W. Steinhoff. 1968. An explanation of geographic variation in litter size. *J. Mammal.* **49**(2): 281–286.

Taber, R. D., and R. F. Dasmann. 1954. A sex difference in mortality in young Columbian black-tailed deer. *J. Wildlife Management* **18**(3): 309–315.

Woodgerd, W. 1964. Population dynamics of bighorn sheep on Wildhorse Island. *J. Wildlife Management* **28**(2): 381–390.

PREDICTING

POPULATION DYNAMICS

After analyzing animal-environment relationships within the two main concepts presented—the concept of homeothermy and the concept of carrying capacity—it is useful to conclude with a consideration of the relative importance of the different factors and forces that have been a part of the analyses. It is hoped that this summary will result in additional in-depth analyses of animal-environment relationships. The complexity of these relationships is so great that a significant amount of progress can only be made if there are many investigators working toward an understanding of the ecology and management of populations of free-ranging animals in their natural habitats.

19-1 ECOLOGICAL PRODUCTIVITY GRADIENT

Members of a population are dynamic entities whose position on an ecological productivity gradient vacillates. The idea of this gradient was first introduced in Chapter 1, in which animals were considered to be either dead or not dead. If they are dead, their contribution to the ecosystem is through the decomposition of body tissue with the resulting release of energy and nutrients. This is a slow process; it may take many years before the energy and matter contained in an animal's body is once again used in living tissue. Further, the net contribution of a dead animal to a subsequent biological process is always less than the gross energy contained in the body at the time of death because some energy is dissipated into nonuseful channels.

Consideration of the productivity of a living animal rather than factors that cause mortality is much more advantageous in a population analysis because of the wide variation in potential productivity. Animals that are "not dead" exhibit a productivity potential varying from less than maintenance to maximum weight gains and full reproductive potential. Nonreproductive members of a population are, in some respects, a detriment to the population because they consume resources without contributing additional numbers to the population. Reproductive members contribute new animals. As the population grows, it may become desirable to harvest a percentage of the total population. *The population growth in each generation is far more dependent on conditions that affect the productivity of the living animals than on the number of animals that are removed by harvest or natural causes.*

19-2 THE RELATIVE IMPORTANCE OF DIFFERENT VARIABLES

ANIMAL REQUIREMENTS.   The requirements for survival, growth, and reproduction vary throughout the biological year, with a deer's requirement for production being highest at the peak of the lactation period in the summer. At that time, the deer population is dispersed and forage quality is high, resulting in conditions that do not appear to cause an appreciable amount of mortality in the adult population. The effect of the nutritional status of females on the amount of milk produced and hence its effect on the growth and subsequent productivity of fawns is a relationship that needs to be investigated further within a total ecological context.

The importance of the growth rate in relation to reproductive capabilities is shown by data on a large captive deer herd in the Seneca Army Depot near Ithaca, New York. There appears to be a threshold weight that fawns must reach before they conceive (Hesselton, personal communication). The position of these deer on the productivity gradient is an important consideration in population ecology inasmuch as a nonreproducing fawn consumes resources without contributing to the population. This may not be bad, however. When range conditions are poor, fawn growth is reduced. At such a time there is little advantage in having a productive fawn class.

For many years, winter survival has been considered a serious problem in deer management. Each spring dead deer are counted in the northern regions, the cause of death usually classified as winter mortality. The deer may indeed have died in the winter, but the causes of death are not necessarily confined to the winter period alone. The condition of the deer at the beginning of winter is an important factor because it determines the amount of reserves an animal has to carry it through difficult periods when food is either unavailable or of low quality.

The timing of the spring dispersal is another factor that is related to the importance of body reserves at the end of winter. A late spring can cause either reduction in productivity or mortality since the body reserves are normally low at the end of winter. Further, the molt begins in March and the insulating value of the hair is reduced in late March and April. Rain is often a part of the spring weather pattern, and heat loss by evaporation from a wet coat may be quite large. These factors can combine to put an animal in a critical hypothermal condition, resulting in the depletion of body reserves at a rate even faster than normal.

There is an indication that the energy requirements of deer rise in the spring. The requirements for gestation are superimposed on this seasonal shift. Thus the timing of the spring dispersal from a wintering area may make the difference between the realization of full reproductive potential, the resorption of some of the fetuses, the mere survival of the deer, or mortality. Since it is difficult to determine the reproductive level of a free-ranging deer herd, it is likely that the effect of variation in reproductive success on population characteristics is not fully appreciated. Population models that include variations in reproductive rates that can be related to animal requirements in relation to the time of spring dispersal and the changes in the quality of the range should be used to analyze the potential importance of these factors.

RANGE SUPPLY. The quantity of forage available on a range is a function of the successional stage of the vegetation and the ingestion rates of the consumers living on that range. The amount of browsing that occurs can affect the total range supply in two different ways. Some plants are stimulated by the removal of part of their current annual growth and forage production is increased by browsing. Too much browsing will cause a decrease in forage production, however, and high populations of wild ruminants can deplete their range supply to the point at which range recovery may take many years.

Canopy characteristics at various stages of plant succession are important determinants of the amount of forage produced within reach of the deer. As the canopy closes, forage production in the understory decreases. A lack of forage due to plant succession, changes in forage quality on a seasonal basis, and the disadvantages of the smaller deer in a population can all combine with severe weather conditions to cause high mortality. It seems apparent that detrimental forces have an additive effect, and if these forces all combine within a single year, both the number and the productivity of deer are noticeably reduced, especially in the younger age classes.

The quality of the forage also influences whether a deer's situation is critical. Low-quality forage is found on some soil types, and deer living on these soils may never reach their full genetic reproductive potential. This situation, coupled with severe winters and short growing seasons, may result in temporary population increases when the successional stage is truly optimum, but later reductions occur more frequently because of the effect of marginal-quality forage on the animal-range relationship. Under these conditions, the relative importance of weather conditions and other stress factors is greater and the rate of mortality is expected to be higher than in situations in which forage quality is generally high.

FACTORS AFFECTING BOTH ANIMAL REQUIREMENTS AND RANGE SUPPLY. The relative importance of different winter shelter and cover conditions for deer and other ruminants has been discussed by biologists for many years. An analysis of the role of cover in relation to the distribution of energy and matter provides a firm basis for the evaluation of different cover types. Analyses of thermal relationships of deer indicate that the most important thermal function of cover is in the reduction of wind. The added benefit from the additional infrared radiation from heavy overhead cover is insignificant at low temperatures. In fact, heavy overhead cover may be detrimental because it obstructs solar radiation, which could reduce the thermal gradients in an animal's hair layer. The thermal benefits from lying in the sun have not been analyzed in a quantitiative way yet, however. Further, forage production is less in a stand with a heavy canopy because of the reduction in radiant energy for photosynthesis.

Snow affects an animal's energy requirements as well as the amount of forage available. Its effect on energy expenditure may be less than expected, however. A deer struggling through deep snow, for example, cannot maintain that high rate of energy expenditure for very long, so movement patterns will be confined to trails. Thus snow serves as a mechanical barrier that regulates the distribution of deer and affects their energy expenditure.

Snow affects the amount of forage available by covering some of it and by preventing animals from reaching that which is located away from the trails. Under these conditions, a deer's intake of forage may be less than if it had an unlimited supply. Thus although snow may cause an ultimate reduction in a deer's energy requirement by confining its movement to trails, it also reduces the food supply so that the deer must depend on its fat reserve.

A crust formed on top of the snow can be a benefit to deer and other wild ruminants if it supports them, for it enables them to reach a new food supply. The occurrence of such crust conditions is beyond the control of the manager, however, so it should not be counted on as a way to compensate for other detrimental conditions. Its benefit is merely supplemental.

The effect of canopy density on snow accumulations can be either beneficial or detrimental to an animal. A heavy coniferous canopy intercepts snow, reducing accumulations on the ground. Reduced wind velocities under the canopy result in lightly falling snow that is less of a mechanical barrier. As the snow ages, it develops a large crystal structure, which is less susceptible to crust formation. Although these effects of snow may be considered a benefit to an animal, the snow also tends to trap the animal if it has sought shelter under the heavy cover during the storm. Deer restrict their movement to trails, reducing the area used and, consequently, the food supply available. When this happens, the fat reserve in relation to the size of a deer becomes very important. The amount of fat reserve is dependent on the condition of the summer and fall range. Thus winter mortality is only partially dependent on the density of deer in a winter concentration area.

The basic principle that underlies any consideration of the thermal benefits of cover in relation to an animal's metabolic processes is that these processes use food as a source of energy. Some of the energy used by an animal may come from body fat, but that fat deposit is the result of food eaten and assimilated earlier in its life. Solar radiation, infrared radiation from cover, heat conserved in the insulative layer—none of these can drive the metabolic machinery of the animal; they can only reduce a thermal gradient between it and its environment.

19-3   FACTORS TO CONSIDER IN POPULATION ANALYSES

Population analyses that merely include numbers representing a field situation have limited usefulness for predictive purposes. Extrapolations can be made from such models, but considerable risk is taken in doing so if the factors that cause changes in mortality or natality are not recognized or understood. Life tables are in this category; they are population models built on characteristics of the population at present and/or at different times in the past.

Population models *designed from* biological relationships *rather than from the effects* of these relationships can be used to predict population trends expected under a variety of conditions. Recognition of biological factors that affect populations and the many possible variations in these factors permits the building of a population model that is a multidimensional series of alternatives. The biologist looks upon this kind of a model as a series of analyses, with each analysis representing a different combination of forces that affect natality or mortality. The probability of occurrence of different situations can then be used to predict long-range trends.

The use of electronic computing equipment is absolutely essential for these kinds of analyses. Consideration of the many possible combinations of factors in relation to the different subgroups in a population might lead to the conclusion that the number of combinations is so great that they could never all be included.

19-4   POPULATION ANALYSES FOR $n = 1, 2, \ldots, n$

A frequent approach to the subject of population ecology is based on the establishment of a given population or cohort, measurement of the number of animals that die within that cohort, measurement of the number of animals in the cohort after the next reproductive period, and calculation of the mortality and natality rates from these figures. This is an oversimplification of this kind of analysis, of course, since many considerations can be made in relation to sex ratios, age ratios, different degrees of vulnerability to loss, and so forth. Many of these were mentioned in Chapter 18.

Many useful conclusions have been drawn from population analyses like those just mentioned. These kinds of data have been used to propose various theories of population control, with the recognition of the effect of some factors that seem to be density dependent and some that are density independent. This implies that the operational environment discussed in Chapter 2 needs to be considered in analyzing population growth. Some factors that affect mortality and natality rates are physical, such as weather. Physical factors are usually density independent, but the effect of physical factors may cause animals to crowd together, resulting in interrelationships that are dependent on the density of the population at that time.

This common approach is based on the numerical representation of a single animal as an entity. Its characteristics are often determined from an examination of the dead members of the species at a checking station. The use of hunter-kill data with respect to numbers alone is often used to calculate the population growth, usually in combination with ground or aerial counts of the remaining population. A problem in using this approach is that it essentially removes the dynamics of an individual in relation to the rest of the members of a population from consideration, treating each animal as if it were equal to others, or, at most, dividing the population into sex and age categories, the two most commonly used population characteristics.

Analyses made with the models of homeothermy and carrying capacity presented in this book indicate clearly that there are distinct differences in thermal and nutritive relationships between animal and environment for deer of different weights. A small deer has been found to be at a disadvantage in just about every situation analyzed so far. The comparisons between small and large deer are nonlinear, indicating that there is a need for analyses of a range of animal sizes to determine the critical weights at which an animal is forced to choose alternatives (e.g., increase ingestion) in order to survive. The number of alternatives available to smaller deer is less than the number available to larger deer (e.g., smaller deer cannot reach as high for forage as larger deer), so smaller deer are usually in a more precarious balance than larger deer.

Recognizing these differences, I propose that students direct their thinking toward the living, productive members of a population, asking questions about these individuals that will permit them to understand the position of each individual on the productivity gradient. Specifically, I suggest that students begin considering the characteristics of a population of *one*, trying to determine its characteristics in time and space, with an understanding of the importance of these characteristics as the individual animal passes from birth through maturity and finally death.

A population of one can be extended to include two organisms that vary in one way only, and the effect of this variation can be studied in relation to productivity, resulting in a prediction of the production of new individuals and of new body tissue of the individuals under consideration. The result then is a model containing several individuals, each generated from the initial one or two under consideration. I call this procedure "building a population model from the inside out." It is a technique that forces consideration of the factors that affect the life of each individual. This has considerable ecological value, for it is each individual that must cope with the factors and forces confronting it each day in the natural world. If the individual is successful, productivity results, including individual gain and the production of a new individual.

MORPHOLOGICAL AND PHYSIOLOGICAL CONSIDERATIONS.   A large number of animals is not necessary for the analyses of population trends in relation to different individual characteristics. Suppose that eight animals were considered, with two weighing 40 kg, two weighing 60, two weighing 80, and two weighing 100. The 40-kg animals are representative of fawns and the higher weights are representative of yearlings and adults. The effect of protein quality on animals of different weights can be extended to the effect of protein quality on population dynamics by considering weight changes and productivity of different weight groups. Data presented in Chapter 17 indicate that smaller animals are affected most by a decline in protein quality. They may be unproductive, and selective mortality could cause the smaller animals to disappear from the population. The effect of this reduction in productivity would be observed for several years.

Another factor that could be analyzed in a population of eight or less is the effect of a late spring on the population. Small deer are in a precarious nutritive balance at that time, with a low safety margin in the form of body protein or energy reserves. Various dates of spring dispersal could be simulated to determine the effect of variation in the time that winter ends and spring begins on the productivity of deer of different sizes. Variations in the reproductive rates of deer of different sizes could then be used to determine the contribution of different weight classes to the population through reproductive processes.

By building a population model from the inside out, the builder is alerted to the effects of a number of possible changes in population structure. An awareness of significant factors allows for a more thorough analysis of population data gathered in the field. The causes for certain changes in field populations can then be understood conceptually, if not numerically, through the use of predictive models.

Analyses of ecological efficiency and natality in relation to body weight suggest that there might be important differences in the stability of populations having different proportions of age or weight classes. It is generally recognized that a younger population has a greater potential for growth because there are more animals approaching their maximum reproductive potential. Conversely, the younger-aged population also exists in a more precarious balance with its food supply. An older-aged population includes more animals that are past their prime, and their productivity will be less. Thus there is a greater potential for variation in numbers of smaller animals because external forces, such as heavy storms, low food quality, late-arriving spring weather, and the like, may have additive or multiplicative effects on the total population number. This general idea has long been recognized in population ecology; smaller animals with high reproductive potentials—insects, for example—exhibit marked population fluctuations over short periods of time. The same principle seems to apply to a single species if different sizes or weights are considered.

BEHAVIORAL CONSIDERATIONS. The disadvantages of being a small deer, in terms of its requirements and the range supply, are further compounded by the subdominant position of smaller deer in a herd. The social structure of a deer herd is generally dominated by the adult doe. Doe dominance has been observed many times in encounters at feeding sites, and often a larger number of does than other members of a herd are trapped during winter trapping because the dominant doe moves into the trap first to feed. A study in Maine, designed to expose the effects of different cover types on the physical condition of deer (Robinson 1960), showed that the final condition of individual deer seemed related to their position of dominance in the pen.

A THEORETICAL AVERAGE DEER. The preceding examples of relationships between biological variables and weight have been presented for discrete populations of one animal each, with each population of one having a different weight characteristic. The different populations have been compared, and it has been shown that the smaller deer are at a distinct disadvantage in each of the relationships shown. The same general pattern emerged in Chapters 13 and 14, inasmuch as the smaller deer were shown to have many disadvantages in the maintenance of homeothermy.

Populations with different weight or metabolic structures can be compared by using a weighted-mean procedure in calculating a theoretical average deer to represent the entire population. This weighted-mean procedure can be used to take into account the number of deer in each weight class, the reproductive condition of the deer, and other factors that affect their requirements for protein, energy, and other essential nutritive elements.

The use of an average deer results in a loss of information and analytical capabilities for a population, however. The averaging of population characteristics tends to mask the dynamic and important relationships that affect the survival of an individual. Ingestion, for example, can be calculated for an average deer, but it has been shown that the relationship between the amount of forage ingested to meet minimum requirements and the size of the rumen is not constant for deer of all weights. The calculation of the rumen size of this theoretical deer would not take into account the disadvantage of being a smaller deer in a population. Thus it might be said that "on the average" theoretical deer will survive, but the smaller deer that constitute a portion of a population may well be at such a disadvantage that they will die.

The importance of considering the individual rather than the average for a population was illustrated further in the relationships between weight loss and initial body weight. An average weight loss could be calculated for all members of a population—0.5 kg day^{-1} was used—but this quantity is far more crucial to a small deer than to a big deer. A small deer cannot tolerate as great a weight loss as a large deer. The importance of weight loss must be analyzed in relation to initial and minimum body weight.

The amount of heat lost from a small deer is greater in proportion to its metabolic rate than the heat-loss–metabolic-rate proportion for a large deer. Convection is relatively greater from a small deer, it has a higher surface area per unit of body weight, and in general is at a disadvantage for survival throughout the winter period. The use of a theoretical mean deer for an analysis of the ability of deer to withstand cold conditions would result in the loss of recognition of basic principles of thermal exchange in relation to the geometry of the animal. Students are cautioned to use average values in any biological analysis *only if there is no loss of significant information when the individual animal is ignored.*

19-5 TIME IN RELATION TO BIOLOGICAL EVENTS

SEASONAL CHANGES IN ANIMALS AND PLANTS. The concept of time was defined in Chapter 2 as a "measure of the intensity of life." This seems applicable to the wild ruminant; the requirements for protein and energy follow a sequence that tends to distribute the requirements throughout the year. White-tailed deer gain weight after lactation ceases in late summer or early fall. Conception occurs most often in November, but the requirements for pregnancy do not accelerate until the last one-third to one-fourth of the gestation period. A deer has little more than survival requirements during the winter season. The winter coat is shed from March through May, and the cost of growing the summer coat coincides with maximum requirements for pregnancy. Lactation is a costly biological process, but lactating deer do little more than move about in the summer. They do not gain weight, the summer hair is retained until almost the end of the lactation period, movements are restricted to a home range of less than a mile, and there is foliage to hide in. In late summer and early fall, the summer coat is shed, lactation ceases, and weight gain begins again.

The biomass of a population becomes more stable at each successive trophic level. This is discussed in Jordan, Botkin, and Wolfe (1970) for wolf-moose relationships on Isle Royal. The factors that affect the survival of individuals within a population are age or weight specific. The moose population analyzed by Jordan, Botkin, and Wolfe has a high annual transfer of biomass from the first-year class, with another peak at about 12 years of age. This is to be expected inasmuch as the analyses discussed in earlier chapters reveal that younger, smaller animals are in a more precarious balance than are mature animals. As old age approaches, physiological efficiency again decreases, but its relationship to age is quite variable.

The range supply follows a sequence that coincides with the sequence of animal requirements. The weight-gaining period in the fall is simultaneous with the availability of seeds and fruits, which are usually rich in fats and oils. The falling of leaves is preceded by a redistribution of nutrients from leaf to stem, increasing the food value of current annual growth that will be used as browse later in the winter. The quality of the forage remains fairly constant throughout the period of dormancy, and it may not be high enough to support the daily requirements of deer, especially during periods of severe weather. The quantity of forage available is often limited, especially in areas with deep snows, which cause the deer to remain concentrated. Under these conditions, they are often forced to eat inadequate quantities of low-quality food. If their needs are not met by the range supply, their stored nutrients must be used.

THE IMPORTANCE OF SEASONAL CHANGES. The physiological efficiency of animals of different weights within a population depends on the impact of environmental forces on the individual animal. The timing of plant growth and the storage of nutrients by the animal is a function of seasonal shifts in the energy balance of the earth's atmosphere, causing changes in weather patterns. The temporal occurrence of these weather patterns and their subsequent effect on both the animal and the range is most critical when body reserves are low, the animal is molting, plant growth is beginning, or some other change in either animal or plant is about to take place.

The analyses of homeothermy and carrying capacity lend considerable support to the idea that the time at which deer disperse at the onset of spring is especially important. Deer show an increase in metabolic rate in the spring, they molt, pregnancy begins to increase nutritive requirements, and body reserves are at a low point. If deer cannot leave the winter concentration area because of continued cold temperatures that prevent the snow from melting as well as vegetative growth, they are trapped with accelerating requirements and a deteriorating range. This may cause many deaths. A more subtle effect on population growth may be in the reduction of fawn numbers due to resorption and an increase in the number of stillbirths. Fawn survival, especially during the first month or two, may also be related to the condition of the doe. Alexander (1962) has shown that lambs from well-fed ewes had twice as much fat reserve as those from poorly fed ewes. Those with a higher energy reserve were able to survive longer during starvation since body fat was their largest source of energy. Cold weather also affects the behavior of lambs, with less teat-seeking activity if the lambs are cold (Alexander and Williams 1966). Thus a cold, wet spring may affect both the doe prior to parturition and the fawn in the early stages of suckling.

Continued cold weather in the spring may have a greater effect on the energy requirements for homeothermy than the same kind of weather would have had earlier because of the reduced insulation quality of the summer coat. Further, spring rains may cause an increase in evaporative heat loss. This can become critical inasmuch as the amount of energy dissipated by evaporation is large. Thus a cold spring rain can have a relatively greater effect on energy requirements than severe cold in the winter.

VARIATIONS IN THE TIME OF BREEDING.   Another temporal variation of importance in population dynamics is the distribution of times of conception and parturition. An early fawn is more likely to be exposed to cold weather than a late fawn. Its mother may have a depleted energy and protein reserve because there has been less time between spring dispersal and parturition than there would be for a later birth. The early fawn has more time to grow to a maximum fall weight, however, so it may be in a better position to survive the following winter.

A doe that gives birth to a fawn early in the spring may terminate lactation at an earlier date than one fawning in July. This gives the doe a longer time in which to gain weight, resulting in a larger energy reserve for winter survival. The early-born fawns face the possible disadvantage of cold spring weather. The compensatory benefits of being born earlier compared with the disadvantages of cold weather have not been measured. These conditions can be simulated by computer analysis, but the limiting factor in such simulation is the lack of biological information about not only fawn growth and reproduction, but also the duration of lactation and subsequent growth of the females on the summer and fall range.

A prediction of the importance of the date of breeding, given information on birth weight and growth rate, is shown in the examples below.

EXAMPLE 1

Given: One adult female, bred November 15, gestation period 200 days, one fetus. Fawn birthweight = 3 kg. Fawn growth rate = 200 g day^{-1}, minimum weight for breeding, 38 kg.

Prediction: Productivity pattern of offspring in relation to time of birth.

Year 1: Female fawn born to adult doe on June 3.
Breeding weight reached on November 25.

Year 2: Female fawn born to yearling doe (drop the adult doe from consideration now; we are testing for the effect of birth date on the productivity of the offspring) on June 13.
Breeding weight reached on December 5.

Year 3: Female fawn born to yearling doe (year 2 fawn) on June 23.
Breeding weight reached on December 15.

Year 4: Female fawn born to yearling doe (year 3 fawn) on July 3.
Breeding weight reached on December 25.

Year 5: Female fawn born to yearling doe (year 4 fawn) on July 13.
Breeding weight reached on January 4.

Year 6: Female fawn born to yearling doe (year 5 fawn) on July 23.
Breeding weight reached on January 14.

The pattern described above continues in a linear fashion because the gestation period is a constant length and the growth rate is given as linear (0.2 kg day^{-1}). The fawns born in Year 5 and later may reach the minimum breeding weight after the bucks have lost their reproductive potential, resulting in a lack of conception even if the fawns could conceive.

There are many unnatural conditions in this example. Only female fawns were born each year; a 1:1 sex ratio might have been used. A female fawn was successfully bred each year as long as it reached a minimum weight; however, a fawn may not necessarily conceive when a threshold weight is reached. The linear growth rate shown may not persist into December and January. The model works, however, subject to the limitation of the set of conditions stated. Let us use the conditions given in this model to test the effect of variation in the minimum breeding weight.

EXAMPLE 2

Given: See Example 1, except for minimum weight for breeding.

Prediction: Productivity pattern of offspring in relation to minimum breeding weights of 30, 32, 34, 36, 38, and 40 kg.

Year 1: Female fawn born to adult doe on June 3.
Breeding weight of 30 kg reached on October 16.
Breeding weight of 32 kg reached on October 26.
Breeding weight of 34 kg reached on November 5.
Breeding weight of 36 kg reached on November 15.
Breeding weight of 38 kg reached on November 25.
Breeding weight of 40 kg reached on December 5.

Year 2: Female fawns born to yearling does of minimum breeding weight of x kg, on
May 4: Breeding weight of 30 kg reached September 16.
May 14: Breeding weight of 32 kg reached October 6.

> May 24: Breeding weight of 34 kg reached October 26.
> June 3: Breeding weight of 36 kg reached November 15.
> June 13: Breeding weight of 38 kg reached December 5.
> June 23: Breeding weight of 40 kg reached December 25.

The productivity pattern, given different breeding weights, shows that the threshold weight for breeding without the delay and subsequent parturition is 36 kg, resulting in breeding on November 15 each year. If deer were to breed at weights less than 36 kg, the conception date would arrive earlier each year, with a 30-kg minimum breeding weight reached on September 16 of year 2, which may be before the bucks are able to service them. Thus there is no particular advantage to a lighter minimum breeding weight if the reproductive capability of the males is controlled by a constant seasonal factor such as day length.

The conditions given in this example are unnatural, too, of course, so the results are applicable only if confined to the stated model. The results indicate that variation in the minimum breeding weight can affect subsequent productivity, and this raises further questions. For example, the observed minimum breeding weight of fawns in one area in western New York is 65 pounds, or about 30 kg. The results from this example indicate that 30 kg is well below the predicted threshold of 36 kg. This raises additional questions, such as the relative importance of weight in relation to age. Maybe a large fawn, born early in May and well nourished, will not come into breeding condition until November because of other influences. In other words, its weight is sufficient, but the hormone balance necessary for reproduction may not be reached. Indeed, the dates of conception by fawns were shown to be confined to December in one of the displays of population dimensions in Chapter 18. Thus there should be further consideration of the factors regulating the reproductive biology of white-tailed fawns, including nutritional factors.

The birth date of a fawn also has an effect on the doe since a late-born fawn will most likely not be weaned as early as an early-born fawn. Since the lowest weights of the annual cycle are often reported for lactating females, an early fawn, weaned for a longer period of time prior to the onset of winter, releases the doe, enabling her to gain weight in the fall. A simple example of weight gains necessary to reach a given late fall weight follows.

EXAMPLE 3

Given: Birth dates of Year 2 in Example 2. Lactation period = 100 days. Weight of doe at end of lactation = 50 kg. Prewinter (December 1) weight desired = 70 kg.

Prediction: Weight gains needed to reach 70 kg by December 1.

May 4 birth of fawn: 0.18 kg day^{-1} = 0.40 lb day^{-1}
May 14 birth of fawn: 0.20 kg day^{-1} = 0.44 lb day^{-1}
May 24 birth of fawn: 0.22 kg day^{-1} = 0.48 lb day^{-1}

$$\text{June 3 birth of fawn:} \quad 0.25 \text{ kg day}^{-1} = 0.55 \text{ lb day}^{-1}$$
$$\text{June 13 birth of fawn:} \quad 0.27 \text{ kg day}^{-1} = 0.59 \text{ lb day}^{-1}$$
$$\text{June 23 birth of fawn:} \quad 0.29 \text{ kg day}^{-1} = 0.64 \text{ lb day}^{-1}$$

Now suppose that 0.25 kg day^{-1} (0.55 lb day^{-1}) was the maximum weight increase possible. The doe bearing a fawn on June 13 would reach a prewinter weight of 68 kg. A subsequent 30% loss in weight that winter would put the doe and her fawn 3.4 and 4.5 kg below the previous summer's minimum, respectively. This may result in a loss of reproduction, although the doe might survive.

These examples illustrate an approach to an analysis of productivity that starts with a single animal in a simple model and progresses to an array of values that might be present in a population of many animals. The next step in building this type of population model is to determine the distribution of these values in a population. There are data in the literature for many of these. Conception dates were given in Chapter 18 for deer in northern and southern New York and for deer in adult and fawn age classes. The distribution of deer in these classes can be related to the distribution of birth dates of the fawn, the distribution of weaning, and the distribution of weight gains necessary to reach a given prewinter weight by a given date. In other words, the examples given can be applied to real populations. Combinations of conditions that result in predicted distributions that are similar to observed distributions stimulate thought concerning the feasibility of these conditions in the wild. If the combinations are unrealistic, the model must be modified to correct for the disparities between the theoretical and the real. If the combinations are realistic, the next step is to make the analysis less descriptive and more analytical. Logical questions follow, centered on such things as the constraints affecting the biological potential for weight gains, the factors regulating the phenology of breeding, and other factors that regulate productivity.

If all were known about the factors that affect productivity, predictions could be made about numbers, mortality rates, natality rates, trophic levels, and all of the other characteristics that are considered in looking at a population as a whole. In other words, the analytical ecologist who starts with a population of one should eventually meet the population ecologist who works with large numbers of animals. At that time, the analytical ecologist will have explanations for the many changes observed by the population ecologist, and the population ecologist will be in a position to present comprehensive equations that represent the effects of these biological functions over appropriate time periods.

This approach relies heavily on a knowledge of the biological factors affecting productivity. Biological systems are always complex, so many alternatives to given conditions or constraints can be expected, and the perfect biological model will probably never be built. The process described is of value, however, since it forces consideration of the processes involved in organism-environment relationships. If these processes are not understood, then the management of natural resources can continue only on the basis of opinion. Chapter 20 includes a discussion of the need for establishing a firm biological base for the decision-making process.

LITERATURE CITED IN CHAPTER 19

Alexander, G. 1962. Energy metabolism in the starved new-born lamb. *Australian J. Agr. Res.* **13**(1): 144–164.

Alexander, G., and D. Williams. 1966. Teat-seeking activity in new-born lambs; the effects of cold. *J. Agr. Sci.* **67**: 181–189.

Jordan, P. A., D. B. Botkin, and M. L. Wolfe. 1971. Biomass dynamics in a moose population. *Ecology* **52**(1): 147–152.

Robinson, W. L. 1960. Test of shelter requirements of penned white-tailed deer. *J. Wildlife Management* **24**(4): 364–371.

IDEAS FOR CONSIDERATION

Develop very simple models of very small populations of any species, adding factors only if the effect of each factor can be predicted. Remember that the main purpose of this exercise is the demonstration of mechanisms, and not the display of real numbers for large populations.

SELECTED REFERENCES

Burger, G. V., and J. P. Linduska. 1967. Habitat management related to bobwhite populations at Remington Farms. *J. Wildlife Management* **31**(1): 1–12.

Gross, J. E. 1969. Optimum yield in deer and elk populations. *Trans. North Am. Wildlife Nat. Resources Conf.* **34**: 372–387.

Klein, D. R. 1968. The introduction, increase, and crash of reindeer on St. Matthew Island. *J. Wildlife Management* **32**(2): 350–367.

Silliman, R. P. 1969. Population models and test populations as research tools. *BioScience* **19**(6): 524–528.

ECOLOGICAL ANALYSES AND
DECISION-MAKING
PROCEDURES

How are ecological concepts applied to management decisions? In the past, decisions were made concerning field-management practices and population changes were then noted. This is useful, but present-day technology permits the addition of intermediate steps that utilize more information and permit the testing of effects of different management decisions by simulation.

20-1 THE SIMULATION OF MANAGEMENT PRACTICES

Biologists and resource managers have spent many hours measuring characteristics of populations before and after a hunting season. They measure the characteristics of animals that are brought to check stations and debate the relative merits of different kinds of seasons. Once a season is set, the biologist or resource manager usually has to wait until it is over to see what effect it has had on a population and whether or not the hunt accomplished the purposes for which it was designed. Often there are circumstances that upset the plans, such as a snow storm or other natural event that affects the success of the hunters.

The simulation of different management practices prior to their implementation in the field provides the resource manager with an opportunity to see the effects of his decisions and of natural events before they occur. This use of simulations or models for testing the effects of different hunting seasons adds an exciting dimension to this phase of resource management. Hunting success can be superimposed on the basic biological foundations within a population model. The effect of variation in the success of hunters can be simulated. This technique permits the resource manager to demonstrate the possible effects of different hunting regulations in different biological situations. At that time, the establishment of regulations becomes more analytical and less emotional.

The need for *realistic* biological models cannot be emphasized too strongly. Models used in population dynamics and resource management must be built on a sound theoretical base rather than on an assemblage of facts, figures, and fiction. Computers programed realistically can aid rational analyses, but unrealistic programs will only compound experimental errors and faulty conclusions.

BIOLOGICAL CONSIDERATIONS. A realistic model can be built only if functional biological relationships are considered. In such a model, it becomes readily apparent that some variables can be controlled by the resource manager whereas others are completely beyond his control. Food quality, for example, may be increased by various management practices such as fertilization or burning. As food quality is analyzed in relation to animal requirements, it may be found that a more important variable is the time of spring dispersal. This depends on snow conditions and weather in late winter and early spring, quite beyond the control of the resource manager.

It is the responsibility of the manager to understand the effects of these natural forces on a population or on the ecosystem. He must also explain these relationships to the public. This can do much toward changing the public's attitudes toward the management of natural resources, although the changing of attitudes through educational processes is always a gradual one. It is vital that the professional resource manager has a firm understanding of the biological system, so that both his management suggestions and his explanations of the effects of variables deserve the public's confidence.

SOCIAL AND ECONOMIC CONSIDERATIONS. The wildlife biologist is interested academically and professionally in the same things that interest many other persons for recreation. Just as every baseball fan in the stand is in a position to second-guess the team's manager, the various publics that are interested in natural resources are in a position to second-guess the resource manager. The manager and publics are both in a position to be heard through the simple expediency of writing to political representatives, assuming positions of leadership in local organizations, giving money to causes of their own choosing, and other similar activities. *However, no amount of democratic action, no amount of money, no amount of sympathy for one cause or another can be productive if the basic characteristics of a biological system are ignored.* The resource manager appears to be in a difficult spot when he must reconcile differences of opinion between himself and the various publics, but he also has the opportunity to establish himself on the firmest base of all—the biological one.

The public has a keen interest in wild animals, including both hunters and those that simply enjoy watching the animals in the field. Sometimes it does not matter much to those watching deer whether they see a small, underweight deer or one in its prime. As more small deer than large deer can be supported on a given quantity of resources, a program aimed at a higher population of small deer may be more desirable from a human sociological point of view.

The relationships between deer and their environment are such that small deer are at some distinct biological disadvantages, however. The impact of severe weather, low food quality, and other environmental forces on high populations of small deer is greater than on lower populations of larger deer. Populations of small deer have a greater potential for variation. An understanding of these fundamental biological relationships is necessary for the decision-making process since social considerations should not be made beyond the biological limits present in a particular habitat.

Many social functions are dependent on the elastic unit of trade called the dollar. The amount of hunting that is done, the proceeds from hunting, the amount of research that is conducted, the quality of the research done, the number of publications that are made available, the quantity of information and educational programs, and many other important interactions between humans and their environment all depend to a large extent on the amount of money that is available.

Money is an elastic kind of analog to energy and matter; its elasticity with time is its greatest liability since it is difficult to predict just what its value will be at some future date. Resource economists build models based on the dollar, and these models are more complex than the models of an ecosystem simply because, ideally at least, the economist's model must contain not only the biological components of the ecosystem—the resources themselves—but their representative, the dollar, as well.

The important thing for the resource manager to realize is that dollars cannot successfully buy a violation of basic natural laws. For example, the rumen capacity of a deer is genetically determined. Low-quality forage, limited by the fertility of the soil, might cause the physical capacity of the rumen to be exceeded before the nutrient requirements of the animal can be met. Under these conditions it may be virtually impossible to improve a population regardless of the amount of money spent on management practices.

POLITICAL CONSIDERATIONS.   Political considerations in resource management have the greatest potential for rapid changes and may even result in complete reversal of policy. Political considerations interact with social and economic ones, making it difficult to plan resource management on a long-term basis. *In the final analysis, the only hope for long-term resource planning depends on the quality of the biological foundation that has been established. This quality must be high enough so that political considerations are not in a position to affect the important functional characteristics of the biological system.* Thus the resource manager must direct his responsibility toward the biological system first. It is essential that every effort be made to establish realistic analyses of the biological system.

20-2   CONCLUSION

Technological capabilities available today make it possible for the resource manager to assume an analytical role in the management of natural resources. Through the use of models, the effects of a variety of management decisions can be analyzed before the decisions are made. These models are of value only if they are biologically realistic, so the resource manager must understand biological relationships as well as be able to synthesize them.

These demands extend a person's abilities to the limit. In fact, an individual cannot comprehend the entire ecosystem in all of its detail. Teams of scientists are necessary for the formation of meaningful and detailed representations of the ecosystem. Resource managers can then use these models as a base for decision-making. The manager should have a working relationship with the analysts so that the results from the use of models can be discussed in preparation for further refinements and improvements.

If strong efforts are not made to develop such biological models, the management of natural resources will be even more subject to social, economic, and political considerations without an understanding of the basic biological functions. Although this approach may be partially successful for short periods of time, the long range use of natural resources in a wise manner can be accomplished only with an understanding of the basic mechanisms involved.

The analyses in this book of the relationships between animals (primarily wild ruminants) and their environments are examples of the kind of analyses that can be made for any organism. Life, in an ecological sense, can be analyzed successfully within the framework of energy expenditure and the redistribution of matter. Both are limited and all life is subject to natural laws describing energy and matter relationships. The focal point for these analyses is the individual organism that is subject to the problems of daily existence; its success in meeting those problems determines whether it remains a productive member of its own population and of the ecosystem as a whole or whether the nutrients in its own body are recycled through the ecosystem.

SELECTED REFERENCES

Giles, R. H., Jr., and R. F. Scott. 1969. A systems approach to refuge management. *Trans. North Am. Wildlife Nat. Resources Conf.* **34:** 103–115.

Walters, C. J., and F. Bunnel. 1971. A computer management game of land use in British Columbia. *J. Wildlife Management* **35**(4): 644–657.

Walters, C. J., and J. E. Gross. 1972. Development of big game management plans through simulation modeling. *J. Wildlife Management* **36**(1): 119–128.

Watt, K. E. F., ed. 1966. *Systems analysis in ecology.* New York: Academic Press, 276 pp.

APPENDIXES

The reference materials in the Appendixes are intended to furnish more detailed information than is practical in the text proper. The items will be of help in converting units of measurement, making more detailed calculations, and selecting additional reading material for classroom and research use.

APPENDIX 1

WEIGHTS AND MEASUREMENTS

A-1-1 CONVERSION OF UNITS OF WEIGHT

Units Given	Units Wanted	For Conversion Multiply by	Units Given	Units Wanted	For Conversion Multiply by
lb	g	453.6	μg/kg	μg/lb	0.4536
lb	kg	0.4536	kcal/kg	kcal/lb	0.4536
oz	g	28.35	kcal/lb	kcal/kg	2.2046
kg	lb	2.2046	ppm	μg/g	1.
kg	mg	1,000,000.	ppm	mg/kg	1.
kg	g	1,000.	ppm	mg/lb	0.4536
g	mg	1,000.	mg/kg	%	0.0001
g	μg	1,000,000.	ppm	%	0.0001
mg	μg	1,000.	mg/g	%	0.1
mg/g	mg/lb	453.6	g/kg	%	0.1
mg/kg	mg/lb	0.4536			

SOURCE: All tables in Appendix 1 are from *Applied Animal Nutrition*, 2d ed., by E. W. Crampton and L. E. Harris. W. H. Freeman and Company. Copyright © 1969.

A-1-2 UNITS OF LENGTH, BRITISH OR U.S. SYSTEM

Inches (in)	Feet (ft)	Yards (yd)	Rods (rd)	Miles (mi)	Metric Equivalent
1	0.08333	0.2778	0.005051	0.00001578	2.5400 cm
12	1	0.3333	0.06061	0.0001894	0.3048 m
36	3	1	0.1818	0.0005682	0.9144 m
198	16.5	5.5	1	0.003125	5.0292 m
63,360	5,280	1,760	320	1	1.6094 km

Note: This system is used in the United States and most of the British Commonwealth Countries.

A-1-3 UNITS OF LENGTH, METRIC SYSTEM

Millimeters (mm)	Centimeters (cm)	Decimeters (dm)	Meters (m)	British or U.S. Equivalent
1	0.1	0.01	0.001	0.03937 in
10	1	0.1	0.01	0.3937 in
100	10	1	0.1	3.9370 in
100	10	1	0.1	0.3281 ft
1,000	100	10	1	39.370 in
1,000	100	10	1	3.2808 ft

A-1-4 UNITS OF AREA, BRITISH OR U.S. SYSTEM

Square Inches (sq in)	Square Feet (sq ft)	Square Yards (sq yd)	Square Rods (sq rd)	Acres (A)	Square Miles (sq mi)	Metric Equivalent
1	0.006944	—	—	—	—	6.4516 sq cm
144	1	0.1111	—	—	—	0.0929 sq m
1,296	9	1	0.0331	—	—	0.8361 sq m
	272.25	30.25	1	0.00625	—	25.2930 sq m
	43,560	4,840	160	1	0.001563	40.4687 sq dkm
	27,878,400	3,097,600	102,400	640	1	2.5900 sq km

Note: This system is used in the United States and most of the British Commonwealth Countries.

A-1-5 UNITS OF AREA, METRIC SYSTEM

Square Meters or Centares (m^2, ca)	Square Dekameters or Ares (dkm^2, a)	Square Hectometers or Hectares (hm^2, ha)	Square Kilometers (km^2)	British or U.S. Equivalent
1	0.01	0.0001	0.000001	0.3954 sq rod
100	1	0.01	0.0001	0.02471 acre
10,000	100	1	0.01	2.4710 acres
1,000,000	10,000	100	1	0.3861 sq mile

A-1-6 UNITS OF CAPACITY (liquid measure), BRITISH OR U.S. SYSTEM

Gills (gi)	Pints (pt)	Quarts (qt)	Gallons (gal)	Metric Equivalent U.S.	Metric Equivalent British
1	0.25	0.125	0.03125	118.292 ml	142.06 ml
4	1	0.5	0.125	0.4732 l	0.5682 l
8	2	1	0.25	0.9463 l	1.1365 l
32	8	4	1	3.7853 l	4.5460 l

A-1-7 UNITS OF CAPACITY, METRIC SYSTEM

Milliliters (ml)	Centiliters (cl)	Deciliters (dl)	Liters (l)	U.S. Equivalent
1	0.1	0.01	0.001	16.2311 minims
10	1	0.1	0.01	2.7052 fl drams
100	10	1	0.1	3.3815 fl ounces
1,000	100	10	1	270.518 fl drams
				33.815 fl ounces

Liters (l)	Dekaliters (dkl)	Hectoliters (hl)	Kiloliters (kl)	U.S. Equivalent
1	0.1	0.01	0.001	1.05671 liq quarts
				0.264178 gallon
				1.81620 dry pints
				0.908102 dry quart
10	1	0.1	0.01	18.1620 dry pints
				9.08102 dry quarts
				1.13513 pecks
100	10	1	0.1	2.83782 bushels
1,000	100	10	1	(no equivalent)

Note: One liter is the volume of pure water at 4°C and 760 mm pressure that weighs one kilogram. 1 liter = 1.000027 cubic decimeters = 1000.027 cubic centimeters.

APPENDIX 2

WEATHER, THERMAL FACTORS, AND THE JULIAN CALENDAR

A-2-1   TEMPERATURE, TRANSMISSION, AND ENERGY

Temperature

Degrees Centigrade ($^\circ$C) = 5/9 ($^\circ$F − 32)

Degrees Fahrenheit ($^\circ$F) = (9/5 $^\circ$C) + 32

Degrees Kelvin ($^\circ$K) = $^\circ$C + 273.16

Degrees Rankine ($^\circ$R) = $^\circ$F + 459.69

Transmission

1 BTU ft^{-2} hr^{-1} $^\circ$F^{-1} = 1.355 cal m^{-2} sec^{-1} $^\circ$C^{-1} = 4.88 kcal m^{-2} hr^{-1} $^\circ$C^{-1}

1 cal m^{-2} sec^{-1} $^\circ$C^{-1} = 0.738 BTU ft^{-2} hr^{-1} $^\circ$F^{-1} = 3.6 kcal m^{-2} hr^{-1} $^\circ$C^{-1}

1 kcal m^{-2} hr^{-1} $^\circ$C^{-1} = 0.278 cal m^{-2} sec^{-1} $^\circ$C^{-1} = 0.205 BTU ft^{-2} hr^{-1} $^\circ$F^{-1}

Energy

BTU = 0.252 kcal

kcal = 1000 cal

langley = 1 cal cm^{-2}

A-2-2 STEFAN-BOLTZMANN CONSTANT $(°K) = 5.6697 \times 10^{-5}$ erg cm^{-2} sec^{-1} $°K^{-4}$

Unit	Area	Sec^{-1}	Min^{-1}	Hr^{-1}	Day^{-1}
kcal	m^{-2}	1.354×10^{-11}	8.127×10^{-10}	4.876×10^{-8}	1.170×10^{-6}
	cm^{-2}	1.354×10^{-15}	8.127×10^{-14}	4.876×10^{-12}	1.170×10^{-10}
	ft^{-2}	1.458×10^{-10}	8.747×10^{-9}	5.248×10^{-7}	1.260×10^{-5}
	in^{-2}	2.100×10^{-8}	1.260×10^{-6}	7.558×10^{-5}	1.814×10^{-3}
ly		1.354×10^{-12}	8.127×10^{-11}	4.876×10^{-9}	1.170×10^{-7}
BTU	m^{-2}	5.375×10^{-11}	3.225×10^{-9}	1.935×10^{-7}	4.644×10^{-6}
	cm^{-2}	5.375×10^{-15}	3.225×10^{-13}	1.935×10^{-11}	4.644×10^{-10}
	ft^{-2}	5.786×10^{-10}	3.472×10^{-8}	2.083×10^{-6}	4.999×10^{-5}
	in^{-2}	8.332×10^{-12}	4.999×10^{-10}	2.999×10^{-8}	7.199×10^{-7}

A-2-3 EQUATIONS FOR USE OF THE ECONOMICAL RADIOMETER

The following equations are satisfactory first approximations. More complete equations, including such things as the thermal characteristics of the polyethylene, increase the accuracy of the instrument by 5%–10% (Moen, unpublished data).

$$Q_{r\downarrow} = [\epsilon\sigma T_t^4 + (T_t - T_b)k_i + (T_t - T_a)k_a]$$
$$Q_{r\uparrow} = [\epsilon\sigma T_b^4 + (T_b - T_t)k_i + (T_b - T_a)k_a]$$
$$Q_{r\uparrow} + Q_{r\downarrow} = \text{total flux}$$
$$Q_r - Q_r = \text{net flux}$$

where

Q_r = infrared radiation flux in kcal m^{-2} hr^{-1}

ϵ = emissivity (1.0 is a satisfactory approximation)

σ = 4.93×10^{-8}

T = absolute temperature (add 273 to °C)

$\quad t$ = top sensing element

$\quad b$ = bottom sensing element

k_i = conductivity of the insulation for the depth used in your own instrument

k_a = conductivity of air [see equation (6-8)]

Day	Jan	Feb	Mar	Apr	May	June	July	Aug	Sep	Oct	Nov	Dec	Day
1	001	032	060	091	121	152	182	213	244	274	305	335	1
2	002	033	061	092	122	153	183	214	245	275	306	336	2
3	003	034	062	093	123	154	184	215	246	276	307	337	3
4	004	035	063	094	124	155	185	216	247	277	308	338	4
5	005	036	064	095	125	156	186	217	248	278	309	339	5
6	006	037	065	096	126	157	187	218	249	279	310	340	6
7	007	038	066	097	127	158	188	219	250	280	311	341	7
8	008	039	067	098	128	159	189	220	251	281	312	342	8
9	009	040	068	099	129	160	190	221	252	282	313	343	9
10	010	041	069	100	130	161	191	222	253	283	314	344	10
11	011	042	070	101	131	162	192	223	254	284	315	345	11
12	012	043	071	102	132	163	193	224	255	285	316	346	12
13	013	044	072	103	133	164	194	225	256	286	317	347	13
14	014	045	073	104	134	165	195	226	257	287	318	348	14
15	015	046	074	105	135	166	196	227	258	288	319	349	15
16	016	047	075	106	136	167	197	228	259	289	320	350	16
17	017	048	076	107	137	168	198	229	260	290	321	351	17
18	018	049	077	108	138	169	199	230	261	291	322	352	18
19	019	050	078	109	139	170	200	231	262	292	323	353	19
20	020	051	079	110	140	171	201	232	263	293	324	354	20
21	021	052	080	111	141	172	202	233	264	294	325	355	21
22	022	053	081	112	142	173	203	234	265	295	326	356	22
23	023	054	082	113	143	174	204	235	266	296	327	357	23
24	024	055	083	114	144	175	205	236	267	297	328	358	24
25	025	056	084	115	145	176	206	237	268	298	329	359	25
26	026	057	085	116	146	177	207	238	269	299	330	360	26
27	027	058	086	117	147	178	208	239	270	300	331	361	27
28	028	059	087	118	148	179	209	240	271	301	332	362	28
29	029		088	119	149	180	210	241	272	302	333	363	29
30	030		089	120	150	181	211	242	273	303	334	364	30
31	031		090		151		212	243		304		365	31

Day	Jan	Feb	Mar	Apr	May	June	July	Aug	Sep	Oct	Nov	Dec	Day
1	001	032	061	092	122	153	183	214	245	275	306	336	1
2	002	033	062	093	123	154	184	215	246	276	307	337	2
3	003	034	063	094	124	155	185	216	247	277	308	338	3
4	004	035	064	095	125	156	186	217	248	278	309	339	4
5	005	036	065	096	126	157	187	218	249	279	310	340	5
6	006	037	066	097	127	158	188	219	250	280	311	341	6
7	007	038	067	098	128	159	189	220	251	281	312	342	7
8	008	039	068	099	129	160	190	221	252	282	313	343	8
9	009	040	069	100	130	161	191	222	253	283	314	344	9
10	010	041	070	101	131	162	192	223	254	284	315	345	10
11	011	042	071	102	132	163	193	224	255	285	316	346	11
12	012	043	072	103	133	164	194	225	256	286	317	347	12
13	013	044	073	104	134	165	195	226	257	287	318	348	13
14	014	045	074	105	135	166	196	227	258	288	319	349	14
15	015	046	075	106	136	167	197	228	259	289	320	350	15
16	016	047	076	107	137	168	198	229	260	290	321	351	16
17	017	048	077	108	138	169	199	230	261	291	322	352	17
18	018	049	078	109	139	170	200	231	262	292	323	353	18
19	019	050	079	110	140	171	201	232	263	293	324	354	19
20	020	051	080	111	141	172	202	233	264	294	325	355	20
21	021	052	081	112	142	173	203	234	265	295	326	356	21
22	022	053	082	113	143	174	204	235	266	296	327	357	22
23	023	054	083	114	144	175	205	236	267	297	328	358	23
24	024	055	084	115	145	176	206	237	268	298	329	359	24
25	025	056	085	116	146	177	207	238	269	299	330	360	25
26	026	057	086	117	147	178	208	239	270	300	331	361	26
27	027	058	087	118	148	179	209	240	271	301	332	362	27
28	028	059	088	119	149	180	210	241	272	302	333	363	28
29	029	060	089	120	150	181	211	242	273	303	334	364	29
30	030		090	121	151	182	212	243	274	304	335	365	30
31	031		091		152		213	244		305		366	31

APPENDIX 3

WEIGHT AND
METABOLIC WEIGHT

A-3-1 METABOLIC SIZE FOR LIVE BODY WEIGHT ($W_{kg}^{0.75}$)

W	$W^{0.75}$	W	$W^{0.75}$	W	$W^{0.75}$
0.5	0.60	130	38.50	480	102.55
1.0	1.00	140	40.70	490	104.15
1.5	1.36	150	42.86	500	105.74
2.0	1.68	160	44.99	510	107.32
2.5	1.99	170	47.08	520	108.89
3.0	2.28	180	49.14	530	110.47
3.5	2.56	190	51.17	540	112.02
4.0	2.83	200	53.18	550	113.57
4.5	3.09	210	55.16	560	115.12
5.0	3.34	220	57.12	570	116.65
5.5	3.59	230	59.06	580	118.19
6.0	3.83	240	60.98	590	119.71
6.5	4.07	250	62.87	600	121.23
7.0	4.30	260	64.75	620	124.2
7.5	4.53	270	66.61	640	127.2
8.0	4.76	280	68.45	660	130.2
8.5	4.98	290	70.28	680	133.2
9.0	5.20	300	72.08	700	136.1
9.5	5.41	310	73.88	720	139.0
10.0	5.62	320	75.66	740	141.9
15.0	7.62	330	77.42	760	144.7
20.0	9.46	340	79.18	780	147.6
25.0	11.18	350	80.92	800	150.4
30.0	12.82	360	82.65	820	153.2
35.0	14.39	370	84.36	840	156.0
40.0	15.91	380	86.07	860	158.8
45.0	17.37	390	87.76	880	161.6
50.0	18.80	400	89.44	900	164.3
60.0	21.56	410	91.11	920	167.0
70.0	24.20	420	92.78	940	169.8
80.0	26.75	430	94.43	960	172.5
90.0	29.22	440	96.07	980	175.2
100	31.62	450	97.70	1,000	177.8
110	33.97	460	99.33		
120	36.26	470	100.94		

SOURCE: From *Applied Animal Nutrition*, 2d ed., by E. W. Crampton and L. E. Harris. W. H. Freeman and Company. Copyright © 1969.

APPENDIX 4

RADIANT TEMPERATURE
IN RELATION TO
AIR TEMPERATURE

A-4-1 LINEAR REGRESSIONS FOR RADIANT SURFACE TEMPERATURE IN RELATION TO AIR TEMPERATURE FOR WHITE-TAILED DEER (based on measurements on a simulator in uniform radiation in the TEST)

	Hide Orientation					
Wind Speed (mi hr^{-1})	*With Wind*			*Against Wind*		
	Regression	*n*	*r*	*Regression*	*n*	*r*
0	$Y = 9.12 + 0.76X$	19	.99	$Y = 10.26 + 0.73X$	16	.97
1	$Y = 9.12 + 0.76X$	22	.97	$Y = 11.42 + 0.70X$	21	.99
2	$Y = 7.68 + 0.80X$	22	.99	$Y = 11.42 + 0.71X$	20	.99
3	$Y = 7.68 + 0.80X$	24	.98	$Y = 7.98 + 0.79X$	21	.98
4	$Y = 6.84 + 0.82X$	24	.99	$Y = 6.46 + 0.83X$	21	.98
6	$Y = 6.46 + 0.83X$	25	.99	$Y = 6.46 + 0.83X$	19	.97
8	$Y = 4.94 + 0.87X$	24	.97	$Y = 6.84 + 0.82X$	21	.99
10	$Y = 3.80 + 0.90X$	23	.99	$Y = 4.56 + 0.88X$	21	.99
14	$Y = 2.67 + 0.92X$	17	.98	$Y = 4.36 + 0.89X$	12	.98

SOURCE: Data from Deborah S. Stevens, Thermal energy exchange and the maintenance of homeothermy in white-tailed deer (Ph.D. dissertation, Cornell University, 1972).

Note: Y = radiant temperature; X = air temperature.

A-4-2 LINEAR REGRESSIONS FOR RADIANT SURFACE TEMPERATURE IN RELATION TO AIR TEMPERATURE FOR WHITE-TAILED DEER (based on measurements on a simulator in the TMST at low radiant temperatures)

Wind Speed (mi hr^{-1})	With Wind			Against Wind		
	Regression	*n*	*r*	*Regression*	*n*	*r*
0	$Y = 4.0 + 0.76X$	14	.99			
1	$Y = 9.12 + 0.76X$	13	.96	$Y = 9.57 + 0.75X$	12	.97
2	$Y = 8.11 + 0.78X$	14	.98	$Y = 9.57 + 0.75X$	15	.96
4	$Y = 7.98 + 0.79X$	13	.97	$Y = 7.40 + 0.81X$	15	.97
6	$Y = 5.42 + 0.86X$	15	.96	$Y = 6.46 + 0.83X$	16	.98
10	$Y = 4.36 + 0.89X$	14	.97	$Y = 5.42 + 0.86X$	16	.97
14	$Y = 4.36 + 0.89X$	9	.98	$Y = 4.94 + 0.87X$	14	.97

SOURCE: Data from Deborah S. Stevens, Thermal energy exchange and the maintenance of homeothermy in white-tailed deer (Ph.D. Dissertation, Cornell University, 1972).

Note: Y = radiant temperature; X = air temperature.

A-4-3 LINEAR REGRESSIONS FOR RADIANT SURFACE TEMPERATURE IN RELATION TO AIR-TEMPERATURE FOR WHITE-TAILED DEER (based on measurements on a simulator)

Mi Hr^{-1}	*Combined Regressions*	*n*	*r*
0	$Y = 9.49 + 0.75X$	51	.98
1	$Y = 9.18 + 0.76X$	68	.98
2	$Y = 8.60 + 0.78X$	71	.98
3	$Y = 7.88 + 0.79X$	45	.98
4	$Y = 7.53 + 0.80X$	74	.97
6	$Y = 6.45 + 0.83X$	75	.98
8	$Y = 5.63 + 0.85X$	45	.98
10	$Y = 4.88 + 0.87X$	74	.99
14	$Y = 4.05 + 0.89X$	52	.98

SOURCE: Data from Deborah S. Stevens, Thermal energy exchange and the maintenance of homeothermy in white-tailed deer (Ph.D. dissertation, Cornell University, 1972).

Note: Y = radiant surface temperature; X = air temperature.

A-4-4 LINEAR REGRESSIONS FOR PREDICTING RADIANT SURFACE TEMPERATURE FOR VARIOUS BODY PARTS OF LIVE WHITE-TAILED DEER IN STILL AIR

Body Part		Regression	n	r
Head	Max:	$Y = 12.34 + .64X$	40	.96
	Min:	$Y = 9.00 + .68X$	40	.95
	Avg:	$Y = 10.67 + .66X$	80	.95
Neck	Max:	$Y = 10.08 + .73X$	67	.93
	Min:	$Y = 7.06 + .78X$	67	.96
	Avg:	$Y = 8.57 + .76X$	134	.95
Body trunk	Max:	$Y = 10.37 + .77X$	179	.94
	Min:	$Y = 7.66 + .79X$	179	.94
	Avg:	$Y = 9.01 + .78X$	358	.94
Upper front leg	Max:	$Y = 12.26 + .70X$	41	.97
	Min:	$Y = 9.24 + .78X$	41	.97
	Avg:	$Y = 10.75 + .74X$	42	.97
Lower front leg	Max:	$Y = 10.06 + .87X$	38	.93
	Min:	$Y = 7.26 + .87X$	38	.95
	Avg:	$Y = 8.66 + .87X$	76	.94
Upper hind leg	Max:	$Y = 10.07 + .74X$	44	.94
	Min:	$Y = 8.34 + .74X$	44	.97
	Avg:	$Y = 9.20 + .74X$	88	.96
Lower hind leg	Max:	$Y = 9.51 + .86X$	38	.99
	Min:	$Y = 7.64 + .80X$	38	.98
	Avg:	$Y = 8.58 + .83X$	76	.98

SOURCE: Data from Deborah S. Stevens, Thermal energy exchange and the maintenance of homeothermy in white-tailed deer (Ph.D. dissertation, Cornell University, 1972).

Note: Y = radiant surface temperature; X = air temperature.

A-4-5 FORMULAS FOR PREDICTING FEATHER SURFACE TEMPERATURES (Y) OF SHARP-TAILED GROUSE FROM AIR TEMPERATURE (X) AT VARIOUS WIND VELOCITIES

Wind Velocity ($mi\ hr^{-1}$)	Regression	n	r
0	$Y = 5.978 + 0.855X$	34	0.995
1	$Y = 9.256 + 0.776X$	34	0.993
2	$Y = 0.726 + 0.765X$	30	0.995
4	$Y = 7.707 + 0.814X$	29	0.979
6	$Y = 7.472 + 0.819X$	27	0.998
10	$Y = 7.113 + 0.828X$	41	0.998

SOURCE: Data from K. E. Evans, Energetics of sharp-tailed grouse (*Pedioecetis phasianellus*) during winter in western South Dakota (Ph.D. dissertation, Cornell University, 1971).

APPENDIX 5

SURFACE AREA
IN RELATION TO WEIGHT

A-5-1 FORM SHEET FOR SURFACE AREA MEASUREMENTS OF DEER
(the circumference, length, and width measurements are in inches;
k is used to convert to the metric system, $k = .0006452$.)

Date _____ County killed in _____ Measured at _____ by _____
Live weight _____ lb; _____ kg Field dressed weight _____ lb; _____ kg
Sex _____ Age _____ Comments _____

Area

1. C: nose C_1 _____
2. C: face C_2 _____ $A = [(C_1 + C_2)/2]L_1 k =$ _____ m^2
3. L: nose (to eyes) L_1 _____
4. L: head L_2 _____ $A = [(C_2 + C_3)/2]L_2 k =$ _____ m^2
5. C: upper neck C_3 _____
6. C: lower neck C_4 _____ $A = [(C_3 + C_4)/2]L_3 k =$ _____ m^2
7. L: neck L_3 _____
8. L: ear L_4 _____ $A = 3\pi[(W_1 L_4)/4]k =$ _____ m^2
9. W: ear W_1 _____
10. C: front body C_5 _____
11. C: rear body C_6 _____ $A = [(C_5 + C_6)/2]L_5 k =$ _____ m^2
12. L: body L_5 _____
13. C: front thigh C_7 _____
14. C: front knee C_8 _____ $A = (C_7 + C_8)L_6 k =$ _____ m^2
15. L: upper front leg L_6 _____
16. L: lower front leg L_7 _____ $A = (C_8 + C_9)L_7 k =$ _____ m^2
17. C: front hoof C_9 _____
18. C: rear thigh C_{10} _____
19. C: hock C_{11} _____ $A = (C_{10} + C_{11})L_8 k =$ _____ m^2
20. L: upper rear leg L_8 _____
21. L: lower rear leg L_9 _____ $A = (C_{11} + C_{12})L_9 k =$ _____ m^2
22. C: rear hoof C_{12} _____

Total _____ m^2

Note: See diagram of deer in Figure 13-17.

A-5-2 SURFACE AREA IN RELATION TO WEIGHT FOR FEMALE WHITE-TAILED DEER

Body Part	Equation for Surface Area*	n	r
Head (two parts)	$0.0023\ W_{kg}^{0.68}$	135	.97
	$0.0083\ W_{kg}^{0.57}$	138	.95
Neck	$0.0078\ W_{kg}^{0.73}$	142	.97
Ears	$0.0092\ W_{kg}^{0.40}$	139	.91
Body trunk	$0.050\ W_{kg}^{0.75}$	140	.99
Upper front legs	$0.013\ W_{kg}^{0.47}$	139	.94
Lower front legs	$0.016\ W_{kg}^{0.41}$	139	.97
Upper hind legs	$0.022\ W_{kg}^{0.51}$	138	.95
Lower hind legs	$0.024\ W_{kg}^{0.44}$	140	.98
Whole body	$0.142\ W_{kg}^{0.635}$	135	.99

*Surface area in square meters.

A-5-3 SURFACE AREA IN RELATION TO WEIGHT FOR PHEASANT

Body Part	Equation	r
Beak	$Y = 1.76 + 0.0057X$	0.916
Head	$Y = 16.01 + 0.0513X$	0.876
Neck	$Y = 29.64 + 0.625X$	0.748
Body	$Y = 74.91 + 0.7914X$	0.973
Upper leg	$Y = 19.36 + 0.0718X$	0.829
Metatarsus	$Y = 12.00 + 0.0285X$	0.820
Toes	$Y = 15.05 + 0.0491X$	0.905

Note: Y = area in cm²; X = weight in g.

A-5-4 WEEKLY WEIGHT LOSS OF WHITE-TAILED DEER FROM NOVEMBER THROUGH THE FIRST WEEK OF JANUARY (see Figure 15-7)

Age (years)	Sex	Equations	
		Y(lbs) X(weeks)	Y(kg); X(weeks)
$4\frac{1}{2}$	M	$Y = 200 - 5.7X$	$91 - 2.59X$
$3\frac{1}{2}$	M	$Y = 191 - 5.4X$	$87 - 2.45X$
$2\frac{1}{2}$	M	$Y = 161 - 4.7X$	$73 - 2.14X$
$3\frac{1}{2}$ & $4\frac{1}{2}$	F	$Y = 121 - 1.2X$	$55 - 0.55X$
$2\frac{1}{2}$	F	$Y = 115 - 1.3X$	$52 - 0.59X$
$1\frac{1}{2}$	M	$Y = 116 - 2.0X$	$53 - 0.91X$
$1\frac{1}{2}$	F	$Y = 101 - 1.6X$	$46 - 0.73X$

SOURCE: Data from H. R. Siegler, ed., *The white-tailed deer of New Hampshire* (Concord: New Hampshire Fish and Game Dept., 1968).

APPENDIX 6

SYMBOLS

The use of meaningful symbols becomes increasingly more difficult as ecological models become more complex. Further, some symbols and notations have been used in several fields of science for many years and are so widely known that it is difficult to change them to more meaningful ones. It is desirable, however, to standardize the symbols used as much as possible. The following list is a combination of traditional symbols and newer notations that are used in the formulas in this book. They are used to communicate ideas and may be replaced by any symbol that communicates the ideas as well or better.

A-6-1 SYMBOLS USED IN METEROLOGY AND THERMAL ANALYSES

d_p = physical depth

d_t = thermal depth

h_c = convection coefficient

Q_c = heat exchange by convection

Q_k = heat exchange by conduction

Q_r = radiant energy emitted

Q_v = heat exchange through vaporization

T = temperature

$\quad T_a$ = air temperature

$\quad T_b$ = body temperature

$\quad T_r$ = radiant temperature

$\quad T_s$ = surface temperature

U = velocity or rate of speed

VP = vapor pressure

VPD = vapor pressure deficit

Δ = difference between two values

ϵ = emissivity

λ = wavelength

σ = Stefan-Boltzmann constant

μ = micron

A-6-2 SYMBOLS USED TO EXPRESS METABOLIC RELATIONSHIPS

I_m = metabolic increment expressed as a multiple of A_{mb}

$\quad I_{ma}$ = metabolic increment for activity

$\quad I_{md}$ = metabolic increment for bedding

$\quad I_{me}$ = metabolic increment for breeding

$\quad I_{mf}$ = metabolic increment for feeding or foraging

$\quad I_{mp}$ = metabolic increment for milk production

$\quad I_{mr}$ = metabolic increment for running

$\quad I_{ms}$ = metabolic increment for standing

$\quad I_{mu}$ = metabolic increment for ruminating

$\quad I_{mw}$ = metabolic increment for walking

Q_e = quantity of energy required

$\quad Q_{ep}$ = energy required for milk production

$\quad Q_{er}$ = energy required for running

$\quad Q_{es}$ = energy required for standing

$\quad Q_{ew}$ = energy required for walking

Q_{eun} = endogenous urinary nitrogen

Q_{mb} = energy expenditure for basal metabolism

Q_{mp} = quantity of milk produced

Q_{mfn} = metabolic fecal nitrogen

Q_n = quantity nitrogen required

$\quad Q_{ng}$ = nitrogen required for gain

$\quad Q_{nh}$ = nitrogen required for hair growth

$\quad Q_{nl}$ = nitrogen required for lactation

$\quad Q_{np}$ = nitrogen required for pregnancy

Q_p = quantity protein required ($Q_p \cong 6.25\ Q_n$)

$\quad Q_{pf}$ = protein required by the fetus

$\quad Q_{pp}$ = protein required for pregnancy

t = time

$\quad t_d$ = time in days

$\quad t_y$ = time in years

XBMR = multiple of the basal metabolic rate—an alternative expression for $(I)Q_{mb}$

APPENDIX 7

REFERENCE BOOKS

The number of reference books available is large, and it is often difficult to select those that are most pertinent to the many facets of ecological investigation. The following lists of statistics references, basic biology references, and general references contain entries that cover a broad range of topics within the major subject areas. Most of these are not cited in any of the chapters; the chapter lists and these lists together provide a good starting point for library work. None of the lists are exhaustive, of course, and the student is encouraged to use the library card catalogs to add additional references to his own file.

STATISTICS REFERENCES

Bishop, O. N. 1971. *Statistics for biology.* The principles of modern biology series. New York: Houghton Mifflin, 216 pp.

Bliss, C. I. 1967. *Statistics in biology.* New York: McGraw-Hill, 576 pp.

Bradley, J. V. 1968. *Distribution-free statistical tests.* Englewood Cliffs, New Jersey: Prentice-Hall, 388 pp.

Noether, G. E. 1967. *Elements of nonparametric statistics.* New York: Wiley, 104 pp.

Siegel, S. 1956. *Nonparametric statistics for the behavioral sciences.* New York: McGraw-Hill.

Sokal, R. R., and F. J. Rohlf. 1969. *Biometry: the principles and practice of statistics in biological research.* San Francisco: W. H. Freeman and Company, 776 pp.

Steel, R. G. D., and J. H. Torrie. 1960. *Principles and procedures of statistics.* New York: McGraw-Hill, 481 pp.

Wyatt, W. W., and C. M. Bridges, Jr. 1967. *Statistics for the behavioral sciences.* Boston: Heath, 389 pp.

BASIC BIOLOGY REFERENCES

Altman, P. L., and D. S. Dittmer, ed. 1966. *Environmental biology.* Bethesda, Maryland: Federation of American Societies for Experimental Biology, 694 pp.

Davis, D. E., and F. B. Golley. 1963. *Principles in mammalogy.* New York: Reinhold, 335 pp.

Devlin, R. M. 1966. *Plant physiology.* New York: Reinhold, 564 pp.

Frieden, E., and H. Lipner. 1971. *Biochemical endocrinology of the vertebrates.* Englewood Cliffs, New Jersey: Prentice-Hall, 164 pp.

Kalmus, H., ed. 1967. *Regulation and control in living systems.* New York: Wiley, 468 pp.

McLaren, A., ed. 1966. *Advances in reproductive physiology.* Vol. 1. New York: Academic Press, 295 pp.

Morowitz, H. J. 1970. *Entropy for biologists: introduction to thermal dynamics.* New York: Academic Press, 195 pp.

Nalbandov, A. V. 1964. *Reproductive physiology: comparative reproductive physiology of domestic animals, laboratory animals, and man.* 2d ed. San Francisco: W. H. Freeman and Company, 316 pp.

Prosser, C. L., and F. A. Brown, Jr. 1961. *Comparative animal physiology.* Philadelphia: Saunders, 688 pp.

Riggs, D. S. 1963. *The mathematical approach to physiological problems.* Baltimore: Williams & Wilkins, 445 pp.

Van Tienhoven, A. 1968. *Reproductive physiology of vertebrates.* Philadelphia: Saunders, 498 pp.

Waterman, T. H., and H. J. Morowitz. 1965. *Theoretical and mathematical biology.* New York: Blaisdell, 426 pp.

Wilson, J. A. 1972. *Principles of animal physiology.* New York: Macmillan, 842 pp.

GENERAL REFERENCES

Benton, A. H., and W. E. Werner, Jr. 1966. *Principles of field biology and ecology.* New York: McGraw-Hill, 499 pp.

Cragg, J. B. 1962. *Advances in ecological research.* Vol. 1–3. London: Academic Press.

Danbenmire, R. F. 1959. *Plants and environment.* New York: Wiley, 422 pp.

Dice, L. R. 1952. *Natural communities.* Ann Arbor: University of Michigan Press, 547 pp.

Giles, R. H., Jr., ed. 1969. *Wildlife management techniques.* Washington, D.C.: The Wildlife Society, 623 pp.

Hansen, H. P., ed. 1967. *Arctic biology.* Corvallis: Oregon State University Press, 318 pp.

Hanson, H. C., and E. D. Churchill. 1961. *The plant community.* New York: Reinhold.

Hochbaum, H. A. 1967. *Travels and traditions of waterfowl.* Minneapolis: University of Minnesota Press, 313 pp.

Humphrey, R. R. 1962. *Range ecology.* New York: Ronald Press, 234 pp.

Jackson, H. H. T. 1961. *Mammals of Wisconsin.* Madison: University of Wisconsin Press, 504 pp.

Keith, L. B. 1963. *Wildlife's ten-year cycle.* Madison: University of Wisconsin Press, 201 pp.

Kendeigh, S. C. 1961. *Animal ecology.* Englewood Cliffs, New Jersey: Prentice-Hall, 468 pp.

Kershaw, K. A. 1964. *Quantitative and dynamic ecology.* London: Edward Arnold, 183 pp.

MacFadyen, A. 1963. *Animal ecology, aims and methods.* London: Pitman, 244 pp.

Owen, D. F. 1966. *Animal ecology in tropical Africa.* San Francisco: W. H. Freeman and Company, 122 pp.

Phillipson, J. 1966. *Ecological energetics.* New York: St. Martin's Press, 57 pp.

Rosene, W. 1970. *The bobwhite quail: its life and management.* New Brunswick, New Jersey: Rutgers University Press, 418 pp.

Smith, R. L. 1968. *Ecology and field biology.* New York: Harper & Row, 687 pp.

Sondheimer, E. S., and J. B. Simeone. 1970. *Chemical ecology.* New York: Academic Press, 336 pp.

Weaver, J. E. 1954. *North American prairie.* Lincoln, Nebraska: Johnson, 348 pp.

Welty, J. C. 1962. *The life of birds.* Philadelphia: Saunders, 546 pp.

APPENDIX 8

INSTRUCTIONS FOR CONTRIBUTORS TO THE PROFESSIONAL LITERATURE

Scientific investigations should be completed with the preparation of a manuscript that will be a valuable addition to the scientific literature. Students should regard their "project reports" as additions to the scientific literature, too, although usually without circulation or publication. Since it is about as easy to write a paper according to an acceptable scientific format as it is according to an unacceptable one, students are urged to prepare their manuscripts at the conclusion of an investigation according to the instructions found in a scientific journal. Because the instructions are quite similar for most journals, the following general summary may serve as a model.

1. Write with originality, clarity, and scientific accuracy.

2. Type manuscripts on $8\frac{1}{2}'' \times 11''$ paper of good quality.

3. Double space throughout, and leave $1\frac{1}{2}''$ margins on all sides.

4. Have a logical sequence of contents, including a title page (follow a thesis format for this), abstract (no more than one page), introductory statement and objectives, methods, results, interpretation and discussion, literature cited, tables with captions, figure legends, and figures.

5. Follow the general rules for preparation of copy described in the *Council of Biology Editors Style Manual* (3d ed., 1972). This is available from the American Institute of Biological Sciences, 3900 Wisconsin Ave., N.W., Washington, D.C. 20016.

6. Select a title that is concise and descriptive (no more than ten to twelve words), with three to five key words that relate to the specific contents of the paper.

7. The common name of an organism is to be accompanied by the scientific name the first time it appears in both the abstract and the whole manuscript.

8. Note all references to the work of others in the manuscript by giving the name of the author(s) and date of publication. Be sure the "Literature Cited" section is complete.

9. Tables should be short, double spaced, and have at least one inch margins. Omit vertical lines.

10. Illustrations should be neatly sketched, labeled, and identified clearly by the caption.

11. Proofread the paper before submitting it to your instructor. It is helpful to read the manuscript from the end to the beginning in looking for typing errors.

12. Retain a copy of the manuscript for your own file.

INDEX